Cybersecurity, Ethics, and Collective Responsibility

Cybersecurity, Ethics, and Collective Responsibility

SEUMAS MILLER AND TERRY BOSSOMAIER

OXFORD
UNIVERSITY PRESS

OXFORD
UNIVERSITY PRESS

Oxford University Press is a department of the University of Oxford. It furthers
the University's objective of excellence in research, scholarship, and education
by publishing worldwide. Oxford is a registered trade mark of Oxford University
Press in the UK and certain other countries.

Published in the United States of America by Oxford University Press
198 Madison Avenue, New York, NY 10016, United States of America.

Library of Congress Control Number: 2024931144

ISBN 978–0–19–005813–5

DOI: 10.1093/oso/9780190058135.001.0001

Printed by Sheridan Books, Inc., United States of America

Contents

Acknowledgements

Seumas Miller would like to gratefully acknowledge the Australian Graduate School in Policing and Security Studies and the Cooperative Research Centre in Cybersecurity at Charles Sturt University, the Uehiro Centre for Practical Ethics at the University of Oxford, and the Digital Ethics Centre at Delft University of Technology for supporting this research.

Seumas Miller would also like to gratefully acknowledge the European Research Council for supporting this research, through the ERC grant GTCMR. No. 670172, *Global Terrorism and Collective Moral Responsibility: Redesigning Military, Police and Intelligence Institutions in Liberal Democracies.*

Terry Bossomaier would like to thank Steve d'Alessandro, Morgan Miles, and Roger Bradbury for helpful discussions and Charles Sturt University for support.

Introduction

Cybersecurity has recently become a strategic priority for governments worldwide. Individual citizens, businesses, and public-sector agencies are increasingly subject to cyberattacks from criminal organizations and malevolent individuals. Many of these attacks are essentially cases of theft and can result in large financial losses, as in the case of identity theft of credit card information or theft of intellectual property (e.g., industrial secrets). Other attacks destroy data and result in lost productivity, equipment damage, and so on. Indeed, cybercrime is evidently now to be measured in terms of trillions of US dollars.[1] Moreover, critical infrastructure, such as electricity grids, are at risk. Further, cyberattacks have emerged as a new form of interstate conflict and, in doing so, the traditional distinction between crimes and acts of war has become increasingly blurred. China, Russia, and North Korea have engaged in cyberattacks that constitute threats to economic interests (e.g., large-scale cybertheft by Chinese state-based agencies of the intellectual property of US corporations), as well as to national security (e.g., Russian attacks on Ukrainian infrastructure). Indeed, in some cases their attacks target liberal democratic institutions (e.g., Russian interference in the 2020 US Presidential election). In early 2022, Russia invaded the Ukraine, and the two countries are, as of the time of writing, engaged in what is essentially a conventional war, waged on Ukrainian soil, over control of the Ukraine. But the war has an important cyber dimension, too, such as Russian cyberattacks on Ukrainian banks and government agencies, and an ongoing Russian disinformation and propaganda campaign in cyberspace—or, what is now referred to as "cognitive warfare" (Chapter 6, Section 6.2). As individuals, businesses, government agencies, and indeed nation states, become increasingly cyber connected, there is a need to ensure that essential technologies and their underpinning infrastructure are secure. More

---

[1] Steve Morgan, "Cybercrime to Cost the World $10.5 Trillion Annually by 2025," *Cybercrime Magazine*, 20 November 2020. https://cybersecurityventures.com/hackerpocalypse-cybercrime-report-2016/. Accessed 24/10/23.

Cybersecurity, Ethics, and Collective Responsibility. Seumas Miller and Terry Bossomaier, Oxford University Press.
© Oxford University Press 2024. DOI: 10.1093/oso/9780190058135.003.0001

generally, there is a need to regulate an increasingly insecure cyberspace in which there is a veritable tsunami of crime that is threatening to undermine (directly or indirectly) legally enshrined sociomoral norms against theft, fraud, and violence. There is also an exponential increase in disinformation, propaganda, and hate speech, which threaten to undermine sociomoral norms of evidence-based truth telling and, in some cases, even democratic institutions. As if this were not enough, cyberspace has become an arena for state-based conflict, in which international legal norms, embodying moral principles, are often flouted by malevolent state actors who can credibly deny their illegal actions due to the problem of reliable attribution in cyberspace. Here we note that fundamental moral rights and principles are not only enshrined in, and protected by, legal systems, but, even more importantly, concretized and maintained in the attitudes and behaviour of individual human beings through their compliance with social norms.[2] Unfortunately, cyberspace is increasingly insecure not only by virtue of the absence of legally enforced regulation, but by virtue of the lack of a commitment to, and compliance with, sociomoral norms. Cyberspace has proven to be an arena in which noncompliance with laws and sociomoral norms alike has to a considerable extent gone unchecked, and thereby generated a host of security problems.

Given these developments, definitions of cybersecurity in terms of the confidentiality, integrity, and availability of information are too narrow (i.e., definitions in terms of data security are too narrow). A more expansive characterization of cybersecurity is called for, one which does not exclude the informational use of the internet for purposes of incitement, defamation, cyberbullying, cognitive warfare, undermining of democratic processes, and so on. Here we need to distinguish between the mode of attack from the nature of the target. By definition, cyberattacks use cyber means (e.g., malware, bots and the internet). However, the target of the attack (e.g., a person being bullied on-line), might be a human being (rather than, say, computer data or computer software) and the means of security might not be a cyberdefence mechanism per se but rather one of the many longstanding, precyber, human defences (e.g., to verbally attack the would-be cyberbully, albeit using the internet to do so). The more expansive characterization of cybersecurity might be couched in terms of the security of cyberspace, including of the

[2] Seumas Miller, *Social Action: A Teleological Account* (New York: Cambridge University Press, 2001) Chapters 4 and 6; Seumas Miller, *The Moral Foundations of Social Institutions: A Philosophical Study* (New York: Cambridge University Press, 2010) Chapters 2 and 3.

informational, communicative, and financial activities conducted in cyberspace. Moreover, cybersecurity is evidently a necessary aspect of the overall security of many potential targets of cyber- and other modes of attack. Consider the overall security of nuclear power plants generating electricity. The security of a nuclear power plant could be threatened by a cyberattack and may require, therefore, extensive cybersecurity measures to be in place. However, the security of the nuclear power station could also be threatened by non-cyber-based means, such as an explosive device placed in the reactor. Our concern in this work is with cybersecurity, but the potential object of a cyberattack is not restricted to computerized data, software, or other forms of cybertechnology. Moreover, those conducting cyberattacks can in so doing have as their primary intention to cause harms to persons and institutions, as well as to access, steal, or destroy computer data, programs, or other forms of cybertechnology. At any rate, in this work we will use a more expansive characterization of cybersecurity (see Chapter 1, Section 1.1.2).

As with security more generally, cybersecurity has given rise to a plethora of ethical concerns, the addressing of which in each case involves not only the identification of the facts but also the analysis and application of ethical principles and values, notably core principles and values of liberal democracy, such as privacy, autonomy, and freedom of communication.[3] Who is responsible for what in institutional and technological contexts in which the weakest link can be exploited in a manner that puts everyone at risk, such as in the context of the Internet of Things (IoT)? Increased security that relies on, for instance, bulk metadata collection and analysis may threaten core liberal democratic values, such as control of personal data and privacy. On the other hand, protection of these same core values, such as privacy protection afforded by encryption, may undermine security by, for instance, impeding legitimate criminal investigations by law enforcement agencies, including (but not restricted to) cybercrimes, such as cybertheft, identity theft, cyberfraud, cybersabotage, and online child pornography. Again, cybertechnologies, notably social media platforms, have an important new role to play in relation to public communication. Thus far, however, this has come hand in glove with an explosion in disinformation, propaganda, and hate speech that is undermining norms of evidence-based truth-telling, and even the response to emergencies, like the COVID-19 pandemic, thereby

---

[3] For a useful introduction to a range of these issues, see Markus Christen, Bert Gordijn and Michele Loi, ed., *The Ethics of Cybersecurity* (Dordrecht: Springer, 2020).

giving rise to security concerns.[4] The rise of so-called big tech—that is, companies like Meta (the parent company of Facebook), Apple, Alphabet (the parent company of Google), and Amazon—has revolutionized marketing and advertising, but at a significant cost to individual privacy and, relatedly, autonomy. New and emerging forms of cyberconflict give rise to questions about justified self-defence and deterrence. What principles ought to govern cyberconflict? Presumably, the principles of necessity and of proportionate response that apply to the waging of conventional wars ought also to apply to cyberconflict. If so, in what forms and under what conditions are they to be applied?

Revelations in 2018 that the data firm Cambridge Analytica had acquired Facebook data and utilized machine-learning (ML) techniques to build psychological profiles of voters, which could potentially be used to influence elections in the US and elsewhere, illustrate a further ethical dimension of certain breaches of cybersecurity—namely, the threat to democratic processes posed by so-called computational propaganda (Chapter 3).[5] In this instance, the security of personal information was compromised, and privacy rights violated. However, it was the subsequent misuse of this information to manipulate voters, and thereby influence democratic processes that is perhaps of greatest concern.

This work is concerned with cyberethics and, specifically, ethical issues that arise in the context of cybersecurity. The nature and extent of cybersecurity threats depends in part on the institutional context and, crucially, on the cybertechnology in play. Moreover, here as elsewhere, addressing practical ethical problems is not simply a matter of the mechanical application of high-level ethical principles (e.g., the application of the principle of utility relied on by utilitarian theories of ethics). A further assumption is that public policy in this area needs to be informed not only by technical, legal, and political input, but also by ethical input. For instance, there is a need for analyses and moral weighing of values, such as privacy and confidentiality. Again, the catch-all notion of security needs to be conceptually unpacked, and its components analysed and given appropriate moral weight. Accordingly, unlike many other works in cybersecurity, this book spends a good deal of time analysing the underlying ethical principles and

---

[4] Dean Cocking and Jeroen van den Hoven, *Evil On-line* (Hoboken: Wiley-Blackwell, 2018); Michael Lynch, *The Internet of Us* (New York: Liveright, 2016).

[5] Samuel Woolley and Philip Howard, ed., *Computational Propaganda* (Oxford: Oxford University Press, 2019).

values that are in need of being secured against cyberattacks and those in play in the response to cyberattacks. For instance, as will be argued in this work, the response to cybersecurity concerns is in large part the discharging of collective moral responsibilities. Accordingly, there are questions as to how these collective responsibilities are to be understood and how they can be institutionally embedded, utilizing cybertechnology as required. In this work, we operate with a theory of collective moral responsibility, and of institutionally embedded collective responsibilities, that has been developed by Miller elsewhere.[6] The notion of institutionally embedded collective responsibilities presupposes a normative theory of institutions and of institutions that in pursuing their institutional purposes are mediated, shaped, and to some extent constituted by technology. On this normative conception of technology, cybertechnology is to be primarily understood in terms of its contribution to institutional purposes, which in turn directly or indirectly serve prior human purposes. The normative theory of institutions favoured here is Miller's normative teleological theory,[7] according to which institutions have the provision of collective goods as their raison d'être, which includes the collective good of cybersecurity. Moreover, the concept of collective responsibility that follows from this theory is an individualist one. Collective moral responsibilities are *jointly held*, individual moral responsibilities, as opposed to responsibilities that might attach to the collective entity (e.g., the Australian government) per se or to each individual (e.g., each member of the Australian government) independently of being attached to the other individuals[8] (see Chapter 4, Section 4.3.5; Chapter 5, Section 5.3; and Chapter 7, Section 7.1.1 for elaboration of the concept of collective responsibility as joint responsibility).

Because practical ethical problems, including many public policy issues, have an ethical—or moral (we use these terms interchangeably in this work)—dimension, as well as a scientific or technological dimension, it has become increasingly clear that professional philosophers have a major role to play in the clarification and resolution of these problems.[9] The ethical dimension is in need of systematic analysis and illumination by means

---

[6] Seumas Miller, "Collective Moral Responsibility: An Individualist Account," in *Midwest Studies in Philosophy*, vol. XXX, ed. Peter French (2006): 176–93; Miller, *Moral Foundations of Social Institutions*.

[7] Ibid.

[8] Miller, "Collective Moral Responsibility: An Individualist Account."

[9] See Jeroen van den Hoven, Seumas Miller, and Thomas Pogge, ed., *Designing-in-Ethics* (New York: Cambridge University Press, 2017) for examples of this.

of philosophically-based moral theories and perspectives. This is not to say that moral philosophers are moral experts who can be left in charge of moral problems. Far from it. Individually and collectively, whether in the private realm or the public arena, we must all take responsibility for the resolution of the moral problems that confront us. However, in relation to many of the more complex of these problems, moral philosophers have an important and distinctive intellectual contribution to make.

Here, to reiterate, it is not simply a matter of philosophical theory being mechanically applied to specific problems. Rather, there is a complex interplay between theoretical perspectives, on the one hand, and specific ethical intuitions and concrete scientific data, on the other. Whether or not ML processes applied to electronic databases constitute an infringement of the right to privacy or undermine democratic processes is partly a matter of figuring out what is important about privacy (i.e., the ethical theory of privacy) and democracy (i.e., the political philosophy of democracy), as well as knowing the scientific facts about the particular ML processes and databases in question. Of course, it might be a matter of banning certain practices (e.g., Cambridge Analytica's voter-profiling practices). On the other hand, in relation to other practices (e.g., targeted advertising of consumer products based on consumer preferences that have been disclosed, let us assume somewhat hypothetically given current practices, under conditions of informed consent) it may well be a matter of balancing the moral weight to be given to privacy against the benefits delivered by these databases in the specific contexts in question.

Thus, philosophical theory operates at several levels of abstraction. There are high-level theoretical claims, such as the principle of maximizing the satisfaction of the greatest number or seeking to benefit the least advantaged. But there are also lower-level philosophical theories of specific values, such as an ethical theory of freedom of the press, or of a specific professional role, such as the moral purposes and characteristic virtues of a national security intelligence analyst. These lower-level normative or value theories operate within specific institutional, occupational, and technological settings. They are context dependent. As such, they grow out of, and are highly sensitive to, context-dependent considerations.

The book has seven substantive chapters and a conclusion. Each substantive chapter is largely self-contained to enable readers to go directly to the ethical issues of interest to them without having to read material on other issues. Chapter 1 provides an overview of cybersecurity threats, countermeasures to cyberattacks, and the institutional landscape in which such threats and countermeasures exist. It can be skipped by those readers who already have a reasonable knowledge of these matters. In the book's conclusion, the ethical guidelines developed in each of the chapters are set forth in summary form. These guidelines are intended not only to assist those interested in solutions to the various ethical problems discussed, but more specifically to provide potential direction to cybersecurity policymakers.

Chapter 1 begins with a discussion of the definition of cybersecurity. It goes on to describe some of the main cyberthreats (e.g., viruses, ransomware) and cybersecurity responses to those threats (e.g., firewalls, encryption), and outlines the institutional landscape (e.g., internet governance bodies such as ICANN, global technology companies, security agencies) in which these threats, and the countermeasures taken in response to them, take place.[10] In relation to the definition of cybersecurity, it distinguishes the narrow notion of data security from a broader notion that includes security issues consisting of, or arising from, disinformation on social media platforms, incitement, and so on. This chapter also introduces a key normative theoretical framework to be used throughout this work—namely, that of joint (cooperative) action in the service of a collective good. An example of a collective good is security, in the sense of law and order. Security in this sense is realized by cooperative or joint action by citizens and police officers. Thus, the police work cooperatively with one another, but they also rely on the citizens (e.g., to report crime). Moreover, this cooperative or joint action is performed in accordance with a collective responsibility of police officers and citizens to maintain law and order. Further, this collective responsibility is discharged in a particular institutional and technological setting.

Chapter 2 addresses privacy and confidentiality issues arising from bulk data, surveillance, and encryption. It provides analyses of the key ethical notion of privacy (and related notions of autonomy, confidentiality, anonymity, and secrecy) and of security. It looks at bulk databases, and the associated use of ML techniques, by governments and their security agencies, on the one hand, and by market-based global technology companies, such as Google, on the other hand. It also addresses the ethical issues arising from the use of high-level or strong encryption (i.e., encryption that is difficult or impossible

---

[10] See Terry Bossomaier, Steven D'Alessandro, and Roger Bradbury, *Human Dimensions of Cybersecurity* (Boca Raton, FL: CRC Press, 2019) for details of measures to assure cybersecurity in relation to data—measures that emphasize the human factor as a risk factor.

to break). In considering privacy rights and security needs in these contexts, and in offering some ethical guidelines to inform policies in this area, it makes use of a range of ethical principles, such as that of necessity and proportionality. Is the use of an intrusive surveillance technology, such as facial recognition technology, necessary and proportionate in relation to the criminal activity or other circumstances in question?

The focus of Chapter 3 is freedom of public communication and, more specifically, freedom of political communication in the context of the exponential increase in disinformation, propaganda, and hate speech on the internet, which is turbo charged by a largely unregulated social media in which social bots, AI, and the like are used (i.e., computational propaganda). On the one hand, there is a presumption in favour of freedom of communication and of political communication, in particular. Indeed, it is a fundamental human right and a cornerstone of liberal democracy. On the other hand, the right to freedom of communication brings with it a moral responsibility to comply with principles of truth telling and evidence-based reasoning. How can disinformation, propaganda, and hate speech be countered without compromising the right to freedom of communication? This chapter provides ethical analysis of key concepts, such as freedom of communication, disinformation, and ideology. It also makes various practical recommendations for policymakers and regulators with an emphasis on the role to be played by epistemic institutions, such as news media organizations and universities, in the context of a more effectively regulated cyberspace.[11]

Chapter 4 concerns a range of practical ethical issues that have arisen in the criminal justice context. In Chapter 1, we emphasize the extent of the crime and related security problems that have arisen in large part because of the advent of cybertechnology and related technologies. However, these new technologies have also resulted in the development of a whole range of new law enforcement tools. In this chapter, we address some of the key ethical problems that have arises as a result of the use by law enforcement of bulk databases, ML techniques, and other related technologies. It looks at these issues through the lens of the moral values, rights and principles constitutive of liberal democracy, such as privacy/autonomy, but also less familiar ones such as personal identity. One's DNA and face are constitutive of one's personal identity and, therefore, other things being equal, DNA profiles and facial

---

[11] Seumas Miller, "Freedom of Political Communication, Propaganda and the Role of Epistemic Institutions in Cyberspace," in *The Ethics of Cybersecurity*, ed. M. Christen, B. Gordjin, and M. Loi (Dordrecht: Springer, 2020).

images ought to be under one's control. The ethical issues addressed include ones in predictive policing, ML-based legal adjudications, universal DNA databases, and facial recognition technology. The chapter concludes with a set of ethical guidelines for resolving some of the main issues in this area.

In Chapter 5, we turn to security issues in the health sector. These include ransomware attacks, such as the WannaCry attack in 2017 on the UK National Health Scheme, that have the potential to cripple hospitals and other medical facilities, thereby causing deaths. The chapter has a particular focus on pandemics, such as COVID-19, and specifically on the health and related security benefits, but also on the moral costs, arising from the use of cybertechnology in combating pandemics. Thus, cybertechnology, such as phone apps and bulk health databases utilizing ML techniques, can mitigate health and related security problems arising from COVID-19. However, such use of cybertechnology has moral costs, notably in terms of infringements of the privacy, autonomy, and 'ownership' rights of individuals. In this chapter, these issues are framed in terms of collective goods (e.g., such as public health and security) versus individual rights. It is argued that the collective goods in questions often override the infringements of the individual rights involved. However, the chapter also makes some suggestions as to how to mitigate such infringements, including in the light of new technological developments such as privacy-preserving health mining software.

Chapter 6 discusses cyberconflict between nation-states and the use of cyber weapons, specifically so-called autonomous weapons.[12] It argues that cyberconflict, including cognitive warfare, is characteristically—although not necessarily—a species of covert political action.[13] Moreover, cyberconflict can be conducted in tandem with conventional warfare, as is taking place in Russia's invasion of Ukraine. If so, it might or might not constitute covert political action. As with other forms of conflict, cyberconflict should be conducted in accordance with moral principles, such as discrimination (according to which one should avoid the deliberate harming of innocents) and necessity and proportionality—although the application of these principles with respect to nonkinetic action (e.g., cognitive warfare) and in

---

[12] George Lucas, *Ethics and Cyber War* (Oxford: Oxford University Press, 2017); David Sloss, *Tyrants on Twitter* (Stanford: Stanford University Press, 2022); Seumas Miller, *Shooting to Kill: The Ethics of Police and Military Use of Lethal Force* (New York: Oxford University Press, 2106), Chapter 10, "Autonomous Weapons and Moral Responsibility."

[13] Seumas Miller, "Cyber-Attacks and 'Dirty Hands': Cyberwar, Cyber-Crimes or Covert Political Action?," in *Binary Bullets: The Ethics of Cyberwarfare*, ed. F. Allfhoff, A Henschke, and B J. Strawser (Oxford: Oxford University Press, 2016), 228–50.

peacetime differs from their application to kinetic action in wartime (i.e., "killing people and breaking things"). Moreover, in this chapter, it is argued that a principle of reciprocity in the service of the collective good of global cybersecurity has application to cyberconflict. Regarding the use of an important category of so-called autonomous weapons, namely human out of the loop weapons, it is argued that they should be prohibited in most, if not all, circumstances.

In Chapter 7, the final chapter of this work, it is argued that the ethical issues in cybersecurity can in large part be framed as dual-use issues. Roughly speaking, cybertechnology can be used to achieve great benefits, but in the hands of malevolent state and nonstate actors, it can cause great harm. The satisfactory resolution of these issues is, therefore, typically complex. Moreover, it involves collective moral responsibilities that need to be institutionally embedded. The chapter elaborates conceptual underpinnings of these claims, including the notions of dual-use technology[14] and of an institutionally embodied collective moral responsibility.[15] Institutionally embodied collective moral responsibilities are manifest in what we refer to as webs of prevention.[16] It identifies computer viruses, autonomous robots, encryption, blockchain, and facial recognition technology as potential dual-use technologies. Following on Chapter 7, there is the conclusion which, as mentioned above, consists in a summary of the main ethical guidelines argued for in this work.

---

[14] Seumas Miller, *Dual Use Science and Technology, Ethics and Weapons of Mass Destruction* (Dordrecht: Springer, 2018).

[15] Seumas Miller, *Institutional Corruption: A Study in Applied Philosophy* (New York: Cambridge University Press, 2017) Chapter 6.

[16] A term we have borrowed from the ethics of biosecurity. See, for instance, Brian Rappert and Caitriona McLeish, *A Web of Prevention: Biological Weapons, Life Sciences and the Governance of Research* (London: Routledge, 2012).

1

Cybersecurity

Threats, Countermeasures, and the Institutional Landscape

1.1 What is Cybersecurity?

1.1.1 Cyberspace and the Internet

Cybersecurity pertains to security in cyberspace. According to the US government cyberspace is defined as follows:

> Cyberspace means the interdependent network of IT infrastructures, and includes the Internet, telecom networks, computer systems, and embedded processors and controllers in critical industries. Common usage of the term also refers to the virtual environment of information and interactions between people.[1]

Klimburg usefully distinguishes between four generic layers of cyberspace:[2] (1) the physical or hardware layer, consisting of computers, cables, and communicating equipment, as well as the switching technology that connects them, and the like; (2) the logic layer, consisting of the computer protocols, software programs, and the like; (3) the data layer, consisting of the emails, documents, and so on that are understandable by the average human being possessed of a relevant natural language, such as English or Mandarin; (4) the social layer, consisting of the human persons and their actions in cyberspace. Each of these four generic layers admits of further analysis. Indeed, each of these generic layers itself consists of more fine-grained, specific layers. Interestingly, layers (2) and (3) loosely correspond

---

[1] Quoted in Alexander Klimburg, *The Darkening Web: The War for Cyberspace* (New York: Penguin Press, 2017), 27.
[2] Ibid, 28–29.

Cybersecurity, Ethics, and Collective Responsibility. Seumas Miller and Terry Bossomaier, Oxford University Press.
© Oxford University Press 2024. DOI: 10.1093/oso/9780190058135.003.0002

to what Karl Popper refers to as the third world of objects—the first world consisting of physical objects and the second world of mental objects.[3] That is, the third world consists of abstract objects, albeit (at least in Klimburg's taxonomy) abstract objects that are physically embodied and invested with semantic properties that are interpretable by humans possessed of the relevant formal or informal language (e.g., a book in a library, a mathematical formula, or an email). We do not need to endorse any specific philosophical ontological theory here, let alone that of Popper, since the intention is merely to make a prima facie useful set of distinctions. At any rate, cyberattacks can directly target all four generic layers. However, we suggest that unless layer (2) (logic) or layer (3) (data) are directly or indirectly attacked (or perhaps are 'collateral damage') then it is not a cyberattack per se. Thus, hacking into an organization's web server to steal intellectual property is a cyberattack, but so is posting hate speech online (albeit, it is a verbal attack). However, beating up a 'geek' for fun is not a cyberattack, nor is stealing a defunct computer to sell its parts for noncomputing purposes.

We can think of the internet as a physical structure (generic layer 1) of computers and connectors. *Computers* here include mobile phones and other specialized devices and connectors, including radio (e.g., Wi-Fi). Or we can think of the internet in terms of its abstract character (generic layers 2 and 3)—or in terms of both. Whichever way one looks at it, the internet is hugely complex. It comprises a vast number of *nodes*, each with its own unique IP (Internet Protocol address). For the purposes of this book, a node can be thought of as any machine connected to the internet—from phone to supercomputer—although this is a slight simplification.

A fundamental concept needed to understand the internet is that of a *protocol*, a procedure or specification for how something is done. Communication then involves multiple *specific layers* (situated within, for instance, generic layer 2), each with its own control protocols. Thus, there are protocols for sending an email (e.g., SMTP, or Simple Mail Transfer Protocol), transferring a file (e.g., FTP, or File Transfer Protocol), or for encryption (e.g., TLS, or Transport Layer Security Protocol—formerly known as Secure Socket Layer).

The number of internet nodes is already around a trillion and is set to increase dramatically with the growth of the Internet of Things (IoT). The *things* in question are televisions, fridges, solar panels, and the like. Information

---

[3] See Karl Popper, *Objective Knowledge* (Oxford: Clarendon Press, 1972), Chapter 4.

in the form of data packets can be routed from any one of these nodes to any other.

The internet with almost a trillion nodes is not only hugely complex; it is a remarkable informational and communicative architecture. From a remote location in Africa or Australia, web sites in New York, US, or Tokyo, Japan, can be accessed, and vice-versa. This can be done because each node has a unique, multicomponent, numerical address, such as 137.166.37.171. The internet needs a way of labelling every node on the network. It does this through a hierarchical structure of IP addresses.

The basic activity of the internet is the transfer of *packets* of data between nodes, using a specification known as the IP, the Internet Protocol. Hence such addresses are known as IP addresses. Remembering such addresses is hard work, which is, fortunately, unnecessary. Domain Name Servers (DNSs) map names (e.g., "google.com") to internet addresses, making access to websites much easier.

The last part of the domain address name indicates the country (e.g., *.au*), or some other special domain (e.g., *.biz* for business and *.org* for nonprofits). Domains are placed before the country code (e.g., *.edu* for universities and *.gov* for government). Routing a packet involves essentially working backwards from the country or special domain.

There is a downside to this remarkable global connectedness. In October 2016, several huge companies had their web services brought to a standstill by a major hack of the service DYN, which provides Domain Name Services (DNS). Companies affected included: media companies such as the Guardian, CNN, and Spotify; social media providers, such as Twitter; and GitHub, the world's major platform for software archiving. All these services were disrupted because the attack was launched on DYN (a company which provided domain name services), which manages domain-name servers for many companies. Attacking DYN effectively blocked access to all the domain names (websites) that it hosts.

The attack on DYN was a Distributed Denial of Service (DDoS) attack using the worm Mirai.[4] A DoS attack disrupts a website by bombarding it with a huge number of requests, too many for it to handle, thus causing it to crash. The source code of Mirai became available on the web, allowing

---

[4] Worms are a form of computer virus. Viruses are self-replicating computer programs that install themselves in computers without necessarily having the consent of the user. Worms look for other computers onto which to copy themselves. A worm is not necessarily harmful; it depends on what it is programmed to do.

further such attacks to occur with little knowledge or effort. The source code contains various strings in Russian, which suggests a possible geographic source. The Mirai and other high profile cyberattacks graphically illustrate the need for cybersecurity measures.

1.1.2 Characterizing Cybersecurity

An initial distinction must be made between cybersecurity and cybersafety. Our concern in this volume is with cybersecurity. Security and safety are not the same thing, and therefore cybersecurity and cybersafety are not the same thing, although they are sometimes conflated. *Security* presupposes an individual or organization with an *intention* to do harm. *Safety* does not presuppose such an intention. Thus, the potential for accidental harm is a safety issue, whereas the potential for intended harm is a security issue. We note, however, a grey area—namely, culpable negligence. Arguably, culpable negligence should be regarded as a security issue, notwithstanding the absence of an intention to do harm.[5] At any rate, in this work we will treat culpable negligence as a security issue, rather than merely a safety issue.

Another initial distinction—or, rather set of distinctions—pertains to the concept of security. For our purposes here, it is useful to distinguish the following (overlapping) categories of security:[6] (1) information security (e.g., of credit card numbers and personal data) in respect of confidentiality, integrity (e.g., compromised by unauthorized modification of data), and availability (e.g., lost because of unauthorized destruction of data); (2) cyber systems (software) security (e.g., of authentication procedures); (3) physical security (e.g., of critical infrastructure that relies on cybertechnology, or devices which could endanger life such as driverless cars); (4) institutional security (e.g., protection against foreign political interference); (5) moral rights security (e.g., protection against violation of individual rights to privacy and autonomy); (6) psychological security (e.g., freedom from fear of cyberbullying); (7) security of security agencies, such as computer emergency response teams (CERTs), police, intelligence, and military agencies

---

[5] Miller, *Dual Use Science and Technology, Ethics and Weapons of Mass Destruction*, 9. To simplify matters, we include *culpable recklessness* as a species of culpable negligence.

[6] For instance, a breach of information security might also be a breach of moral rights security, if the information in question is sensitive personal data.

(e.g., compromised by infiltration of police, or identity theft of names/ locations of spies or undercover operatives).

The term *cybersecurity* has been variously defined. For instance, Kosseff[7] offers this useful, if somewhat familiar, definition from a legal standpoint:

> Cybersecurity law promotes the confidentiality, integrity, and availability of public and private information, systems, and networks, through the use of forward-looking regulations and incentives, with the goal of protecting individual rights and privacy, economic interests, and national security.

This definition is couched, as many are, in terms of the confidentiality, integrity, and availability of information. However, it has a teleological or goal-focused dimension in that it defines cybersecurity in part by reference to the goals or ends of protecting individual rights, economic interests, and national security. Its invocation of some of the main goals of cybersecurity is, we suggest, a strength. It is a strength in that it allows cybersecurity to consist in part in, for instance, protecting privacy, while, nevertheless anchoring it in the protection of information transferred by, or stored, in computers, like devices, or cyberspace, more broadly. So, protecting property rights might or might not be an example of cybersecurity. Thus, a person protecting their wad of cash from being stolen by hiding it under a mattress is not an example of cybersecurity, whereas using a password to ensure that their bank account is not hacked and the money stolen is an example of cybersecurity. Moreover, it should be noted that these goals (i.e., individual *rights*, economic *interests*, and national *security*) are moral or ethical in character—which is not to say that instances of legalized property rights, economic interests, and proffered national security justifications are necessarily morally legitimate. And, of course, one could supplement this definition by recourse to the assemblage of technical devices and related practices (e.g., firewalls and password secrecy) that promote informational security but might not be subject to laws or regulations.

Definitions of cybersecurity purely in terms of information security and information systems security—and, therefore, couched in terms of the confidentiality of information (e.g., potentially breached by data theft), integrity of information (e.g., potentially breached by unauthorized data changes), and availability of information (e.g., potentially breached by unauthorized

---

[7] Jeff Kosseff, "Defining Cybersecurity Law," *Iowa Law Review* 103 (2018): 1010.

data destruction)[8]—are correct insofar as they exclude the type of cases mentioned above; they exclude beating up 'geeks' and stealing defunct computers. However, these definitions are too narrow in two main respects.

Firstly, the definitions in question are too narrow insofar as they exclude information possessed by someone who is not entitled to that information but who, nevertheless, wants to exclude others from accessing, corrupting, or destroying it. The possessor of such information may not have privacy or ownership rights to it; nor is his or her possession of this information necessarily governed by any other relevant moral principle (e.g., of confidentiality). Rather the possessor simply wants to retain the information in its intact form and keep it *secret* from others. Here we need to distinguish confidentiality from secrecy (for further discussion of the distinction between confidentiality and secrecy, see Chapter 2, Section 2.1.2). Moreover, by preventing others from accessing, corrupting, or destroying the information, its possessor is *securing* the information. The possessor is engaged in an exercise of cybersecurity. Accordingly, we need to broaden the notion of cybersecurity to include the mere prevention of access to possessed information. We do so in order that cybersecurity (and security, more generally) does not in all circumstances connote a moral good (e.g., in terms of confidentiality or theft), which is does in narrower definitions. This is not to say that the notion of cybersecurity (and security, more generally) is not necessarily a moral good in the minimal and derivative sense that it implies defence against an attack, since the harm or damage inflicted by a successful attack is morally problematic.

Secondly, these revised definitions are still too narrow insofar as they exclude the informational use of the internet for purposes of incitement, defamation, cyberbullying, cognitive warfare, undermining of democratic processes, and so on—as opposed to, for instance, purposes of legitimate communication between friends or colleagues. Accordingly, it might be best to distinguish between the revised, narrow notion of data security, defined in terms of prevention of access, confidentiality, integrity, and availability of information, and the broader notion of cybersecurity that helps itself to the notion of cyberspace and, more specifically, the security of cyberspace.

A further distinction to be made is that between cybersecurity per se and closely related forms of security—namely, forms of security dependent on cybersecurity. Thus, cybersecurity consists of the security of the

---

[8] Christen et al., *The Ethics of Cybersecurity*, 12–13.

informational, communicative, financial, and so on activities conducted in cyberspace. Accordingly, the security of a nuclear power plant, for instance, is not necessarily an instance of cybersecurity since its security could also be threatened by a non-cyber-based attack, such as by exploding a bomb in it, thereby releasing radioactive material into the environment. That said, the security of a nuclear power plant could be threatened by a cyberattack on its control system, which might intentionally or foreseeably result in the release of radioactive material into the environment. If so, the security of the nuclear power plant's control system would be a cybersecurity issue. However, the security of the nuclear plant in respect of its release of radioactive material into the environment would in this instance presumably be dependent on a cyberattack initially focussed on its computer-based control system and, specifically, on the logic and/or data layer of that system (see above).

A further point pertains to the so-called weaponization of data. The term *weaponization* used in relation to data is problematic insofar as it conflates epistemic concepts with kinetic concepts. Epistemic concepts pertain to knowledge, beliefs, statements (including false ones) and the like, whereas kinetic concepts pertain to essentially physical actions and events. Of course, many kinetic actions have an epistemic aspect (e.g., intentional actions involve beliefs) and many epistemic actions have a physical manifestation (e.g., many beliefs are vocalized). However, this does not extinguish the conceptual distinction. Moreover, here as elsewhere, conflation may well obscure rather than assist. More specifically, it may be unhelpful to conflate the use of weapons that kill people and destroy buildings with communications that seek to change beliefs or otherwise induce changes in psychological attitudes—even if the latter are reasonably described as verbal attacks or are done so as to indirectly (via multiple persons) cause kinetic changes, as when disinformation is deployed in a kinetic war. From the perspective of ethics, the conflation is especially unhelpful since the ethical principles pertaining to epistemic actions (e.g., Don't tell lies) are importantly different from those pertaining to kinetic actions (e.g., Don't kill innocent persons). From the perspective of politics—and therefore, ethics, indirectly—this conflation often serves ideological purposes. For instance, the claim that one's political opponents are weaponizing data tends to assimilate their activity to kinetic activity such as bombing raids, thereby inflating the seriousness of their moral offence and potentially justifying a kinetic response. Indeed, some theorists have, in our view, unhelpfully sought to apply Just War Theory—a normative theory developed in relation to kinetic wars—to the essentially epistemic activity of intelligence agencies.[9] But surely information theft or disinformation per se does not typically justify waging a kinetic war, although in theory it might (Chapter 6, Sections 6.1.2 and 6.2).

An important initial task in relation to cybersecurity is to identify and characterize the main cyberthreats, some of which are obvious (e.g., malicious viruses, denial-of-service attacks), some of which might not be (e.g., disinformation campaigns—see Chapter 3.1). Another task is to identify the main sources of these threats (e.g., cybercriminals or nation-states). A third and final task is to develop cybersecurity responses to those threats. Obviously, cybersecurity measures need to be responsive to cyberthreats. However, security is often presented as essentially a technical issue, one requiring passwords, firewalls, encryption, blockchain technology, and so on. Certainly, technology lies at the heart of cybersecurity concerns. However, as the Edward Snowden disclosures graphically illustrated,[10] the human factor[11] is crucial to cybersecurity, and cybersecurity has a fundamental moral dimension. This moral dimension is reflected in criminal law. In cybersecurity, as elsewhere, criminal law typically expresses moral beliefs.[12] Yet in respect of cybersecurity, criminal law and regulation have often lagged the moral concerns (see Chapter 4). Importantly, cybersecurity also has an institutional dimension, as well as its technical and ethical dimensions. Indeed, the ethical issues arising from cybersecurity are embedded in sociotechnical systems in which technology is embedded in institutional and, more broadly, social arrangements.[13] It is these ethical issues that are the main focus of this work.

1.1.3 Cybersecurity: The Key Actors

In relation to the key actors, there are several points to be made. Firstly, there is a need to outline these key actors, as well as the regulatory and governance

---

[9] Ross Bellaby, *The Ethics of Intelligence* (London: Routledge, 2014). For criticisms, see Seumas Miller, "Rethinking the Just Intelligence Theory of National Security Intelligence Collection and Analysis," *Social Epistemology* 35, no. 3 (2021): 211–31.

[10] Patrick Walsh and Seumas Miller, "Rethinking 'Five-Eyes' Security Intelligence Collection Policies and Practices Post 9/11/Post-Snowden," *Intelligence and National Security* 31, no. 3 (2016): 345–68.

[11] Bossomaier et al., *Human Dimensions of Cybersecurity*.

[12] Seumas Miller and Ian Gordon, *Investigative Ethics: Ethics for Police Detectives and Criminal Investigators* (Oxford: Wiley-Blackwell, 2014), Chapter 1.

[13] Seumas Miller, "Design for Values in Institutions," in *Handbook of Ethics, Values & Technological Design*, ed. I. Poel, J. Van den Hoven, and P. Vermaas (Dordrecht: Springer, 2015), 769–81.

arrangements, relevant to the ethical issues in cybersecurity addressed in this volume. (We consider regulatory and governance arrangements in 1.5 and 1.6 below, as well as at various point throughout this work.) Notable such actors are national governments and their security agencies, such as the National Security Agency (NSA) in the US, the Government Communications Headquarters (GCHQ) in the UK, and the People's Liberation Army (PLA) of China's Unit 61398; global technology companies, such as Apple, Alphabet (parent company of Google), and Meta (formerly Facebook); and civil society organizations, such as universities, NGOs, and loosely organized social media groups and movements that use the internet's public communication platforms to propagate a common viewpoint or ideology (e.g., the Black Lives Matter movement). Moreover, there are the commercial and other organizations that might utilize (with or without consent) the vast amounts of data collected by the above-mentioned companies.

In addition, there are those actors who are the sources of cyberthreats to individuals, private companies, public sector agencies, critical infrastructure, and so on. These sources include, most obviously, criminals and criminal organizations, as well as terrorist groups and other extremist political organizations. As noted in the introduction, cybercrime is evidently now to be measured in terms of trillions of US dollars.[14] These sources of cyberthreats also include malevolent individuals and groups whose harmful actions might not constitute crimes. Importantly, and worryingly, one of the main sources of cyberthreats are national governments, such as Russia in the context of its February 2022 invasion of Ukraine. Moreover, they may include private companies themselves, such as Cambridge Analytica. In some cases companies have pursued profits at the expense of ensuring appropriate security or safety measures, leading to loss of life as a result of, for instance, flawed software. Consider the example of the Boeing 737MAX crashes. The first Boeing 737MAX crashed in Indonesia in October 2018, killing 189, and another crashed five months later in Ethiopia, killing 157. All Boeing 737MAX jets were grounded worldwide for nearly two years. One of the causes was badly designed software.[15]

In relation to the regulatory and governance arrangements, we can distinguish between two sets of institutions, regulations, and norms. On the

---

[14] Steve Morgan, "Cybercrime to Cost the World $10.5 Trillion Annually by 2025.".

[15] Charles Bramesco, "'All Those Agencies Failed Us': Inside the Terrifying Downfall of Boeing," *The Guardian*, 23 February 2022. https://www.theguardian.com/film/2022/feb/22/downfall-the-case-against-boeing-netflix-documentary-737-max.

one hand, there are the institutions, regulations, and norms that govern the technical infrastructure of cyberspace, including but not restricted to the internet, such as the numerous organizations which develop protocols and standards—notably, IEEE (Institute of Electrical and Electronic Engineers) and ACM (Association for Computing Machinery), as well as internet-specific entities such as ICANN (Internet Corporation for Assigned Names and Numbers), and IETF (Internet Engineering Task Force). On the other hand, there are the institutions, regulations, and norms that govern the informational uses of the technical infrastructure for economic, human rights, law enforcement, and other legitimate purposes. Thus, the EU has recently introduced legislation to protect data privacy (*inter alia*)—notably, the GDPR (General Data Protection Regulation) (see Chapter 2, Section 2.2.2, for discussion of these issues).[16]

The concept of security and, therefore, cybersecurity implies a distinction between attackers (or other perpetrators), those attacked or threatened (victims), and defenders (security personnel). (In addition, there are, of course (see below), the 'weapons' used in the attack (e.g., malware), the direct objects of the attack (e.g., the software controlling infrastructure), and the defences against an attack (e.g., AI-driven software). However, the lines between attackers and defenders are not necessarily clear-cut. For instance, an offensive measure might be a form of defence. Accordingly, the lines between the perpetrators and the victims are not necessarily clear-cut. National governments of both an authoritarian and a liberal democratic persuasion evidently engage in cyberattacks. Consider, on the one hand, the above-mentioned Russian cyberattacks on Ukraine and China's cyberespionage (including industrial espionage), as reported by the cybersecurity firm Mandiant; and, on the other hand, the US's alleged Stuxnet attack on Iranian nuclear facilities (Chapter 6, Section 6.1). Moreover, there are grey areas in relation to issues of culpability. For instance, the UK Security Minister called technology companies "ruthless profiteers" for refusing to allow access to encrypted messages, thereby forcing the UK to spend millions of pounds on extra human surveillance.[17] These companies have hosted terrorist propaganda and been

---

[16] Regulation (EU) 2016/679 (General Data Protection Regulation) (GDPR). https://gdpr-info.eu/.
[17] H. Mance and A. Ram, "British Minister Calls Silicon Valley Giants 'Ruthless Profiteers,'" *Financial Review*, 2 January 2018. https://www.afr.com/technology/british-minister-calls-silicon-val ley-giants-ruthless-profiteers-20180101-h0byun.Accessed 25/10/2023.

accused of collateral damage to media and democracy for allowing "fake news" to circulate on social media—including "news" that may constitute foreign interference in elections in the US and elsewhere. Further, ideologically motivated groups of various persuasions have engaged in widespread cyberbullying and attempted censorship of those who express views antithetical to their own. In some cases, the goals of these groups might in themselves be morally worthy. However, there often remains a moral problem with the means used to realize their ends.

Another important issue is the relationship between nation-states, technology companies, and other institutional actors in in respect of internet infrastructure, electronic communication. and data flows. For instance, a huge proportion of internet activity is routed through the US, partly because the five big tech companies—namely, Alphabet, Meta, Amazon, Apple, and Microsoft—are domiciled there. This gives US security agencies, like the NSA, unprecedented access to metadata, for instance. Again, the so-called Great Firewall of China enables the Chinese government to limit the internet access of Chinese citizens.

In a book on cybersecurity and ethics it is crucial to consider the normative dimension of the institutional actors in cyberspace, including governments, internet providers, and so on, and in particular to explore in general terms the institutional relationships that ought to exist in cyberspace in relation to cybersecurity. Thus, we need to address the question: Who ought to be responsible for what security measures in cyberspace? A salient normative concept—or, rather, set of related concepts—here is that of public goods, collective goods, the so-called commons and common pool resources.[18] Arguably, the internet is or, at least, ought to be a collective good, at least in the minimal sense that it ought to be designed and regulated in a manner that ensures that it benefits everyone and does so more or less equitably. We return to this notion of a collective good (and related notions) in relation to cybersecurity throughout this volume (see Chapter 5, Section 5.2). Another salient concept, applicable at a more general level, is that of a social institution since, as we have seen above, many of the key actors are in fact institutions (see 1.6 and 1.7 below, and Chapter 7.1).[19]

---

[18] See, for instance, Elinor Ostrom, *Governing the Commons: The Evolution of Institutions for Collective Action* (Cambridge: Cambridge University Press, 1990).

[19] Miller, *Moral Foundations of Social Institutions.*

1.2 Cyberthreats

1.2.1 Cyberthreats and Cyberattacks

It is hard to overestimate the dependence of individuals and institutions on cybertechnology in modern societies. Consider the centrality and ubiquity of cybertechnology in the technological systems that run everything in a modern society, from transport (e.g., trains and planes) to energy (e.g., electricity grids and nuclear power stations), to security (e.g., criminal records and submarines), to finance (e.g., payment systems and stock exchanges), to health (e.g., medical records and infusion pumps). Now consider the interconnectedness of computers with one another via the internet, and especially the exponential increase of computing devices embedded in everyday things—from fridges to pacemakers—sending and receiving data over the internet (e.g., the IoT, or Internet of Things). As a consequence of this dependence of individuals and institutions on cybertechnology, successful cyberattacks have the potential to cause an extremely wide variety of direct (e.g., destruction of data on one's computer) and indirect (e.g., financial loss as a result of cybertheft of one's credit card number) harms—and, crucially, cause a very great magnitude of harm. At one end of the scale, a cyberattack might simply cause inconvenience by infecting an unimportant file with a virus that can be isolated and removed. At the other end of the scale, a cyberattack might cripple critical infrastructure (e.g., the electricity grid during a severe winter), causing the deaths of thousands. Moreover, it is useful at this point to distinguish between harms and damage, although these concepts overlap. An attack on physical infrastructure, such as severing a cable, does damage. However, it would be infelicitous to say that the cable was harmed. On the other hand, an attack of cyberbullying typically harms the victim. Human beings can be harmed in cyberattacks, but physical infrastructure can only be damaged. What of attacks on software and data per se? Strictly speaking, if data and software are abstract in character, then to that extent they are neither damaged nor harmed, although their physical embodiments can be damaged. However, they can presumably be corrupted (e.g., if tampered with via their physical embodiment), and, of course, this can result in their human users and recipients being harmed.[20]

---

[20] That is, as mentioned above, they belong to what Popper referredto as the third world of objects (i.e., abstract objects, albeit abstract objects that are physically embodied but also humanly

Here it might be helpful to distinguish a few different categories of malware (i.e., the malicious software frequently involved in cyberattacks).

A virus is a computer program that causes damage of some kind, such as a ransomware virus that encrypts a user's hard drive. Like a biological virus, it may use some of the system software on the host computer, allowing its own code to be very small and inconspicuous. The most common way a virus gets into a machine is by hitchhiking on the back of some other software—very commonly email attachments. It might arrive over the internet or be delivered on software provided on a physical device. One danger, not always recognized, is the risk of viruses on USB sticks handed out with demo and promotional material at events. In fact, the Stuxnet virus is thought to have been delivered by a rogue USB stick.

A virus may not cause any damage immediately. A variant, the malicious bot, is a software agent (program) with a particular role, which lurks on the infected computer until activated by some signal. Sometimes such bots are connected to other bots to form a botnet, which is often comprising hundreds of thousands of bots. These bots may not cause damage to the host machine but be used for attack on other machines. One frequent and very costly attack is a DDoS, a Distributed Denial of Service, in which bots flood a website with data packets that overload it—first reducing its response time and, often, ultimately crashing it completely.

Sometimes a virus might arrive on a worm, which is a software program that spreads autonomously throughout a network. Once it arrives on a machine it looks for other machines to which it is connected, and then replicates itself on them.

Obviously, cyberattacks can harm individual human persons and collective entities, such as groups, private- and public-sector institutions, nation-states. However, collective entities are presumably harmed (as opposed to damaged) by virtue of the harm done (directly or indirectly) to the human persons who comprise them, rather than to their physical or abstract properties.

Kosseff usefully uses the Sony attack to illustrate the wide range of harms potentially resulting from a cyberattack.[21] According to Vox, in late November 2014, Sony Pictures Entertainment was hacked by a group calling

constructed and invested with semantic properties). In what follows, we shall not always adhere to this trichotomy of damage, harm, and corruption.

[21] Kosseff, "Defining Cybersecurity Law," 989.

itself the Guardians of Peace.[22] The hackers, who were widely believed to be working in at least some capacity with North Korea, stole huge amounts of information from Sony's network. They leaked the information to journalists, who wrote about various embarrassing things Sony employees had said to one other. Then the hackers, using one of their near-daily communiqués on the website Pastebin, threatened to commit acts of terrorism against movie theatres, demanding that Sony cancel the planned release of the film *The Interview*.

The Sony hack caused a wide range of harms to the United States, its companies, and its citizens. Among the most prominent harms were: (1) privacy harms to Sony employees; (2) embarrassment of Sony executives and celebrities; (3) reduced market value of leaked films; (4) internal operations slowdown at Sony; (5) harm to Sony's business reputation; (6) reduced public confidence in the security of electronic communications; (7) a chilling effect on free speech and press; and (8) a symbolic victory of the North Korean government over the United States.

The variety and seriousness of the harms and damage that cyberattacks can directly or indirectly cause reflects in large part the human goods that these attacks destroy or diminish. These include lives and livelihoods, individual physical and mental well-being, individual freedoms and political autonomy, law and order, the integrity of economic and political institutions, the integrity of epistemic institutions (e.g., universities, the media), and public trust—to name some of the main ones. We note that the indirect harms might be the unintended consequences of cyberattacks. For instance, an unintended consequence of numerous and well-publicized, successful cyberattacks on a government's databases of confidential personal information might be a loss of public trust in the government's ability to ensure information security, and ultimately result in the refusal of many citizens to provide the personal information required for an efficient and effective government provision of the services in question.

It is important to distinguish between cyberattacks and cyberthreats, which might not be cyberattacks per se. Let us distinguish between, for instance, hate speech and a newsworthy factual claim posted on the internet that is not couched in offensive language, and appears on its face to be true, yet which is known by its speaker to be false (i.e., it is fake news). Hate speech is an attack—and, if communicated on the internet, is a

²² https://www.vox.com/2015/1/20/18089084/sony-hack-north-korea. Accessed 4/7/2023.

cyberattack—that might have devastating effects (e.g., hate speech posted by the Myanmar military on Facebook about the Rohingya had the effect of inciting murder).[23] However, the newsworthy factual claim mentioned above is an instance of fake news but is evidently not in itself an attack, even a verbal attack, and so is not a cyberattack. However, it might be a cyberthreat. Fake news posted on social media, even if it is presents as an inoffensive factual claim, might in each context have very harmful effects, especially if tweeted and retweeted. Consider, for instance, the claim that COVID-19 is no more harmful than the flu. In this instance, fake news is a cyberthreat, but not a cyberattack as such, since it only consists in some purportedly factual content that is not in itself offensive. Again, there are undetected or undeterred threats arising from failure to penetrate the cybersecurity of malevolent actors. Such threats include encrypted terrorist communications (Chapter 2, Section 2.3), authoritarian state actors using machine-learning (ML) techniques (Chapter 4, Section 4.1; and Chapter 6, Section 6.2) to interfere in democratic elections, and criminals (including some state actors) utilizing blockchain technology (e.g., used by Bitcoin) to engage in transfers of the proceeds of crime (or, in the case of North Korea, to steal Bitcoins to fund nuclear weapons). While these activities are not cyberattacks per se, they are nevertheless cyberthreats since, firstly, they utilize cybertechnology to engage in criminal and moral offences and, secondly, combating them typically requires using cybertechnology.

In this connection, it is also important to distinguish between, on the one hand, a single, discrete cyber*attack* (e.g., Stuxnet, or a number of unconnected, discrete cyberattacks) and, on the other hand, a cyberwar comprised of multiple, interconnected cyberattacks in the service of a strategic outcome.[24] With respect to wars (as opposed to covert political action—see Chapter 6, Section 6.1), we need to distinguish cyberwars per se from kinetic wars in which there is a cyber dimension comprised of cyberattacks on, for example, enemy military communication and other facilities. An important recent instance of an essentially kinetic war with a cyber dimension is Russia's invasion of Ukraine. Indeed, it is highly unlikely that in waging a modern kinetic war there will be no recourse to cyberattacks.

[23] Paul Mozur, "A Genocide Incited on Facebook, with Posts from Myanmar's Military," *New York Times*, 15 October 2018. https://www.nytimes.com/2018/10/15/technology/myanmar-facebook-genocide.html.

[24] Lucas, *Ethics and Cyber War*.

1.2.2 Cyberthreats: Disinformation, Propaganda, and Hate Speech

It is also important to distinguish cyberattacks—and, indeed, cyberwars—from a generalized cyberthreat that consists of an ongoing, widespread process, involving numerous, connected cyberattacks and/or disinformation/propaganda/hate speech communications, and multiple actors—whether individual attackers (e.g., hackers), propagandists or their targets (e.g., electoral offices, voters) or all of these. Consider, for instance, the deliberate, systematic and ongoing campaign of cognitive warfare (Chapter 6, Section 6.2) conducted by foreign powers—notably, Russia but also Iran and others—to influence the outcome of various elections, including the US Presidential elections in 2016 and 2020.[25]

At this point, a more fine-grained or nuanced approach seems called for. Thus, the distinction between the above-described generalized cyberthreat and a cyberwar, including cognitive warfare, between identifiable protagonists is one of degree. For instance, arguably, Russia and the US are engaged in a cyberwar/cognitive war of sorts—or, at the very least, a form of cyberconflict short of war (and perhaps appropriately characterized as covert political action) (Chapter 6, Sections 6.1 and 6.2).[26] Again, the cumulative effect of numerous discrete cyberattacks—or disinformation communications and campaigns—might constitute a process with discernible corrosive and/or corruptive effects on institutions or social norms. Consider the widespread and ongoing dissemination of fake news on social media by disparate actors (e.g., Russian disinformation campaigns, Donald Trump's numerous falsehoods on Twitter,[27] left-wing and right-wing propagandists, antivaxxers, and so on), which is known as computational propaganda (Chapter 3).

---

[25] National Intelligence Council, "Foreign Threats to the US 2020 Federal Elections" (Washington, DC: National Intelligence Council, 2021), 1–15, https://www.dni.gov/files/ODNI/documents/asse ssments/ICA-declass-16MAR21.pdf. Accessed: 26/10/2023. Regarding the UK, see Nigel Gould-Davies, "The Russia Report: Key Points and Implications" (London: IISS, 2020), https://www.iiss. org/blogs/analysis/2020/07/isc-russia-report-key-points-and-implications. Accessed 26/10/2023.

[26] Michael Gross and Tamar Meisels, eds., *Soft War: The Ethics of Unarmed Conflict* (Cambridge: Cambridge University Press, 2017); J. Galliott, ed., *Force Short of War in Modern Conflict: Jus ad Vim* (Edinburgh: Edinburgh University Press, 2019); Allhoff et al., *Binary Bullets*; Sloss, *Tyrants on Twitter*.

[27] N. Rattner, "Trump's Election Lies Were Among His Most Popular Tweets," *CNBC (International)*, 13 January 2021, https://www.cnbc.com/2021/01/13/trump-tweets-legacy-of-lies-misinformation-distrust.html, accessed 26/10/2023; G. Kessler, S. Rizzo, and M. Kelly, "Trump's False or Misleading Claims Total 30,573 over Four Years," *Washington Post*. https://www.washing tonpost.com/politics/2021/01/24/trumps-false-or-misleading-claims-total-30573-over-four-years/. Accessed 26/10/2023.

Evidently, this ongoing 'tsunami' of disinformation is now undermining social norms of truth telling and trust in cyberspace.[28] Moreover, the integrity of institutions, such as liberal democratic governments, relies in part on
widespread compliance with norms of truth telling and trust. Accordingly,
disinformation poses a cyberthreat to these institutions, a threat that is a security threat because it threatens the integrity, and therefore existence, of
institutions. This was graphically illustrated by the 6 January 2021 attack on
Capitol Hill in the US, which was carried out by Trump supporters who were
fuelled by disinformation campaigns based on Trump's false claims of election fraud.

The disinformation threat is greatly facilitated by social media platforms
(Chapter 3, Sections 3.1.1 and 3.3). The social media platforms can be split
into three categories: (1) the tech giants; (2) the well-established specialized platforms; and (3) the many start-ups and smaller enterprises. The
tech giants (e.g., Meta (including Facebook), Alphabet (including Google),
Apple, Microsoft, and Amazon—all of which are trillion-dollar companies)
are the largest of *any* type of company. At $60 billion, General Motors is a
much smaller company. Indeed, Apple is fifty times its size. These companies
are *so* big that they have a huge influence on the media landscape, including
the legal framework in which they operate. Their deep pockets enable them
to absorb smaller, competing companies. Thus, Google owns YouTube, while
Facebook (or, its parent company, Meta) owns Instagram, WhatsApp and
others. Just beneath the giants, we have a range of platforms, with a billion or
more users: YouTube (2.6 billion); WhatsApp (2 billion); Instagram (1.5 billion); WeChat (1.3 billion); and TikTok (1 billion). Then there are many other
companies just below 1 billion users, like Twitter with slightly over 400 million. There are also many smaller ones, such as Flick and Tumblr, which are
made possible by a surfeit of venture capital funding. The mantra, made popular by Amazon, is to go for growth, regardless of profitability. Many of these
smaller tech companies are debt ridden but rely on a voracious market to
fund their inflated IPOs.

Such a growth model has the unfortunate side effect of prioritizing attention over truth. Fake news, if it is sensational (and perhaps accompanied by
realistic, but fake, photos), often gets a bigger audience than the cold truth.

---

[28] Cocking and van den Hoven, *Evil On-line*; Michael Lynch, *The Internet of Us* (New York: Liveright,
2016); Miller, "Freedom of Political Communication, Propaganda and the Role of Epistemic
Institutions in Cyberspace."

A case in point in early 2022 was Spotify. It reputedly paid one-hundred million dollars for the podcast The Joe Rogan Experience, which was at the time the highest rated podcast. Unfortunately, Rogan turned out to have scant regard for the truth, propagating misleading statements about COVID-19 vaccination. The company decried his statements, but nevertheless claimed that silencing him was not the answer.[29] Evidently, facilitating his propaganda was the answer.

1.2.3 Cyberthreats and Privacy

A somewhat different process that is not a prima facie instance of even a generalized a cyberthreat is that of an emerging so-called surveillance society. Arguably, the authoritarian government in China is in the process of creating such a society, at least in Xinjiang and Tibet. Moreover, even in liberal democracies there are concerns about governments and security agencies, like the National Security Agency (NSA) in the US. The Snowden revelations disclosed that the NSA was collecting, storing, and analysing bulk data—notably, metadata consisting of caller/receiver numbers, location and duration of calls, and so on, but increasingly also biometric data, such as fingerprints and facial images—on their citizens (see Chapters 2 and 4). It is also claimed by some, notably Shoshana Zuboff, that a somewhat different version of a surveillance society ("surveillance capitalism," according to Zuboff) is emerging in the liberal democracies at the hands of technology companies like Alphabet, Meta, and Amazon.[30] These global technology companies are engaged *inter alia* in collecting, storing, and analysing vast amounts of personal information to create dossiers on users for commercial purposes—either their own or those of other corporations. For instance, the technology companies use cookies to log a user's activity on websites to build up a comprehensive dossier on the user, which enables targeted advertising. One issue here is the lack of consent given by users to this collection and use of their personal data. Another issue is the potential manipulation of users in a manner that compromises their autonomy[31] (Chapter 2, Section 2.2.2).

---

[29] https://www.cloudflare.com/products/bot-management/. San Francisco. Accessed 3/1/2022.
[30] Shoshana Zuboff, *The Age of Surveillance Capitalism* (London: Profile Books, 2019).
[31] Ibid.

An example of the misuse of personal data that involves public-sector organizations, as well as corporations, is what is referred to as Real-Time Bidding (RTB) for advertising.[32] This essentially involves onselling personal data for advertising. Brave has recently filed complaints under the GDPR against local government councils in the UK using RTB. They note therein that more than half of council websites use RTB ad auctions. RTB faces multiple GDPR investigations for systematic data breaches because it broadcasts people's personal data to countless companies. Ninety-six UK council websites use Google's RTB system. Google's RTB shares data with hundreds of companies, without any assurance of who that data is then shared with or how it will be used. The Irish Data Protection Commission opened an investigation into Google's RTB in May 2019.[33]

In China, the government claims that the widespread use of surveillance is a necessary or, at least, desirable security measure. The measures used include high resolution CCTV cameras in public spaces, facial recognition technology, integrated bulk databases (of metadata, biometric data, genomic data, and so on—see Chapter 4, Sections 4.3 and 4.4), interception of phone and email messages, geolocation devices, and so on. However, the question to be asked is: Who is being secured and against what threats? Arguably, it is primarily the authoritarian state itself that is being secured, and secured against political opposition, rather than securing the citizenry against criminal activity (i.e., activity that is morally justifiably criminal). Indeed, the security of the citizenry, understood in terms of the protection of their basic moral rights, is under threat from the state.

Our concern here in relation to the so-called surveillance society is not so much with what is being secured, but rather with what is being threatened (including, potentially, by security measures) and left unsecured—namely, individual privacy and autonomy. In short, whatever its security benefits might be, the surveillance society is itself also a security threat. For it not only uses cybertechnology in ways that threaten individual privacy and autonomy, but it also fails to introduce institutional and other measures that afford protection to these moral rights—and, in some cases, it undermines existing protective measures.

[32] "RTB Evidence: Selected evidence submitted to data protection authorities to demonstrate RTB's GDPR problems." https://brave.com/rtb-evidence/. Accessed 5/1/2022.

[33] Liam Tung, "Google Accused of Leaking Personal Data to Thousands of Advertisers." ZDNet. https://www.zdnet.com/article/google-accused-of-leaking-personal-data-to-thousands-of-advertisers/. Accessed 6/7/2020.

1.3 Cyberattacks

As we have seen, cyberattacks are a species of cyberthreats. Cyberattacks can be categorized into various types, especially based on their constitutive actions, purposes, and the means and/or vector by which they are launched. The various types of attack will come under scrutiny in the rest of the book. Moreover, we earlier distinguished between cybersecurity in the narrow sense of information security (and information systems security), on the one hand, and cybersecurity in the wider sense that includes security from incitement, computational propaganda, cognitive warfare, and so on, on the other. In this section, we want to give a brief overview of some of the salient types of cyberattack that potentially breach information and information systems security. That is, our concern is only with cybersecurity in the narrow sense. In doing so, we seek to highlight some of the technical issues. Likewise, in 1.4 below we highlight some of the technical issues that arise in countering or preventing cyberattacks (i.e., we highlight salient countermeasures). The technical issues briefly introduced here and in 1.4 will be discussed in more detail throughout the book.

As with any complex phenomenon, there are many perspectives from which one can describe cyberattacks. Moreover, there are numerous elements of the cyberattacks in question that could be addressed. For our purposes here, it is sufficient to limit ourselves to brief discussions of: (1) attacks on data (1.3.1); (2) attacks on computer services (1.3.2); (3) outcomes of these attacks (1.3.3).

1.3.1 Data Attacks

As mentioned above, there are three main ways in which the (physically embodied) data may be the subject of an attack: theft; corruption; and destruction. The twenty-first century has seen data become a valuable commodity, sufficient to warrant the *Economist*'s headline, "Data Is the New Oil,"[34] Zuboff's bleak assessment of the surveillance world in which we now live, as well as the extraordinary level of global cybercrime.[35] According to the FBI's cybercrime statistics, a minimum of 422 million individuals

---

[34] "Data Is the New Oil," *Economist*, 26 February 2019.
[35] Zuboff, *Age of Surveillance Capitalism*.

were impacted in 2022, and nearly 33 billion accounts will be breached in 2023, at a cost of eight trillion dollars.[36] Perhaps this is to be expected given that, throughout history, anything which is valuable tends to be stolen—and, given that cyberspace is under-regulated and, perhaps, even more important—existing regulations are underenforced. Most of the ways data is stolen are very much like the theft of physical objects. First the thief needs to get access (e.g., break the window of a jewellery store), then appropriate the items (e.g., put jewels in a bag) and escape. Just as in the many jewellery thefts, the way-in frequently involves deception or insider cooperation. Sometimes someone just leaves the doors or the safe unlocked. Data theft first involves breaking into a data store, which can be accomplished by guessing or stealing a password, or by bribing somebody to provide authorization.

The most common form of data security is password protection. Guessing a password is like finding the combination of a safe. It's hard work, and the more digits in the combination of the safe, then the more work is required for entry. Thus, passwords need to be long and unguessable. However, a major source of cyber vulnerability, especially for the IoT, is leaving passwords unchanged from their factory settings. If the safe combination has been reset to some unguessable number, rather than somebody's birthday or the fourth of July, an enterprising thief still has some options. For instance, remembering 82428573 requires a concentrated effort, thus our safe owner has written it on a post-it note stuck underneath his desk.

Sometimes people will give away data motivated by greed (e.g., someone gives out their credit or debit card details in response to an email promising a large payment into their account) or even, ironically, by security concerns. Concerning the latter, an email purporting to be from a bank may ask the user to change their password because of alleged unauthorized accessing of their account. However, in fact the email has been sent by a fraudster who has the goal of getting the recipient to enter their password, allowing the fraudster to gain access to their bank account and steal their money. Such attacks are variously called spoofing or phishing. According to the FBI's 2022 statistics, there were more victims of phishing than any other crime category.[37] Other motivations to give away data can be frustration or revenge.

---

[36] Federal Bureau of Investigations, *Internet Crime Report 2022* (Washington, DC: Internet Crime Report Centre, 2022), https://www.ic3.gov/Media/PDF/AnnualReport/2022_IC3Report.pdf.
[37] Ibid.

For instance, disgruntled employee might provide an important password to his employer's business competitor.

The general point to be made here is that access to computer systems may involve human weakness and deception in the same way as scams have done long before digital computers were invented. The more valuable the data, the larger the effort to find vulnerabilities due to human frailty. Such ways of gaining access are the equivalent of trying to get in through the front door.

Other ways involve snooping on the data that goes in and out. Such attacks involve hardware or software interception of data traffic. However, the data might be encrypted. For instance, there are multiple Wi-Fi encryption standards—WPA3 being the encryption standard that should be used at the present time. If the data is encrypted, then interception may need to take place before encryption or after decryption. However, this might not be difficult if there is weak password protection. Sometimes, for example, one finds a hotel Wi-Fi that is not password protected, or else has a generic password for all users.

There are various sorts of hardware snoopers, which have been used, for example, to gain access to cars. Once the attacker has gained access to the data, it needs to be extracted. This could occur via the computer network to which the compromised device is attached, or it could be removed onto some form of removable storage (e.g., a USB stick, portable hard drive, or DVD). It may not be data that is being stolen, though. With the ever-increasing use of internet-enabled computer control, objects can be stolen too. Luxury cars are an excellent target. In this case, the attack exploits the car's own wireless technology for remote unlocking to gain access to the car's computer system.

Gaining access to a computer system, though, may not be about stealing data. It might be, to corrupt or destroy the data. The corruption of data could consist of changing it (e.g., adjusting payments made), or making additions to it (e.g., adding pornographic images to a person's photo album). When flight KAL007 was shot down by the Russians, one of the numerous theories why it strayed into Russian airspace was that the wrong coordinates were fed into the flight control system. It is quite conceivable that a major terrorist attack may be done remotely by corrupting its crucial data, causing the cataclysmic loss of a plane, train, dam, or power station.

The invader might not corrupt but destroy data—or threaten to destroy it. Destruction is effectively cybersabotage, and it has various, different goals. The threat to destroy is often an extortion attempt, referred to as ransomware.

Typically, the victim's data is encrypted by the attacker, rendering it unusable. The victim must pay a ransom for its decryption.

Cyberattacks conducted for the purpose of data corruption or destruction often use malware (malicious software), of which the two primary examples are worms and viruses. Worms propagate autonomously through computer networks and often carry a payload, which is a computer virus that can cause some sort of undesired consequence on the infected machine. The most common way a machine is infected by a virus, though, is not actually a worm, but through an email attachment that the user unwittingly opens.

Finally, a cyberattack might consist of a secret 'invasion' with no immediate corruption or destruction of data. Rather, the malware is a 'sleeper', often referred to as a Trojan Horse, which sits inside a system, perhaps waiting for some trigger before it acts. In 2020, the Australian Government announced a major security breach of just such an invader from a state attacker, revealed in 2021 to be China. Some months before this, one of Australia's most prestigious universities, the Australian National University, also encountered such a low-profile attack. This is what is usually referred to as spyware. It may often take root in personal computer systems and use the camera and/ or microphone to record user activity, or his or her environment. Keyloggers are a variant that record keystrokes and sends them to a hacker, who mines for passwords and other sensitive information. Sometimes user agreements might have hidden in them the right to use spyware. An example is a robot vacuum cleaner that can make a map of its user's house and sent it back to base.

1.3.2 Attacks on Computer Services

Data theft, corruption, or destruction might not be the goal of an attack. Rather, the intention might be to interfere with the actions of an internet service (e.g., the ability to send and receive emails). Doing so does not require access to the server. An obvious example would be cutting the ethernet cables that connect an internet service to the internet. But a common form of remote attack is Denial of Service (DoS) (see Chapter 5, Section 5.5). If a heckler throws a single rotten tomato at a politician at a rally, it doesn't stop him delivering his speech. However, if a hundred protesters throw rotten tomatoes, and keep throwing them, it will become impossible to deliver the speech. As with the Mirai attack, this is what a DoS does. Data packets are sent in huge volumes to the target server (computer), clogging up the network and rendering it unusable. Just as in our tomato example, the attack is often more severe when it comes from multiple sources at the same time, a Distributed Denial of Service attack. Because the DDoS attack simply consists of a flood of arbitrary packets (rotten eggs would work just as well as tomatoes), the packet sources can be very simple machines. Thus, devices on the IoT, such as televisions, refrigerators, and security cameras, are all potential hosts for the attacker. The software sitting on each computer is referred to as a bot, an autonomous software agent, like a robot.

1.3.3 Attack Outcomes

As we have seen, cyberattacks that compromise the security of information and information systems attack, directly or indirectly, the logic layer (e.g., software) or the data layer (e.g., emails), but they frequently do so as a means to the following ends: physical damage, e.g., to equipment controlled by computers; moral, psychological, physical or other harms to individual human beings; institutional damage, e.g., to corporations or governments, and their agencies; or harm to a community or nation-state at large. Regarding physical damage, one of the most well-known and ingenious such attacks was Stuxnet, a virus that attacked the uranium enrichment centrifuges in Iran. By increasing their rotation speed beyond their design limits, it caused major damage. It was thought that a state actor (e.g., Israel and/or the United States) was responsible. As the IoT, which involves computer control of everyday devices, ramps up, we may expect to see more and more such attacks. Moreover, they may do harm to individuals, damage to institutions and, ultimately, harm whole communities. For example, a future terrorist attack might involve implanting malware into General Motors' automotive control systems for a vehicle's fuel pump. Since all fossil fuel vehicles need a fuel pump, the same software might be used for all of them. At some specified day and time, the fuel pumps in the vehicles could be deactivated, causing traffic chaos in cities around the world, including car crashes and injuries to drivers. Moreover, it may do damage to General Motors. Individual harm may be financial, through attacks on bank accounts. It may also be more far reaching by stealing personal data that can later be used for identity theft. It might not even involve theft at all. The attack may be psychological, as with cyberbullying or troll activity, which uses malicious invective against other online users. Cyberattacks on institutions, including corporations, may attack the corporation's reputation or steal data for resale. An attack on an individual person may involve stealing a credit card number, with which the attacker will then go on a shopping spree. In a corporate attack, as for example happened with Sony, millions of credit card numbers may be stolen. The attacker is not going to use all of these. They instead get offered for resale on the Dark Web, which is a large, hidden part of the Web, accessed through the ToR browser.[38]

Finally, damage caused by state actors is the new face of cyberattacks. This has largely consisted of ongoing cyberattacks that, nevertheless, do not rise to the level of war and can be credibly denied by the attacker (e.g., the Russian and Chinese cyberattacks mentioned above). As such, they constitute, for the most part, instances of covert political actions (see Chapter 6, Section 6.1). However, cyberattacks can be an important component of a kinetic war. At any rate, whether it is espionage, covert political action, attacks on communication networks, or automated weapons in the context of a kinetic war, attacks by one state actor against another will increasingly become cyber dominated.

1.4 Cyber Countermeasures and Threats
to Countermeasures

In cybersecurity, as in other security contexts we can distinguish between reactive, preventive, deterrence, deflection, and resilience strategies, although an integrated mix of strategies is generally the most effective. In any case, the different strategies are not entirely separable (e.g., deterrence is a means of prevention). Reactive strategies presuppose an attack has been launched. However, ideally, a reactive response identifies an attack prior to its completion (e.g., by means of an intrusion detection system) and thwarts it. Moreover, again ideally, the attackers are identified during or after the attack, although in cyberspace this can be difficult (i.e., there is a so-called attribution problem). Prevention strategies include technical measures such as firewalls or, in relation to the so-called human factor, refusing to hire past

---

[38] The ToR browser uses onion routing (literally, "The onion Router"). Onion routing is a way of securely routing information. Packets have an onion-like set of addresses and travel over many nodes. No node can see all the layers of the onion. Each node strips off the top layer to find the address of to where it should send the packet.

offenders. Deterrence strategies include, most notably, launching a cyber counterattack or threatening to do so. As with deterrence strategies, deflection strategies are a means of prevention (e.g., the attack is diverted from its target by means of honeypots). Resilience strategies focus on strengthening defences and ensuring there is a viable recovery process, if attacked.

So called defence-in-depth involves several phases: intrusion preparation, monitoring and response, mitigation, recovery, and continuity planning. Moreover, there are several additional features of a successful cyberstrategy and defensive process, as well as principles that should govern them, such as 'trust but verify', 'need to know', 'separation of duties', and 'understand one's attackers' (e.g., Lockheed Martin's Cyber Kill Chain).[39]

1.4.1 Defensive Measures

As discussed in Bossomaier et al., much of cybersecurity is about human factors.[40] Every year huge data compromises occur due to, for instance, carelessness and disaffected employees (e.g., Edward Snowden). Technical solutions must be backed by human solutions, such as good employee vetting, professional reporting mechanisms, and cybersecurity training programs.

There are various basic measures to defend against cyberattacks. These include the following ones: ensuring passwords are strong; using a password safe to store them; keeping software up to date, so that security patches are applied as soon as vulnerabilities are identified; managing cookies (e.g., turning off third-party cookies, if possible). Other basic measures pertain to system logs and erasing files.

System logs are not much of a problem unless a machine is hacked. A system log is a file containing information such as device events, operations, and changes. System logs can be problematic, if hacked, because data may exist there that would otherwise have been encrypted. In modern password systems, the password is never stored anywhere. Only a hashed (one-way encrypted) version is stored. When the password is entered, it is hashed and compared with the stored version. Having the hashed password is of no use to the hacker because it will be hashed again when it is entered, which is why a forgotten password cannot be retrieved. Suppose however that the

[39] https://www.lockheedmartin.com/en-us/capabilities/cyber/cyber-kill-chain.html. Accessed 4/7/2023.
[40] Bossomaier et al., *Human Dimensions of Cybersecurity*.

simple password 'kangaroo' is used, but 'kangeroo' is mistakenly typed instead. This may appear in the system log as is, unencrypted, from which it is easy to guess the password.

Properly erasing files or e-shredding is slightly more complicated. Deleting a file usually only removes it from a table of files that are located on the disk. Nothing happens to the data; it is simply not accessible by normal means. But there are tools used by hackers and in digital forensics that can still locate and read the data. To really be rid of it, the area it occupies on the disk must be overwritten with random data, all zeros or a million repetitions of a word. This can be especially problematic with SSDs, which do not enable the system to write to a specific disk area.

Beyond these basic measures, there is now a plethora of programs that can be installed to guard against malware—often packaged as security suites. These include the following types. Antivirus software detects and destroys viruses that have been seen before. Only these can be guaranteed to be eliminated, and therefore virus writers rely on pre-existing viruses. Firewalls are software that sit on individual computers or servers and determine which IP addresses may send data in and out of a network and which ports they may use to do so. Firewalls can only block traffic that is already known to be unwanted, but that traffic could be from a large domain, such as an entire country. The Great Firewall of China controls access to everything outside of China for those within, and vice versa. Network monitoring software enables programs to be identified and blocked from roaming extensively around the internet, possibly picking up malware or supplying personal information to data brokers.

A fundamental cyber-defensive measure is encryption (Chapter 2, Section 2.3). Virtual Private Networks (VPNs) hide IP addresses and encrypt data in transit. The idea here is that all traffic out of a computer goes not directly to its intended destination but goes via one or more servers that may be in a different city, country, or continent. The server is like a relay station. The destination node only knows the address of the last relay in the chain, thus hiding the original address of the sender. Destination hiding is one of the two primary roles of a VPN. The second is encryption of traffic along the way. VPNs are vital for privacy, but they also allow people in authoritarian regimes to access forbidden, external sites and provide security for opposition to such regimes. The data is encrypted to the remote server, but may be, and sometimes must be, decrypted there and re-encrypted before sending it on to its destination. The gold standard of encryption is end-to-end encryption where data is not, and cannot be, decrypted until it reaches the destination machine. Numerous programs such as WhatsApp, Facetime, and Signal use end-to-end encryption.

The above list applies to individuals as well as to governments and corporations. However, there are constraints in relation to these measures. Whereas ideally all personal data should be encrypted, in fact there are numerous breaches, such as the loss of large quantities of credit card numbers, that show that this is not always the case. Indeed, widescale encryption is difficult to achieve in organizations or groups with a large number of users (e.g., employees) and extensive workforce casualization. Again, some organizations, such as banks, should provide round-the-clock protection. However, this requires maintaining hardware and network infrastructure so that they can keep up with demand and avoid DDoS attacks. Moreover, averting DDoS, in particular, costs money. Doing so requires state-of-the-art intervention software to monitor and block attacks—bearing in mind that the attacks come from hundreds up to tens of millions of machines—and complex backup structures to allow for a switch to some other server when one is under attack. As this book goes to press, there are reports of huge attacks, an order of magnitude greater than earlier.[41]

1.4.2 Authentication

The last topic we consider in this section is that of trust and authentication. The largest source of malware in 2018 was email. There are already procedures in place to make email more secure. However, more can be done.

Sending links in email is ubiquitous. Though typically genuine, these links can be malicious, causing the download of malware. To make a payment for, say, your electrician, you receive an email from an accounting software system that you do not know, with a payment button therein. Hackers have been getting more and more sophisticated at mimicking such emails, with disastrous consequences for those fooled. The hacker simply makes a copy of a genuine email, but changes the link to their own malicious link. Since the email message was copied directly from a genuine email, it is very hard to tell from the email content that it is fake. The only way to tell is to look at

[41] https://www.reuters.com/technology/internet-companies-report-biggest-ever-denial-service-operation-2023-10-11/} Accessed 17/10/2023.

the email sender and headers, something email software should do, but frequently does not.

Ideally any message, be it email or something else, should bear a digital signature. This process uses public-private key cryptography (Chapter 2, Section 2.3.1). In such asymmetric cryptography protocols, a message signed with one key (public or private) can only be decrypted with the other key. Thus, if Zak wants to send a message out to many different people, he signs it with his private key. Anybody can now check that it came from Zak by checking the message against his public key. On the other hand, if Zak wants to send a private message to Zena, he encrypts it with her public key. Only she can decrypt it, using her private key.

Finally, in the situation that many different people are contributing to a distributed ledger, some record of transactions or documents of any kind, blockchains create a series of data units secured cryptographically (see Chapter 7, Section 7.1.5).

1.5 Cybercrimes and Cybermoral Offences

Cyberlaw pertains to the use of the internet, computers, smartphones and the like. It comprises both criminal and civil laws (e.g., disputes about who owns a website). We can also distinguish between, on the one hand, pre-existing crimes that merely utilize cybertechnology, such as child sexual abuse or some forms of online fraud (i.e., crimes that are *qua* crimes merely contingently cybercrimes, sometimes referred to as computer-assisted crimes), and, on the other hand, crimes of which this is not true, such as computer hacking and virus distribution (i.e., crimes that are *qua* crimes essentially cybercrimes).[42]

Cybercrimes include online fraud and theft (e.g., of data, including private or confidential information, like credit card numbers and trade secrets); extortion (e.g., by means of a denial of service attacks or a ransomware attack); e-money laundering; illicit drug selling on the Dark Web; copyright violations, like distribution of pirated software; online stalking; cyberdefamation; *crimen falsi* or crimes of falsehood or deceit (e.g., perjury, false statements, theft by deception, electoral interference); incitement

---

[42] S. L. Soni and C. P. Bhargav, *Cyber Security and Cyber Law* (New Delhi: Prashant Publishing, 2016), 53.

to violence using social media; and child pornography in the form of computer-stored images.

According to Astra's security audit, which collates statistics from a range of sources,[43]

> Over the last 21 years from 2001 to 2021, cyber crime has claimed at least 6.5 million victims with an estimated loss of nearly $26 billion over the same period . . . . In 2020, internet crime victims over the age of 60 experienced $966 million in losses, while victims under 20 experienced almost $71 million in losses . . . . Cyber crime earns cybercriminals $1.5 trillion every year. . . Ransomware and business email compromise (BEC) attacks were the leading cause of losses from the five years of 2017-2021 . . . . Ransomware will cost its victims around $265 billion (USD) annually by 2031.

Cybercrimes in breach of the international laws of war, include sabotage of specific facilities (e.g., Stuxnet attack, Chapter 6, Section 6.1), or cyberattacks on critical infrastructure by a foreign power (e.g., attack on Estonian infrastructure by Russian hackers, Chapter 6, Section 6.1). Espionage is a crime in most jurisdictions if it is conducted by foreign powers against them, but not if they are the ones conducting it against foreign powers. For this reason, it is not enshrined in international law.

Some of the major legislation pertaining to cybercrimes includes, in the US, the National Cybersecurity Protection Act and the Cybersecurity Enhancement Act of 2014 and the Health Insurance Portability and Accountability Act (HIPAA) of 1996, and, in the EU, the various EU directives on cybercrime and GDPR.[44] The Budapest Convention on Cyber Crime 2001 is the only binding treaty that specifically regulates cyberspace behaviour.

There is a paucity of laws and regulations pertaining to cybersecurity, both in domestic jurisdictions and internationally. For instance, according to Kosseff, existing US cybersecurity frameworks focus on protecting the confidentiality of information for the purposes of protecting individual privacy.[45] However, the laws could be greatly improved to focus more other aspects,

---

[43] Nivedita James, "90+ Crime Statistics 2023: Costs, Industries and Trends," *Astra*, 29 April 2023. https://www.getastra.com/blog/security-audit/cyber-crime-statistics/. Accessed 6/7/2023.

[44] The HIPAA was passed to protect the confidentiality, integrity, and availability of all electronic protected health information.

[45] Kosseff, "Defining Cybersecurity Law."

including: (1) integrity and availability; (2) protecting systems and networks; and (3) promoting economic and national security interests.

The EU's General Data Protection Regulation (GDPR) is the most important regulation of cyberspace in respect of privacy rights to date. The GDPR is widely acknowledged to have significantly enhanced individual privacy rights in the digital age. Notably, the GDPR requires that companies who collect and store the personal data of EU citizens must have the informed consent of these citizens in order to do so. However, the market domination of Google, Facebook and other tech giants is such that it is doubtful whether the preconditions for genuine consent can be met (Chapters 2, Section 2.2.2).[46] Moreover, intelligence agencies and national security purposes provide an important restriction on the application of the informed consent principle and the raft of other privacy protecting principles that constitute GDPR. According to article 23 of the GDPR:[47]

> Union or Member State law to which the data controller or processor is subject may restrict by way of a legislative measure the scope of the obligations and rights provided for in Article 12 to 22 and Article 34, as well as Article 5 insofar as its provisions correspond to the rights and obligations provided for in Articles 12 to 22, when such a restriction respects *the essence of the fundamental rights and freedoms and is a necessary and proportionate* measure in a democratic society to safeguard: (a) *national security* [italics added].

This restriction has caused some concern among privacy advocacy groups. For instance, according to Human Rights Watch:

> The EU regulation will not curtail large-scale government surveillance, as it allows for government surveillance under broad exemptions. Government agencies can process personal data without consent if there is a "national security," "defence," or "public security" concern, terms the regulation does not define. As the EU's Court of Justice has established, however, such terms do not provide carte blanche for countries to do whatever they like. International and regional human rights laws (and any national regulations that do not conflict with the EU regulation) still apply to limit

---

[46] Very large fines have been handed out to Google and Facebook by the EU for data breaches, but at the time of writing are still under appeal.
[47] General Data Protection Regulation (GDPR). https://gdpr-info.eu/. Accessed 7/11/2023.

the surveillance and data processing activities of intelligence and law enforcement agencies. However, many European states have expanded their surveillance laws in recent years, undermining protections for privacy and other human rights. In the coming years, the EU Court of Justice is likely to be called on to delineate the regulation's state interest exceptions in the context of EU, European, and international human rights law.[48]

Additional proposed EU legislation includes the Digital Services Act (DSA), which would introduce new obligations on platforms, such as Facebook, YouTube and Twitter, to reveal information and data to regulators about how their algorithms work, how decisions are made to remove content, and how adverts are targeted at users. Failure to comply could result in very substantial fines based on a percentage of annual revenue (e.g., six percent). Another EU legislative proposal is the Digital Market Act (DMA), which targets monopolistic technology companies and anticompetitive practices. It would carry significant sanctions including the possibility of the breaking up of companies such as Alphabet and Meta.

What of cyberwar and cyberweapons? Regarding cyberwar and international law, many hold that international law regulates cyberwar by analogy with conventional war. Thus, according to Duncan Hollis:

. . . international law is widely assumed to govern "cyberwar" by analogies that delimit the boundaries of the jus ad bellum (the set of laws regarding *when* force can be used) and the jus in bello, also known as international humanitarian law or "IHL" (the set of laws regulating *how* states may use force).[49]

However, Hollis goes on to problematize this conception. At the very least there is clearly a lacuna in international law in respect of cyberwar (Chapter 6, Sections 6.1 and 6.2). The second Tallin Manual was intended to address this lacuna but is regarded by many as a failure.[50]

---

[48] Human Rights Watch, "EU General Data Protection Legislation," 6 June 2018. https://www.hrw.org/news/2018/06/06/eu-general-data-protection-regulation.

[49] Duncan B. Hollis, "Rethinking the Boundaries of Law and Cyberspace," in *Cyber War: Law and Ethics for Virtual Conflicts*, ed. Jens David Ohlin, Kevin Govern, and Claire Finkelstein (New York: Oxford University Press, 2015), 131.

[50] Michael M. Schmitt, ed., *Tallinn Manual on The International Law Applicable to Cyberwar* (Cambridge: Cambridge University Press, 2013)). It is believed to have failed by, for instance, George Lucas who expresses this view in his *Ethics and Cyber War*.

Regarding cyberweapons, according to Prunckun,[51] international law or treaties could, but do not, ban certain cyber weapons (Chapter 6, Section 6.3), as is done in the case of biological and chemical weapons. Moreover, legal definitions of hardware and software cyberweapons (e.g., functional definitions) could be provided. Further there could be the prohibition/criminalization of the uses of at least some (e.g., not dual-use 'weapons') of these weapons to conduct attacks, as well as of their possession, sale, transfer, and so on, other than by governments or with government approval (e.g., for use in deterrence against hostile states).

The paucity of laws and regulations is partly due to the transjurisdictional nature of cybercrime. As Prunckun says, "how does a government enforce laws against multiple offenders who operate in jurisdictions that are hard to determine based on the rules of the physical world—e.g., extraterritorial issues"[52] Moreover, "[t]he striking feature of cybersecurity is that these [cybersecurity] measures focus firmly on . . . locks, fences and doors." Further, "[t]hese security measures overlook the fact that existing jurisdictional laws do not provide an adequate legal foundation for control."[53] Indeed, arguably, governments are turning a blind eye to the problem.

Clearly, there is a close relationship between legal and moral offences; indeed serious moral offences are typically enshrined in the criminal law. Moreover, as noted above, there is a need for additional national and international legislation and regulation of cybercrime. Nevertheless it needs to be kept in mind that there are numerous moral offences, including cybermoral offences that should not be criminalized (e.g., minor use of an office computer for private purposes). Further, there is a grey area between moral offences and criminal offences giving rise to the critical questions of which moral offences under what circumstances should be criminalized (e.g., failure of a worker with access to confidential information adequately to protect a password).

There are many moral offences that are not also criminal offences. These include lying, disseminating falsehoods (other than crimes of perjury, defamation, and so on), breaking promises (other than contracts), censorship (e.g., of unfashionable views that are not unlawful), cheating (other than crimes of fraud, and so on), economic coercion, harassment/bullying/

---

[51] Henry Prunckun, "The Rule of Law: Controlling Cyber Weapons," in *Cyber Weaponry: Issues and Implications of Digital Arms*, ed. Henry Prunckun (Dordrecht: Springer, 2018), 94–95.

[52] Ibid, 91.

[53] Ibid, 93.

vilification (short of criminal activity), injustices (short of criminal activity). Many moral offences that are not also criminal offences are committed by organizations (e.g., allowing dissemination of falsehoods of social
media platforms to be protected by the legal status of platforms not being
publishers)[54] and nation-states (e.g., authoritarian states censoring alternative political viewpoints). One of the most important categories of moral
offences that are not necessarily criminal are instances of corruption engaged
in by powerful institutional actors (e.g., political leaders' use of social media
platforms to vilify other institutions and institutional actors, such as the judi-
ciary and media and intelligence agencies, which have the important role in
liberal democracies of ensuring the government power is circumscribed and
its exercise held to account).

Given the necessity not to criminalize all moral offences, there is a need
to have strategies to enhance cybersecurity other than by legislation to reduce moral offences. These include recourse to market mechanisms,
regulations and enhancing the resilience of sociomoral norms. While recourse to regulations is necessary and recourse to market mechanisms
frequently useful, of particular importance is enhancing the resilience of
sociomoral norms (including at times by recourse to regulations and market
mechanisms). Roughly speaking, sociomoral norms are regularities in action
that are also moral norms; members of the social group in question believe
they have a moral duty to conform to these regulaties or that they otherwise morally *ought* to conform. or to which they otherwise morally *ought* to
conform.[55] Such sociomoral norms include ones respecting and enforcing
moral rights. Here the *ought* is not that of mere instrumental rationality;
it is not simply a matter of believing that one ought to conform because it
serves one's purpose. Some conventions and many laws are also sociomoral
norms. For example, the convention and the law to drive on the left is a norm;
people feel that they morally ought to conform. Moral rights and principles
are concretized and maintained by sociomoral norms. Moreover, compliance with such moral rights and principles and, for that matter, the laws that
might enshrine these moral rights and principles, consists in large part in
members of the relevant social group adhering to these sociomoral norms—
morally approving of doing so and morally disapproving of failing to do so.
It is not simply a matter of each individual independently believing in the

---

[54] Recently, the EU has enacted legislation to enable platforms to be held legally liable for unlawful
communications on their platforms.
[55] Miller, *Social Action*, Chapters 4 and 6.

correctness of the moral rights and principles in question, although this is a necessary condition for compliance with the corresponding sociomoral norms. Naturally, it does not follow from this that sociomoral norms necessarily concretize and maintain objectively correct moral rights and principles, as opposed to those falsely believed to be objectively correct. Far from it. Consider, for instance, the sociomoral norm of child sacrifice or racial discrimination. Nevertheless, the serious, ongoing erosion of fundamental, objectively correct, sociomoral norms, such as the norms of truth telling and trust, and the norms of refraining from theft, fraud, and murder, ultimately undermines the possibility of functional societies and their constitutive spheres of communicative, economic activity, including activity undertaken in cyberspace.

Considering the above, the question arises as to the appropriate regulatory model for cybersecurity, given that 'black letter' law, (so to speak) is never going to be adequate on its own and given that regulations need to buttress underlying sociomoral norms.

According to Kosseff, a regulatory model based on coercion and deterrence assumes robust government oversight through extensive government monitoring and inspections, coupled with severe penalties for observed violations.[56] There is an alternative model, based on cooperation between the public and private sectors. Under this model, cybersecurity law should contain a mix of penalty-based regulatory deterrence along with cooperation and incentives. Cooperation is particularly appropriate because companies' goals are often aligned with those of the government. For instance, it is in national interests and Sony's corporate interests to avoid another such attack. The necessity of a forward-looking component of cybersecurity law is also evident in cyber-resilience: the ability not only to prevent cyberattacks but to withstand or quickly recover from them.

1.6 The Institutional Landscape and Internet Governance Arrangements

As described above, some of the key institutional or, more broadly, organized social and economic actors in cyberspace are as follows: governments, security agencies, internet companies, civil society (e.g., NGOs, universities,

---

[56] Kosseff, "Defining Cybersecurity Law."

media companies), corporations, criminal organizations, terrorist groups, paedophile networks, and more loosely organized, social media-based movements (e.g., Black Lives Matter, who use the public communication channels afforded by the internet and social media providers).

These actors operate in various overlapping, interconnected, spheres of cyberspace—notably, the political sphere (e.g., nation-states, terrorist groups, political protest movements such as the pro-democracy Hong Kong protests); the military sphere (e.g., the armed forces of nation-states, NATO, Five Eyes, and the UN General Assembly First Committee); the law enforcement sphere; the economic sphere; the social sphere (e.g., Facebook friends); and the internet governance sphere.[57]

1.6.1 Internet Governance

Obviously, a relatively comprehensive map of the institutional landscape of cyberspace would be an extremely large and complex construct, with myriad details. Accordingly, we shall not attempt to provide such a map here, and instead display a map of the institutional governance arrangements in cyberspace. Unsurprisingly, even the provision of a map of the governance of cyberspace is a complex and problematic undertaking:

> The distinctive challenge of cyber governance is that it simply touches on so many different actors, single institutions, and collections of institutions working together on specific parts of the wider body of cyberspace. This requires a unique approach of cooperation that is fairly unusual for governments. The likelihood of governments legislating themselves out of the present conundrum of achieving technical security and stability while at the same time guaranteeing basic freedoms is slim.[58]

However, Joseph Nye has usefully displayed this complexity in a diagram.[59] See Figure 1.1.

For our purposes here, we do not need to drill down into the details of this complex model. Rather, we need to make two general points. Firstly,

---

[57] Klimburg, "The Darkening Web."
[58] Ibid, 319.
[59] Joseph Nye, "The Regime Complex for Managing Global Cyber Activities" (London: Global Commission on Internet Governance, 2014), 8.

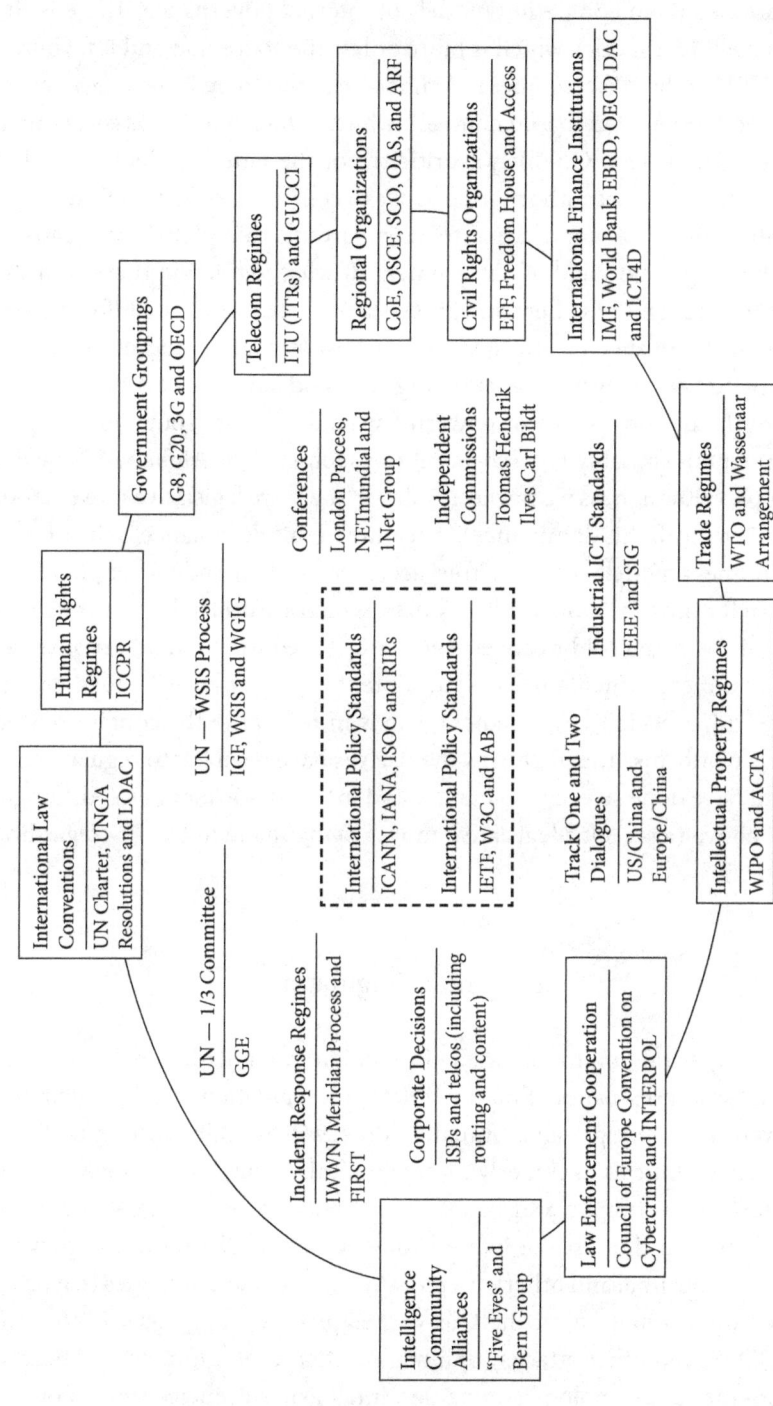

Figure 1.1 Managing Global Cyberactivities

there are two main competing models of internet governance. There is the multistakeholder model, which is more or less the status quo and is favoured by the US and the Western liberal democracies; and there is the cyber sovereignty model (i.e., state-based control), which is favoured by Russia, China, and their supporters. One major criticism of the latter model is that it is driven by the desire of authoritarian states to control information flows.

Another point of importance to us regarding this global landscape is the lack of regulation and of enforcement of what regulation there is in the face of the tsunami of cybercrime, interstate cyberconflict, disinformation, and so on. More specifically, there is a lack of regulation and enforcement of regulation with respect to ensuring free and fair markets of tax arbitrage, of the dominant role of powerful special interest groups (other than governments)—notably, large corporations such as Meta, Alphabet, Amazon, and Apple (although, as we saw in 1.5 above, there are legislative moves afoot in the EU to curb this dominance). A feature of this dominance is the underlying business model—namely, 'free' access to content and use of platforms in return for the provision of privacy data, which are utilized to generate revenue (e.g., by being sold to advertisers, as opposed to the standard business model of direct payment for access to content or for a service). A further regulatory and, especially enforcement, gap is in respect to the content posted on these platforms. The platforms themselves have been left to regulate content and have done so only reluctantly and in an ad hoc manner. (Although, see 1.5 above (again) for legislative moves being made in the EU regarding this issue.)

1.6.2 Smart Regulation

The more general point to be made in relation to regulation is the need for so-called smart regulation. Smart regulation is regulation which contributes to preventing or mitigating criminal, or otherwise morally unacceptable activity, thereby buttresses underlying sociomoral norms, but in a manner that facilitates the legitimate and beneficial purposes of institutions and technology. Thus, in the cases of interest to us, smart regulation would prevent or mitigate criminal and otherwise morally unacceptable activities (e.g., data destruction, corruption, or theft, as well as privacy violations, disinformation, DDoS, and other attacks) in a manner that the legitimate and beneficial activities of technology companies, intelligence agencies, and so on are facilitated rather than obstructed. This, of course, is easier said than done. Given many cyberattacks are forms of epistemic rather than kinetic action, and therefore the harms caused are indirect and incremental (e.g., such as in the case of spreading disinformation), governments are often reluctant to enact stringent regulations, even if such regulation is needed (Chapter 3). Moreover, here, as elsewhere, there are powerful interests opposed to needed regulation and, of course, given the complexity of these issues, the devil is very much in the detail.

One source of this complexity, which is not always recognized or acknowledged, arises from the limitations of new and emerging cybertechnology. Another arises from the dual-use character of cybertechnology (Chapter 7). In relation to the limitations, consider ML techniques. These are hugely beneficial in a wide range of areas, including, health (e.g., distinguishing melanomas from benign moles) and criminal justice (e.g., identifying offenders based on profiles developed using ML). However, ML has limitations; it is subject to error and bias (Chapter 4, Section 4.1). Consider, for instance, the racial bias introduced by the selection of a training data set comprised in large part of white males, as happened with the introduction of facial recognition technology utilising ML. Levels of accuracy were comparatively low in relation to identifying, for instance, black females.[60] Accordingly, regulation needs somehow to be able to respond to potential problems of error and bias, while the ethics of facial recognition are coming under increased scrutiny.[61]

The essence of ML is a training dataset, and thus the larger the better. There are two broad categories: supervised, in which the correct pattern category is specified for each item in the dataset; and unsupervised, in which the computer program has to determine suitable categories. Fairly obviously, supervised ML is subject to human error and bias, given that the 'correct' pattern is specified in advance and, therefore, might be incorrect by virtue of human error. By contrast, unsupervised might seem to be unbiased. However, consider the example of determining the make of cars from a data set of car pictures. Supervised, we assign each image of a car a brand (e.g., Audi, Toyota, and so on) This is objective. In the unsupervised case, we may end up with cars being classified, for example, by colour. To correct this, we would need to do something such as extracting from the car pictures the grill

---

[60] Alex Najibi, "Racial Discrimination in Facial Recognition Technology." https://sitn.hms.harv ard.edu/flash/2020/racial- discrimination-in-face-recognition-technology/. Harvard University, Science Blog: Accessed 27/10/2023.

[61] https://www.nature.com/articles/d41586-020-03187-3.

and badge logo at the front of each car. In other words, we need to *pre-process* the data, a process that can be subject to error or bias.

More generally, the phenomena that ML seeks to identify—and, in some cases, predict—often undergo change over time. To stay with our car example, over the last couple of decades cars have tended to become bigger and more rounded. Recently Rolls Royce did something that surprised quite a few people. It modified the iconic flat radiator grill to a more rounded shape. And herein lies the challenge: if we sample Rolls Royce uniformly over time, then most will have the old radiator grill. The new ones could be dismissed as errors. There are many facets of preparing and using data in ML, like this, which occur across algorithms and methods.[62]

These problems of error and bias in ML are potentially highly morally problematic in ways that need to be accommodated by regulations. Consider, for instance, that driverless cars using ML, which recently treated a stop sign as a speed limit sign after coloured dots were added to the stop sign. We discuss the limitations of ML in criminal justice contexts in Chapter 4, Sections 4.1 and 4.2.

What of the complexities confronting smart regulation that arise from the dual-use character of much cybertechnology? As is the case with, for instance, nuclear technology, cybertechnology can be used to provide enormous benefits to humankind, but in the hands of malevolent actors it can also be used to cause great harm. Consider, for instance, encryption. On the one hand, it protects the right to privacy. However, on the other hand, it is used, as we saw above, in ransomware attacks to potentially cause great harm. We discuss the dual-use problem in relation to cybertechnology in Chapter 7.

1.7 Social Institutions: A Normative Theory

Cyberattacks, even if they are perpetrated by individuals and target individuals, take place in institutional settings, such as the internet, which comprises technology embedded in a complex institutional landscape. The institutional dimension of cyberthreats and, therefore cyberattacks, is perhaps most obvious in the case of cyberattacks launched by hostile nation-states. However, it is also evidenced in the use of cybertechnology by

---

[62] D Heaven "Why Deep Learning AIs Are so Easy to Fool." https://www.nature.com/articles/d41 586-019-03013-5. Accessed 27/10/2023.

corporations, law enforcement agencies, and governments that may threaten individual rights to privacy and autonomy. In relation to corporations, it is especially evidenced in the use of cybertechnology by companies, such as Meta and Alphabet, whose core business is inherently cyber based. Social media, for instance, is inherently cyber based. Hence the threat posed by disinformation, propaganda, and hate speech on social media in the context of regulatory gaps in relation to content on social media platforms.

The institutions that comprise the institutional settings in which cyberthreats exist are institutions the purposes and activities of which are mediated, shaped, and in part constituted by technology, including cybertechnology. This is obvious in the case of social media platforms and security agencies that deploy often very sophisticated cybertechnology, but it is also true of businesses like finance and retail and their customers, all of whom rely on the internet, at the very least. An initial question to be asked regarding any new technology is what its purpose is and, more particularly, what its benefit is to society within the institutional settings in which it plays its mediating role. In terms of our favoured normative institutional theory (of which more below) the benefit to society is to be understood as a collective good, such as efficient, effective, and ethically sustainable public communication, and the role of technology is to assist or enable the institution(s) in which it plays its mediating role to provide or maintain the collective good(s) that are the raison d'être of that institution(s).[63]

As mentioned above, it is therefore crucial to consider the normative dimension of the institutional actors in cyberspace and in particular, to provide answers to questions, such as the following. What is the nature, magnitude, and moral significance of cyberthreats posed *by* various institutional actors (e.g., by hostile foreign governments, criminal organizations, big tech companies, law enforcement agencies)? For instance, are there internal and external threats to the institutions that are constitutive of liberal democracy and, perhaps, to liberal democracy itself? What institutional actors ought to be morally and legally responsible for which security measures in cyberspace (e.g., in respect of preserving privacy and curtailing disinformation)? In answering many of these questions, it is necessary to take into account

---

[63] Miller, *Moral Foundations of Social Institutions*, Chapters 1 and 2; Seumas Miller, "Design for Values in Institutions," in *Handbook of Ethics, Values & Technological Design*, ed. I. Poel, J. Van den Hoven, and P. Vermaas (Dordrecht: Springer, 2015), 769–81; Seumas Miller, "Ignorance, Technology and Collective Responsibility," in *Perspectives on Ignorance from Moral and Social Philosophy*, ed. Rik Peels (London: Routledge, 2017), 217–37.

the epistemic, as opposed to kinetic, nature of many cyberthreats. Moreover, in addressing some of these questions it may be necessary to rethink and redesign some existing institutions—notably, a diverse range of epistemic institutions, such as news media organizations and national security intelligence agencies, which may require reminding ourselves what the fundamental purposes of these institutions are. In relation to new and emerging institutions—notably, those constitutive of cyberspace, such as the internet itself or social media companies—it may be necessary to provide new normative theories. Accordingly, a salient normative concept or set of concepts relevant here is that of a public good in the economists' sense of that term,[64] collective goods,[65] the commons and common-pool resources (e.g., water in rivers, forests) and, relatedly, that of the social institution itself. Thus, one model of the internet that is often advanced is that it is a common pool resource, akin to the world's oceans—although, while the analogy is useful, it can be drawn too far.[66]

An important debate that has arisen in relation to this matter pertains to so-called common carriers and network neutrality. Arguably, the internet should have the status of a common carrier, akin to telephone networks and public highways. Network neutrality means applying common carrier rules to the internet to preserve its freedom and openness. Common carriage prohibits the owner of a network that holds itself out to all comers from discriminating against information by halting, slowing, or otherwise tampering with the transfer of any data, except for legitimate network management purposes, such as easing congestion or blocking spam.

On Miller's normative teleological account of institutions, as on more broadly functionalist accounts, the definition of an institution will typically include a description of the human good or social benefit that it purports to produce and, at least on Miller's theory, that good will be a collective good.[67] For example, universities purport to produce knowledge and understanding. Language enables the communication of truths; marriages and civil unions facilitate the raising and moral development of children; economic systems ought to produce material well-being, and so on. Such goods, or benefits,

---

[64] Public goods in the economists' sense are nonrival and nonexcludable. If a good is nonrival, then my enjoyment of it does not prevent or diminish the possibility of your enjoyment of it (e.g., a street sign is nonrival). Again, a good is nonexcludable if it is such that if anyone is enjoying the good then no-one can be prevented from enjoying it (e.g., national defence).

[65] Goods that are jointly produced (e.g., law and order).

[66] Klimburg, "The Darkening Web," 351.

[67] Miller, *Moral Foundations of Social Institutions*.

are collective in character and are generated by collective action in response to collective responsibilities. Hence, the notion of collective moral responsibility has a central role in the normative-teleological theory (Chapter 5, Section 5.3). Moreover, over the last hundred years or so special normative theories of particular institutions have become widely accepted, at least in liberal democracies (e.g., the normative theory of police organizations and of the so-called Fourth Estate). However, as we have seen, one of the effects of the existence and global utilization of the internet and of information and communication technology more generally, has been to disrupt preexisting institutional structures and call into question the normative theories that explain and justify particular institutions and their constitutive activity. An obvious example of this disruption is the tsunami of disinformation, propaganda, and hate speech (i.e., computational propaganda) that has resulted from the coming into existence of the social media companies, such as Facebook and Twitter. Moreover, this disruption has brought with it a new set of pressing ethico-institutional questions. Some of these pertain, as we have seen, to technology companies.

In relation to normative theories of the institutions constitutive of cyberspace, we need to distinguish the overarching normative theory—or perhaps theories—of cyberspace, including of the internet (e.g., the normative-teleological theory in terms of collective goods) and the governance model (e.g., the multi-stakeholder model). Both the overarching institutional theory and the governance model are normative insofar as both are prescriptive and not merely descriptive. Their respective adherents advocate their implementation, or their continued existence, in the case of the multistakeholder governance model. Moreover, normative institutional theories need to be anchored in descriptive reality. There is little point to a normative theory of institutions if it cannot be implemented in the actual world. However, the overarching teleological theory and the governance model are different, albeit related, normative accounts. Specifically, the governance model presupposes the institutional theory. We need to be clear on the nature and point of an institutional theory before we settle on its governance arrangements. In this respect, the internet, for instance, is no different from the market or from the criminal justice system, to take two obvious examples.

And there is a further point, a point regarding the relationship between normative theories and regulation. If one's institutional theory of, for example, the housing market holds that its ultimate purpose is the provision of a sustainable supply of housing of reasonable quality to a population and at a reasonable price,[68] rather than generating wealth for house owners, then one needs to regulate the housing market accordingly (e.g., by reducing incentives to speculate on house prices). By parity of reasoning, if one's institutional theory of, for instance, social media platforms holds that they are essentially common-pool resources to enable the communication of information (*inter alia*) then they ought to be regulated accordingly.

[68] Miller, *Moral Foundations of Social Institutions*, Chapter 10.

2

Privacy and Confidentiality

Bulk Data, Surveillance, and Encryption

It is agreed, on all hands, that there is an important right to privacy and, relatedly, autonomy, and an important principle of confidentiality. There are also ownership rights, such as the rights of companies to their intellectual property, including some of the data that they possess or use. In addition, there are moral rights to control data concerning features *constitutive* of human, familial, or personal identity (e.g., genetic and DNA data), which are evidently not simply privacy rights or ownership rights—at least in the sense of intellectual or other property rights (see Chapter 4, Section 4.3.1 and Section 4.4.1). Naturally, since these are rights to control data, like one's DNA profile, their exercise is an aspect of individual (or collective) autonomy.[1]

These various moral rights to privacy, autonomy, confidentiality, and ownership have a central role in a work on the ethics of cybersecurity for two main reasons. Firstly, as we saw in Chapter 1, cybersecurity in the narrow sense consists of data security, and the ethical significance of data security depends in large part on the moral rights to privacy, autonomy, confidentiality, or ownership of data. Thus, cyberattacks frequently target personal data (including genetic data), confidential data, or data that is owned by someone else. They do so for a wide variety of criminal or otherwise malevolent purposes—notably, theft. As noted in Chapter 1, global cybercrime can now be measured in terms of trillions of US dollars. However, the violations of the rights to privacy/autonomy or confidentiality (if not ownership) that these crimes typically involve is a moral cost above and beyond the resulting financial costs. Secondly, as we also saw in Chapter 1, ownership, privacy/autonomy, and confidentiality rights to personal data can be violated in

---

[1] Thus, the overlapping DNA profiles of members of a family might be regarded as a matter of collective or joint autonomy—as opposed to individual autonomy. For ease of exposition, we do not always distinguish here between types of personal data (e.g., between one's financial data and one's genetic data). For discussion of issues pertaining to genomic data and their implications for law enforcement, see Chapter 4, Section 4.3 and Section 4.4.

Cybersecurity, Ethics, and Collective Responsibility. Seumas Miller and Terry Bossomaier, Oxford University Press.
© Oxford University Press 2024. DOI: 10.1093/oso/9780190058135.003.0003

ways other than by cyberattacks. For one thing, such personal data can be collected, stored, and analysed in bulk electronic databases without the consent of the owners of that data (e.g., it can have been sold by companies that had the consent to collect and store the data but not to sell it). While this does not involve a cyberattack as such, it is a culpable failure to secure the personal data in question in cyberspace (i.e., it is a breach of cybersecurity). For another thing, personal data that has otherwise been legitimately collected, stored, and analysed can be used for a purpose other than the original legitimizing purpose, and therefore misused, since this would likely involve a violation of privacy, autonomy, confidentiality and/or ownership rights. Again, while this does not involve a cyberattack as such, it is a culpable failure to secure the personal data in question in cyberspace (i.e., it is a breach of cybersecurity). A further related issue in respect of privacy/autonomy pertains to the quantum of aggregated personal data made available to security agencies, even when consent is given (e.g., indirectly via democratically elected legislatures). Beyond a certain point, security agencies' establishment and use of integrated bulk databases of personal and other data may incrementally compromise the privacy/autonomy rights of individuals by virtue of this process, and perhaps over a lengthy period. This could result in an unacceptable power imbalance between a government and its security agencies, on the one hand, and the citizenry of the liberal democratic state, on the other hand. If so, it would be a failure—presumably, a culpable failure, if not addressed—to adequately ensure the *extent* of the security afforded to the personal data in cyberspace, meaning the government and its security agencies have come to have too much access to too much personal data. Therefore, the privacy/autonomy rights of the members of the citizenry (jointly held rights—see 2.1.1 below) have been violated by the government and its security agencies. These rights have been violated by virtue of the unacceptably reduced extent of security that the members of the citizenry now have in respect of their personal data vis-à-vis their own government and security agencies. Since the data in question is in cyberspace the security in question is cybersecurity. So, it is an ongoing, and perhaps escalating, breach of cybersecurity—notwithstanding the absence of a cyberattack.

We distinguish in what follows, as we have above, between privacy, autonomy, confidentiality, and ownership rights. They are not the same things (although privacy, confidentiality, and ownership rights to data all involve rights to control it, and therefore entail autonomy rights). For instance, one can own a business, including its name, and yet the name of the business would not be—indeed, perhaps could not be—private or confidential. So, one can have an ownership right to something to which one does not also have a privacy or a confidentiality right. Moreover, something to which one has a privacy right or a confidentiality right is not necessarily owned. For instance, a doctor might have a right and duty based on the principle of professional confidentiality not to divulge her knowledge of the medical conditions of her patients. However, it is the patient and not the doctor who has prior ownership or autonomy rights with respect to this information.

We discuss the difference, and the relationship, between privacy and confidentiality rights below. Here it suffices to offer the following example to illustrate the distinction. Consider a confidential government report on climate change. Leaking the report would be a breach of the government's right to confidentiality, but it would not necessarily involve a violation of anyone's right to privacy, since it may not contain any personal data.

In relation to privacy, autonomy, confidentiality, and ownership rights, it is important to distinguish between the different categories of 'things' to which these rights attach. On the one hand, there are privacy, autonomy, confidentiality, and ownership rights to data (i.e., abstract objects). On the other hand, there are privacy, autonomy, confidentiality and ownership rights to thoughts, actions, and cars (i.e., mental and physical objects). As we saw in Chapter 1, data are abstract objects with a physical embodiment; data belongs to Popper's third world, as opposed to the first world of physical objects or the second world of thoughts and other mental phenomena. If personal and other data are owned, in the sense that the relevant persons have moral (as opposed to legal) rights to these data, then, as is the case with other abstract objects to which persons have moral rights, they are instances of intellectual property, in the moral sense of that term.[2] Naturally, many abstract objects, including propositions that are known, are not instances of (legal or moral) property, intellectual or otherwise. Matters of common knowledge—for instance, that the world is round, or roses are red—are a case in point, as are some genomic data.

While we need to keep in mind the distinctions between privacy, autonomy, confidentiality, and ownership rights, our primary concern in this chapter is with privacy and confidentiality rights. Moreover, we need to

---

[2] Some corporations have managed to secure *legal* rights to genomic data (i.e., as a form of legal intellectual property). This is highly morally problematic as is argued in David Koepsell, *Who Owns You?: Science, Innovation and the Gene Patent Wars* (Hoboken: Wiley-Blackwell, 2015).

acknowledge that there are other potentially competing moral values, rights, and principles that are also important—notably, security, but also various freedoms that are constitutive of market-based institutions, such as principles of free trade. Since rights to privacy are not absolute, privacy rights can be overridden (as can confidentiality, ownership, and autonomy rights) by, for instance, the need for security from terrorist attacks. Thus, security requirements might dictate that emails between suspected terrorists be intercepted.

Two related sets of moral concerns that arise in relation to privacy and confidentiality in cyberspace are, firstly, the collection, storge, analysis, and use of bulk data (including personal data, such as health and financial information, as well as biometric data, like facial images), and secondly, encryption. Such bulk data collection, storage, analysis, and use is done not only by governments and their security agencies, but also by corporations and other organizations in the private sector. The first set of moral concerns is perhaps best encapsulated by the phenomenon of transnational criminal organizations engaged in all manner of cybercrimes on a vast scale, as well as by the spectre of the so-called surveillance society of which Tibet and Xinjiang in China are increasingly being seen as instantiations. Encryption is evidently part of the solution to these concerns, given that end-to-end encryption (see 2.3 below) affords strong protection of privacy and confidentiality, not only against criminal organizations but also against authoritarian governments. However, encryption gives rise to additional security concerns. For instance, it enables criminals and terrorists to evade detection. Before discussing the privacy and confidentiality concerns arising from bulk data and those arising from encryption, we need to get some clarity on the notions of privacy and confidentiality.

2.1 Privacy, Confidentiality, and Security

2.1.1 Privacy/Autonomy

The notion of privacy has proven difficult to adequately explicate.[3] Nevertheless, there are several general points that can be made. First, privacy

---

[3] Seumas Miller and John Blackler, *Ethical Issues in Policing* (London: Routledge, 2016) Chapter 4—83–110; Miller and Gordon, *Investigative Ethics* Chapter 10— 243–262; Samuel D. Warren and Louis D. Brandeis, "The Right to Privacy," *Harvard Law Review* 4, no. 5 (1890): 193–220; Charles Fried,

is a right that people have in relation to other persons, organizations, and governments with respect to: (a) the possession and/or accessing of personal information[4] about themselves by other persons, organizations, or governments (e.g., data stored in telecommunication company, technology company, or government databases); (b) the observation/perceiving/tracking of themselves—including of their movements, personal relationships and so on—by other persons (including public officials) by means of, for instance, CCTV (perhaps using facial recognition technology) or mapping metadata (e.g., caller number, duration of calls, location of caller/receiver) to determine geolocation history; and (c) the interception of their communications, such as their phone conversations, emails, and/or text messages.

Second, the right to privacy is closely related to the more fundamental moral value of autonomy. So much so that we will often refer to *privacy/autonomy rights*. Roughly speaking, the notion of privacy delimits an informational and observational 'space': the sphere of privacy. However, the right to autonomy consists of a right to decide what to think and do, and the right to control the sphere of privacy, therefore deciding *whom to exclude and whom not to exclude* from it, as well as the extent of that exclusion. Especially in the case of data, this means control of the use to which personal data is put by those granted access. So, the right to privacy consists in large part of the right of an individual to control—that is, a right of autonomy held against organizations and other individuals—access to and uses of personal data, and rights against observation and monitoring of the sphere of privacy.

Naturally, as already mentioned, the right to privacy is not absolute. It can be overridden. Moreover, its precise boundaries are unclear. A person does not have a right not to be casually observed in a public space, but arguably does have a right not to have their movements tracked via their smartphone, even if this right may be overridden under certain circumstances

---

"Privacy," *Yale Law Journal* 77, no. 3 (1969): 475–93; Daniel Solove, *Understanding Privacy* (Harvard: Harvard University Press, 2008); Julie C. Inness, *Privacy, Intimacy, and Isolation* (New York: Oxford University Press, 1992); Helen Nissenbaum, *Privacy in Context: Technology, Policy, and The Integrity of Social Life* (Stanford: Stanford Law Books, 2009).

[4] Here, as elsewhere, we need to distinguish between personal data with respect to properties constitutive of personal identity (e.g., one's DNA profile) and personal data not so constitutive (e.g., one's current salary). Both types of data give rise to privacy and confidentiality issues (and to that extent, as discussed below, to autonomy issues). However, for ease of exposition we will not always make this distinction and, in any case, the identity/autonomy issues particular to the former type of personal data are discussed in Chapter 4, Section 4.3 and Section 4.4.

(e.g., they are suspected of terrorism or carrying an infectious disease, such as COVID-19).

Moreover, whereas the right to privacy/autonomy is in large part a right of the single individual, this is not always the case. For instance, there are joint rights to privacy, such as in the case of an intimate sexual encounter, and the joint right of close relatives to a DNA profile (Chapter 4, Section 4.3.4).[5]

Further, the right to privacy/autonomy is really a set of related rights. Moreover, these constitutive rights are to some extent context dependent. Consider, for instance, a religious community in which everyone shares the same religion, and the religion is central to the community, and given daily public expression in it. Indeed, let us assume, membership in the community is dependent on commitment to this religion. In this community, the right not to divulge one's religious affiliation may not make much sense.

The third general point about privacy is that some degree of it is necessary simply for people to pursue their personal projects, whatever those projects might be. For one thing, reflection is necessary for planning, and reflection requires some degree of freedom from the distracting intrusions of others, including intrusive surveillance. For another, knowledge of someone else's plans can lead to those plans being thwarted, or even lead to the undermining of the institution in which such planned activity takes place. Plans might be thwarted if one's political or business rivals can track one's movements and interactions, and thereby come to know and thwart one's plans prior to their implementation. Democratic institutions, for example, might be undermined if the citizens' votes are not protected by a secret ballot, including the prohibition on cameras in private voting booths. Again, genuine freedom of choice in market-based institutions might be compromised if advertisers can create detailed profiles of potential customers, thereby enabling advertisers to use psychological techniques to manipulate customers' choices. Autonomy—including the exercise of autonomy in the public sphere—requires a measure of privacy.

The fourth general point about privacy is that the stringency of privacy and related autonomy rights to specific forms of personal data varies. The privacy right to personal financial data, such as one's annual salary, might

[5] Seumas Miller, "Joint Rights: Human Beings, Corporations, and Animals," *Journal of Applied Ethics and Philosophy* 12 (2021): 1–7; Seumas Miller, "Joint Political Rights and Obligations," *Phenomenology and Mind* 9 (2015): 136–45; Seumas Miller, "Collective Rights and Minorities," *International Journal of Applied Philosophy* 14, no. 2 (2000): 241–57; Seumas Miller, "Collective Rights," *Public Affairs Quarterly* 1, no. 4 (1999): 331–46.

be less stringent than the privacy right to health data, such as one's terminal illness. Again, the right to control one's DNA profile might be more stringent than the right to control one's fingerprint image. There are various criteria here for determining the stringency of privacy rights. One criterion is the potential harm that one might suffer if a specific privacy right is infringed, and this in turn might vary from one context to another. For instance, a potential employer's knowledge of a person's religious beliefs might threaten a person's career prospects in some circumstances but not in others. Another criterion is the centrality to an individual's personal identity of the specific type of personal data to which there is a privacy right. One's facial image and one's DNA profile, for instance, are central to one's personal identity, whereas one's annual salary is not. Accordingly, an individual has more stringent control rights over access to his or her DNA profile, or to the use of his or her facial image, than overaccess to or use of his or her annual salary data (see Chapter 4, Sections 4.3 and 4.4).

Thus far we have described privacy and autonomy, considered for the most part as the rights of a single individual. However, it is important to consider the implications of the infringement, indeed violation, of the privacy and autonomy rights of the whole citizenry, not only by criminal organizations, but also by the state; and also of the rights of large cohorts of consumers by powerful corporations. It is uncontroversial that violations by criminal organizations are morally and legally unacceptable. The main problem is enforcement of the law, assuming the law reflects prior ethical principles. However, violations by the state or by large corporations are more ethically and legally problematic. Such violations on a large scale can lead to a power imbalance between the state and the citizenry, and thereby undermine liberal democracy itself; or, can lead to a power imbalance between consumers and corporations, and thereby undermine the workings of free and fair markets. The surveillance system imposed on the Uighurs in China, which used a full range of technologies (e.g., bulk metadata collection, facial recognition technology, DNA profiling, and so on), graphically illustrates the risks attached to large-scale violations of privacy and related autonomy rights, if governments use them in a discriminatory manner.[6] Naturally, this

---

[6] John Kleinig, Peter Mameli, Seumas Miller, Douglas Salane, and Adina Schwartz, *Security and Privacy: Global Standards for Ethical Identity Management in Contemporary Liberal Democratic States* (Canberra: ANU Press, 2011), 1–289; Seumas Miller and Patrick Walsh, "NSA, Snowden and the Ethics and Accountability of Intelligence Gathering, " in *Ethics and the Future of Spying: Technology, Intelligence Collection and National Security,* ed. Jai Galliott and W. Reed (Abingdon: Routledge, 2016), 193–204; Adam Henschke, *Ethics in an Age of Surveillance: Virtual Identities and Personal*

is an extreme example but it serves to highlight the problem: where on the spectrum of surveillance options is the line to be drawn?

In light of the above analysis of privacy, and especially its close relationship to autonomy, we are entitled to conclude that some form of privacy is a constitutive human good. As such, infringements of privacy ought to be avoided. That said, privacy can reasonably be overridden by security considerations under some circumstances, as when lives are at risk. After all, the right to life is a weightier moral right than the right to privacy.

Individual privacy is sometimes confused with anonymity, but these are distinct notions. Anonymity is preserved when a person's identity in one context is not known in another. Anonymity can be a means to privacy, confidentiality, or avoidance of harm (e.g., reputational damage to a politician that might arise from the public disclosure of his extramarital affair). Indeed, anonymity is literally a matter of life and death in some situations. For example, facial recognition technology may reveal the real identity of an undercover operative to the criminal organization he has infiltrated, if that organization were to use facial recognition technology to compare the operative's current facial image with a facial image of him when he graduated from a police college some years before, which is available on the internet. Such examples demonstrate that anonymity is sometimes a very important *instrumental* good, which can also be instrumentally harmful, in some contexts (e.g., anonymous online bullies). But these examples do not demonstrate that it is a *constitutive* human good.[7] Indeed, anonymity in cyberspace is frequently relied upon by those engaged in illegal, immoral, or otherwise harmful behaviour (e.g., peddling child pornography, cyberbullying, hate speech, disinformation, and so on). Nor is it simply a matter of the harm done to victims by their anonymous perpetrators. For the perpetrators may ultimately be harming themselves by their actions, at least in the sense of corrupting their own moral character. Moreover, in a context in which anonymity (and, relatedly, secrecy—see below) was not available to them, the malevolent actions of potential perpetrators may be curtailed by the social disapproval, as well as formal sanctions, that would be visited upon them.

Information (New York: Cambridge University Press, 2017); Kevin Macnish, "Government Surveillance and Why Defining Privacy Matters in a Post-Snowden World," *Journal of Applied Philosophy* 35, no. 2 (2018): 417–32; Xiao Qiang, "The Road to Digital Unfreedom: President Xi's Surveillance State," *Journal of Democracy* 30, no. 1 (2019): 53–67.

[7] Miller and Walsh, "NSA, Snowden and the Ethics and Accountability of Intelligence Gathering."

They (as well as, of course, their potential victims) might be the better for this intervention. They might be less morally corrupt, and better socially adjusted. At any rate, we must conclude that anonymity is not a constitutive human good. In this respect, anonymity is quite different from privacy.

2.1.2 Confidentiality

While the concept of privacy is not the same as the concept of confidentiality, they are closely related. The sphere of individual privacy can be widened to include other individuals who stand in a professional relationship to the first individual, for example, a person's doctor. When it is so widened, it gives rise to confidentiality requirements on the part of the members of the profession in question. Moreover, morally legitimate institutional processes give rise to confidentiality requirements with respect to information. For instance, law enforcement operations give rise to stringent confidentiality requirements, given what is often at stake (e.g., the outcome of important investigations that could be compromised by exposure, or, as mentioned above, the risk to an undercover operative if his or her identity is revealed).[8] At least in the case of security agencies, such as police, military, and intelligence agencies, a degree of compliance with principles of confidentiality is a *constitutive institutional* good in the sense that security agencies could not successfully operate without a high degree of confidentiality in respect of their intelligence data, methods ('tradecraft') and operations.[9] That is, confidentiality is a necessary feature of security agencies, which is not to say that a degree of transparency is not required to ensure, for instance, appropriate accountability.

Confidentiality might be referred to as informational security. However, information that is not confidential can be secure, and some confidential information is not secure. Therefore, confidentiality of information and informational security are different concepts. Nevertheless, they are related insofar as, other things being equal, confidential information should be held securely. Confidential information and informational security should go together. Moreover, confidentiality is, as we saw above, often based on privacy (e.g., the confidentiality of personal information). Accordingly, not only is privacy not necessarily in conflict with security, but privacy often depends

---

[8] Miller and Gordon, *Investigative Ethics.*
[9] Miller and Walsh, "NSA, Snowden and the Ethics and Accountability of Intelligence Gathering."

on security. On the other hand, the integration or interlinking of databases of confidential information is potentially problematic from a privacy and autonomy perspective, as the example of China's surveillance system in demonstrates.

Another related notion of interest to us here is secrecy. Secret information should not be confused with, or morally equated to, private or confidential information. For unlike privacy and confidentiality, secrecy is a morally neutral or even pejorative notion. Secrecy is desirable in legitimate institutional settings of conflict and fierce competition, for example in market-based companies. However, high levels of secrecy are often undesirable in both institutional and interpersonal settings. For secrecy can mask incompetence, lying, cheating, corruption, illegality, and, in the case of authoritarian regimes, human rights abuses. Indeed, even in liberal democracies there is the risk that if the use of databases is not closely monitored and appropriately transparent then they will be used for unintended purposes, such as surveillance, and thereby enable function creep (see below). Speaking generally, those engaged in behaviour that is illegal, harmful, immoral, or otherwise generally disapproved of tend to want to do so in secret. Accordingly, secrecy is at odds not only with compliance with laws and regulation, but, perhaps more fundamentally, with adherence to the sociomoral norms that underpin laws and regulations. If secrecy allows those engaged in illegal activity to evade formal sanctions, it also allows those engaged in immoral or otherwise harmful activity to evade informal sanctions, like social disapproval. Accordingly, in contrast with confidentiality, secrecy is not a constitutive institutional good, and in contrast with privacy, it is not a constitutive human good. Indeed, in both institutional and interpersonal settings, transparency and disclosure are typically to be preferred to secrecy. This is obviously the case in institutional settings. However, it is also the case in interpersonal settings. Consider, for instance, the problem of child sexual abuse, a crime that is typically perpetrated by relatives or by family friends.

With respect to specific privacy rights, we have distinguished degrees of stringency. Some privacy rights are more stringent than others. We have also distinguished privacy, autonomy, anonymity, confidentiality, and secrecy, and argued that whereas privacy is a constitutive human good—in part by virtue of its relation to autonomy—and confidentiality, a constitutive institutional good, neither anonymity nor secrecy are constitutive goods.[10]

---

[10] Ibid.

Moreover, given the close relationships between privacy and confidentiality, the sharp contrast often drawn between privacy and security does not necessarily obtain.

2.1.3 Security

The notion of security is somewhat vague. Sometimes it is used to refer to various forms of collective security, notably national security (e.g., harm to the public from a terrorist attack), community security (e.g., disruptions to law and order by violent political demonstrations), and biosecurity (e.g., threats to public health and society caused by bioweapons). At other times, it is used to refer to personal physical security (e.g., from threats to one's life or limbs). Importantly, threats to personal physical security, when scaled up, constitute threats to collective security, as in the case of a terrorist bomb attack.

For our purposes here a critical notion of security is, of course, cybersecurity. In Chapter 1 we distinguished the narrow notion of data security from the wider notion of the security of cyberspace.

As mentioned, data security is the security of a person's or organization's data, software, and the like, from especially theft, corruption, or destruction. Here, again as discussed in Chapter 1, it is important to see that there are multiple different sources of data security threats, and correspondingly multiple different potential victims. Some of these threats might be legally and/ or morally legitimate, although many are not. Consider the personal data of a citizen. It could be accessed without his consent by a fellow citizen, a criminal organization, a corporation, a law enforcement agency, a foreign power, and so on. In each of these cases, there has been a breach of data security. In some cases, such as a morally unjustified and unlawful accessing of personal data by members of a criminal organization, the breach is obviously morally and legally illegitimate. However, in other cases, such as an authorized accessing of personal data by law enforcement because there is a reasonable suspicion of criminal activity, then a breach of data security might well be both morally and legally legitimate. Nevertheless, this latter accessing by law enforcement remains a breach of data security, and is an *infringement* of his right to privacy. However, it is not a *violation* of his right to privacy. More generally, it is a mistake to regard breaches of data security as necessarily consisting only of unlawful, or morally illegitimate, breaches.

The notion of the *scope* of security—including, but not restricted to, cybersecurity—refers to the extent of the persons or things to be secured against a threat (e.g., an individual person or a collective). The scope of security (e.g., the personal, organizational, and national levels) can be distinguished from the *type* of security. There are different types of security because there are different threats to different kinds of things. In respect of cybersecurity, we have already distinguished data security from the broader notion of the security of cyberspace. Data security consists at its core[11] in ensuring that information held in electronic databases and associated software, including private and confidential information, is protected from unauthorized (by the possessors of the databases) or otherwise illegitimate accessing for purposes of theft, corruption or destruction. Encryption (see section 2.3 below) plays a key role in ensuring data security. Clearly, data security is critical in the face of sustained hacking by state and non-state actors that can compromise privacy and confidentiality. Other types of security pertain to physical or psychological harm to human beings, damage to physical objects, and certain forms of harm to institutional processes or purposes (e.g., as a result of widespread corrupt institutional activity such as bribery of public officials). In modern societies, many of these latter types of security are dependent on data security.

Aside from the scope and types of security, there are also various *contexts* of security. These include crime, counterterrorism (CT), war, cyberwar, trade wars, and so on. Moreover, the stringency of privacy rights and confidentiality requirements need to be relativized to context. In wartime, for instance, military intelligence gathering is largely unfettered, and the privacy rights of citizens curtailed by emergency powers. By contrast, in domestic law enforcement there is a strong presumption in favour of the privacy rights of citizens. Moreover, in domestic law enforcement there is likely to be increased accountability when privacy/autonomy rights are overridden. For instance, police might not have the legal right to access personal information other than by means of a judicial warrant, issued by a magistrate, under limited conditions. Counterterrorism in well-ordered jurisdictions is typically a matter of law enforcement. However, in war zones, such as combating the Islamic State in Iraq, Syria, and elsewhere, counter-terrorism operations—including intelligence gathering—are military in character. Below we turn to

---

[11] Notwithstanding cases of morally justified unauthorized accessing.

bulk databases and associated data analytics and their implications for privacy and confidentiality.

A final point pertains to the multiplicity of interdependent means to ensure security. Naturally, enforceable laws and regulations come to mind. However, it is also important to stress the importance of reducing the opportunities for breaches of security. Additionally, the role of sociomoral norms should not be ignored (e.g., in respect of promoting the attitudes and behaviour that reduce security violations in cyberspace as elsewhere). We note that, in contexts of rapid institutional and technological change, there is often a need to develop and internalize new security laws, regulations, and protocols. These may need to be given direction by ethical guidelines. Such guidelines are themselves based upon more fundamental moral rights, responsibilities, and principles, which are hopefully embodied in, and animated by, existing underlying sociomoral norms. Consider, for instance, the moral responsibility to protect the vulnerable, upon which the duty to report instances of child sexual abuse might be based. This latter moral duty may in turn provide guidance to legislators and regulators, focused on the problem of child sexual abuse in a new domain, such as cyberspace. However, compliance with any resulting regulations will ultimately depend on the commitment of members of the relevant social group or online community to their moral responsibility in this regard and, therefore, to the underlying sociomoral norm. Formal sanctions, while necessary, will typically not be sufficient for regulatory compliance.

2.2 Bulk Data and Surveillance

2.2.1 Governments and Security Agencies

Governments, law enforcement, and intelligence agencies (e.g., the US National Security Agency [NSA] and the UK Government Communications Headquarters [GCHQ]) increasingly rely on bulk data and associated data analytics. Regarding bulk databases established and used by governments and security agencies in the service of security, consider the following recent developments.

After 9/11, the US Foreign Intelligence Surveillance Court (FISC) authorized the collection of bulk telephone metadata (e.g., the unique phone number of caller/recipient, the time of calls, their duration, and the location of caller/recipient), allowing the NSA access to all call records.[12] The whistle-blower, Edward Snowden, revealed information about this telephone metadata program. His revelations included material about the NSA's PRISM program, which allows the agency to access a large amount of digital information (e.g., emails, Facebook posts, and instant messages). The difference between telephone metadata collection and PRISM is that the latter also collects the *contents* of those communications. In addition, the so-called Verizon program involved the collection by the NSA of the metadata from the calls made within the US, and between the US and any foreign country, of millions of customers of Verizon and other telecommunication providers. An important difference between the Verizon program and PRISM is that the latter involved agreements between the NSA and various US-based internet companies (e.g., Google, Facebook, Skype and so on) to enable the NSA to monitor the online communications of non-US citizens, based overseas.

More recent developments include the establishment of bulk biometric—including genomic, such as DNA—databases and the use of face-recognition technology.[13] In relation to databases, the UK's National DNA Database (NDNAD) now holds the DNA profiles of 10 percent of its population, and China has collected the DNA profiles of the entire populations of Tibet and Xinjiang.[14] Moreover, tens of millions have submitted their genomic data for testing to commercial providers, such as Ancestry.com. Further, there are now bulk databases of facial images scraped from the internet by, for instance, the company Clearview AI. So, there are now large databases of biometric and genomic data in the private sector. Clearview AI developed facial recognition technology that has been used by law enforcement, and potentially private-sector organizations, to identify unknown individuals from its huge database of billions of facial images.

We need to note three developments that are constitutive of the so-called surveillance society, which raise privacy concerns. Firstly, it is not only a matter of the huge quantity of personal data collected and stored, but also of the wide range of categories that constitute such data (e.g., phone and internet metadata and content, financial and health data, DNA data, CCTV video footage, and facial and other images from social media). Secondly,

---

[12] The FISC was established to provide judicial oversight of intelligence agencies (the NSA and the FBI) seeking interception of communications of suspects.

[13] Marcus Smith and Seumas Miller, "The Ethical Application of Biometric Facial Recognition Technology," *AI and Society* 36 (2021): 167–75..

[14] David Kaye and Michael Smith, "DNA Identification Databases: Legality, Legitimacy, and the Case for Population Wide Coverage," *Wisconsin Law Review* (2003): 413–59.

there is the interlinkage of these databases and the application of powerful analytical tools, such as machine-learning (ML) techniques, to analyse this data and, as in the case of China, to assess citizens in accordance with its social credit system. Thirdly, the use of surveillance technologies, such as CCTV cameras and facial-recognition technology, not only in order to provide data for the profiles of rapists, corrupt officials, terrorists and so on, but in order to track the movements and activities of ordinary citizens.

These privacy and confidentiality concerns regarding the surveillance society are not the only concerns. For instance, there is also the relationship between privacy and autonomy. As already noted, privacy is not an absolute right, and indeed privacy rights (and confidentiality requirements) can be overridden by other considerations, including security.[15] Privacy rights can also be overridden by public health concerns, under certain circumstances, as we shall argue in Chapter 5. In short, as is the case with encryption, while the establishment of bulk databases and associated analytical technologies has significant moral costs (e.g., with respect to privacy), it also provides significant benefits. Hence, the need for empirically informed ethical analysis. The ethical analysis in question needs to identify salient ethical or moral principles.

There are several moral and legal principles that are typically utilized to justify and constrain the collection, analysis, and dissemination of bulk data. These principles include: (1) the degree of stringency of the privacy or confidentiality right to the data in question; (2) the principle requiring a morally legitimate purpose (e.g. detection of criminal activity); (3) the law enforcement principle that is analogous to the principle of discrimination (or distinction) applicable in war proscribing the deliberate killing by combatants of non-combatants, but which proscribes the deliberate targeting by police officers of those who are neither offenders nor suspects; (4) the principle of necessity (e.g. the use of facial-recognition technology needs to be demonstrated to be necessary) and; (5) the principle of proportionality (e.g. universal DNA databases need to be demonstrated to be proportionate to, for instance, the current terrorist threat).[16] However, the overarching principles governing the moral legitimacy of police and intelligence agencies' activities in liberal democracies are also applicable, and in particular the principles of

---

[15] Tom Sorell, "Privacy, Bulk Collection and Operational Utility," in *National Security Intelligence and Ethics*, ed. Seumas Miller, Mitt Regan, and Patrick F Walsh (London: Routledge, 2021).
[16] Miller, "Rethinking the Just Intelligence Theory of National Security Intelligence Collection and Analysis."

democratic consent and responsibility for the security of ordinary citizens; and accordingly, even if the establishment of a particular bulk data collection, accessible by law enforcement, is otherwise morally justified, it must also be consented to by the citizenry. Thus, the citizenry might consent to a police database of the fingerprints of all those who have merely been arrested on suspicion of a crime (as well as, of course, those convicted of a crime). However, the citizenry might not consent to a counterpart DNA database (Chapter 4, Section 4.3)—notwithstanding that the DNA database has more or less the same degree of compliance with these moral and legal principles (i.e., discrimination, necessity, and proportionality). And there is this further point. If the establishment of a particular bulk database (e.g., of fingerprints) is compliant with these moral and legal principles, and is consented to by the citizenry—and is, therefore, morally justified—then the citizens may have a collective moral responsibility to provide their personal data.

Criminal justice and national security intelligence activity exists at both a micro- and a macrolevel (as long as this distinction is understood as being a fairly loose one). The microlevel is the level of specific operations (e.g., intrusive surveillance of the murder suspect John Glover, the 'Granny Killer').[17] This is a level at which the principles of discrimination, necessity, and proportionality are manifestly applied. Thus, consistent with the principle of discrimination or distinction ("Do not deliberately target the innocent") as it applies in criminal justice contexts, there is surveillance of a suspect or known offender; the use of highly intrusive surveillance technology is necessary (i.e., if, for instance, metadata collection would not be sufficient), and; the degree of intrusiveness of the surveillance is proportionate (i.e., murder, for instance, is a very serious crime). But intelligence activity also exists at a macrolevel, and this has implications for the application of the principles of discrimination, necessity, and proportionality.

Consider bulk data collection by national security intelligence agencies. Intelligence activities, ultimately aimed at identifying terrorists and thwarting acts of terrorism, often now involve the application of ML techniques to bulk databases, which consist in the main of the communications and other data of innocent civilians, indeed, frequently innocent fellow citizens (i.e., such data is deliberately collected and accessed). Prima facie this is a violation of the principle of discrimination. In response, it can be argued that while the bulk data of these innocent persons is read by a machine, or perhaps only

---

[17] Miller and Gordon, *Investigative Ethics*, 26–28.

seen by human eyes in an anonymized form, it is for the most part not seen in a manner that constitutes an infringement of privacy. Of course, the particular data items that result from the application of the ML process are de-anonymized and, ultimately, seen by human eyes. However, the argument might continue, such data meets the standard of reasonable suspicion, already applicable to intelligence gathering and investigation by law enforcement agencies, by virtue of being the result of that very process. Whatever the merits of this argument as a justification for the application of ML techniques to bulk databases by way of mitigating the degree and extent of intrusion into the privacy of innocent citizens, this intrusion into the privacy of innocent civilians is nevertheless deliberately done. As such, it remains an infringement of the principle of discrimination, even if a morally justified infringement. Moreover, it is not analogous to the principle of discrimination as it applies to the use of lethal force by combatants since combatants in war are not permitted to *deliberately* kill innocent civilians even as a means to some further legitimate end. The reason for this difference between the principle of discrimination as it applies to intelligence activities and as it applies to the use of lethal force reflects the much greater moral significance that attaches to deliberately overriding an innocent person's right to life than attaches to deliberately overriding their right to privacy. This difference in significance in turn reflects—indeed in large part is derived from—the much greater moral weight that attaches to life than to privacy. Hence there is a (more or less) absolute legal prohibition of deliberately killing the innocent, even in wartime, but not of deliberately overriding their privacy, even in peacetime.

At this point, the principle of consent is relevant, perhaps in the form of implied and/or indirect consent (e.g., via elected representatives). There is also the possibility of hypothetical consent, which according to some philosophers has considerable moral weight, at least under some circumstances.[18] For it might be that *innocent* citizens, appropriately informed, would not object to their personal data being stored for a limited period in a particular bulk database if it were in an anonymized form—and, assuming this was necessary to enable law enforcement to protect them from terrorists, for example. Moreover, it might be that innocent citizens who are, nevertheless, suspected of a crime as the result of a process of analysis data in a particular bulk database would not object to this process and the

---

[18] John Rawls, *A Theory of Justice* (Cambridge, MA: Harvard University Press, 1971) and Michael Skerker, *The Moral Status of Combatants* (London: Routledge, 2020), 126–28.

subsequent investigation of them, if they believed it would be exculpatory—and, again, assuming this was necessary to enable law enforcement to protect them from, say, terrorists. Of course, offenders (e.g., terrorists), would object to such bulk databases and associated analytics since it would enable their detection and arrest. However, *offenders'* objections would be overridden by the threat that they pose to their fellow citizens and the liberal democratic state. More problematically, innocent citizens, especially if they were suspects, might not trust their government or security agencies, whether justifiably or no, and therefore may refuse their consent for this reason. However, if so, this would be a consideration in favour of building trust (e.g., by improving the performance of intelligence officers and investigators), and by implementing accountability measures, rather than abandoning bulk databases. Naturally, there will always be those who will be unmoved by even well-intentioned, significant, trust-building measures. However, in a liberal democracy it is at least in principle possible that such measures will be effective in building trust among the vast majority of innocent citizens, including suspects. Yet, even in such a liberal democracy there will likely be a small minority of innocent citizens who continue to distrust their government or its security agencies. Accordingly, full consent will not be forthcoming. At this point, the consent of the vast majority, taken in conjunction with the need for security, may well override the rights to privacy of this minority.

Key ethical issues at the macrolevel pertain to the necessity and proportionality of the establishment and general uses of the bulk databases themselves.[19] In his influential UK report, David Anderson distinguishes between bulk interception, bulk acquisition, bulk equipment interference (e.g., hacking into computerized devices and copying material), and bulk personal datasets (e.g., electoral roles, passport database, driving license database, national insurance numbers, and passenger name records from flights [PNRs]).[20] He also distinguishes between databases held by the security agencies and their accessing of databases held by other agencies (e.g., private sector firms). His concern was with the former. Regarding the necessity of

[19] David Omand recommends the distinction between laws and their application as being serviceable in the attempt to understand how *jus ad intelligentiam* and *jus in intelligentia* might relate to national security intelligence activities. See David Omand and Mark Phythian, *Principled Spying: The Ethics of Secret Intelligence* (Oxford: Oxford University Press, 2018), 99. See also Scott Robbins, "Bulk Data Collection, National Security and Ethics," in *Counter-Terrorism: The Ethical Issues*, ed. Seumas Miller, Adam Henschke and Jonas Feltes, *Counter-Terrorism: The Ethical Issues* (Cheltenham: Edward Elgar, 2021).

[20] David Anderson, "Report of the Bulk Powers Review," 2016. https://terrorismlegislationrevie wer.independent.gov.uk/wp-content/uploads/2016/08/Bulk-Powers-Review-final-report.pdf.

establishing and utilizing these databases, Anderson said: "The bulk powers play an important part in identifying, understanding, and averting threats in Great Britain, Northern Ireland and further afield. Where alternative methods exist, they are often less effective, more dangerous, more resource-intensive, more intrusive, or slower."[21] Clearly, given, for instance, the existence of alternative methods that are more resource-intensive, this is a relatively weak, and therefore permissive notion of necessity.[22]

Anderson did not address the question of proportionality in his report. To do so, we would need to distinguish between the proportionality of establishing a particular database for national security purposes, as opposed to accessing and analysing (for national security purposes) an existing database created for a purpose other than national security. Moreover, the weight to be accorded to the right to privacy in any such application of the principle of proportionality is a complex matter, not least because of the close, in part conceptual, relationship between privacy and other fundamental rights, such as the right to individual autonomy in the context of the liberal democratic concern not to allow individual autonomy vis-à-vis the state to be compromised. Evidently, the application of the principle of discrimination at this macrolevel is problematic insofar as the databases in question necessarily contain the data of citizens who are innocent of any national security breach—indeed, most of the data in many of the databases pertain to innocent citizens. Nor is this problem necessarily entirely resolved, even if it is considerably mitigated, by virtue of, for instance, the anonymized form in which personal data in these databases exists in the collection and filtering phases of the national security intelligence process. For instance, it might be argued that the personal data is owned, and therefore there is a right to control its collection. On the other hand, even if this is conceded, it might be countered that this right to control is overridden in the contexts in question (assuming the data is anonymized and infringement of the right thereby mitigated).

There is also the question of the relationship of the microlevel to the macrolevel from the perspective of the application of the principles of discrimination, necessity, and proportionality. For instance, in order

---

[21] Ibid., Chapters 5–8.

[22] Assuming, of course, that the principle of necessity is what Anderson had in mind. But if he did not have the principle of necessity in mind, then what principle did he have in mind? See Kevin Macnish, *The Ethics of Surveillance* (London: Routledge, 2017), ch. 5, for an account of the ethical issues in this area.

successfully to undertake a single microlevel CT operation it might not be necessary to establish bulk databases for national security purposes (or, for that matter, access existing bulk databases). However, in order successfully to undertake a large number of microlevel CT operations in a short period it may well be necessary (as per Anderson's report) to establish such bulk databases (or to access existing bulk databases). Again, taken in aggregate, the nature and extent of the infringements of privacy of innocent citizens resulting from the accessing of databases of personal information might be held not to be disproportionate to the aggregated outcomes of successful CT operations that relied on accessing the data in question. Note that compliance with the principles of necessity and proportionality at the macrolevel does not entail compliance with these principles at the microlevel (i.e., does not entail compliance with these principles on each and every specific intelligence collection operation). This is in part because microlevel operations might be justified ultimately in terms of their contribution to macrolevel outcomes. For instance, spreading the intelligence-gathering net wide and over a long period of time might enable insight into an entire terrorist network and its activities by virtue of a detailed process of identifying direct and indirect links between multiple individual terrorists and their associates—associates who might not themselves be terrorists or even terrorist sympathizers (e.g., they might be the unknowing relatives of terrorists). This is so notwithstanding that accessing the personal data of a person who was not a suspect, but merely thought (falsely, as it turns out) to be a potential associate of a terrorist might not—*considered on its own*—be justified by the principles of discrimination, necessity, or proportionality.

The principle of proportionality needs to take into account not only the somewhat vague character of the end or goal of national security (definitive of the principle of necessity) and the obstacles faced by intelligence officers (e.g., high level encryption), but also the potential harms arising from national security intelligence activities, and in particular from the utilization of bulk data. To reiterate, privacy concerns in this area are somewhat mitigated by the fact that the bulk data collected and analysed is typically in an anonymized form (e.g., by means of ML techniques), and therefore arguably only the privacy rights of genuine suspects are infringed or, perhaps, seriously infringed (i.e., the individuals identified upon completion of the analysis). However, these harms, such as the weakening of individual autonomy vis-à-vis the state, arising from extensive privacy infringements by intelligence agencies, and a diminution in public trust (a collective good—see

Chapter 5, Section 5.2) as a consequence of the secret nature of national security intelligence activities, may be incremental, difficult to quantify, and cumulative.[23] Please also note that, considered at the macrolevel, the harms in question are potentially various in terms of our previously mentioned taxonomy of harms. For instance, since intelligence officers are themselves citizens, their intelligence activities might turn out to be (indirectly and incrementally) a form of collective *self*-harm, given their membership in the harmed collective.

Accordingly, it can be difficult to know exactly where to draw the line between proportionate and disproportionate intelligence activities when it comes to the utilization of bulk data for national security purposes. Consider in this connection the potential utilization of integrated biometric and nonbiometric databases. One prominent concern about the inadequacy of privacy protections is the potential for 'function creep', where the use of information taken for a particular purpose is used for purposes other than those for which consent was obtained. The underlying concern in relation to function creep is the threat to individual autonomy posed by comprehensive, integrated biometric and nonbiometric databases utilized by governments and their security agencies in the service of ill-defined notions of necessity and national security without appropriate regulatory constraints and democratic accountability.

2.2.2 Corporations

What of the collection and use of bulk personal data by corporations? We have already mentioned the examples of Ancestry.com and Clearview AI, but only in terms of the use of this data by governments and security agencies. However, there are also individual privacy concerns in relation to private sector organizations—hence, the potential appropriateness of Zuboff's notion of a surveillance society.[24] As mentioned in Chapter 1, a particular development is the business model according to which individuals provide their personal data in return for free use of internet services. As also mentioned in Chapter 1, an egregious example of privacy violations in the overall context

---

[23] The notion of a collective good is elaborated in Miller, *Moral Foundations of Social Institutions*, Chapter 2.
[24] Zuboff, *Age of Surveillance Capitalism*.

of this business model is what is referred to as Real-Time Bidding (RTB) for advertising.[25] This essentially involves onselling personal data for advertising purposes. More than half of UK council websites use RTB ad auctions. RTB faces multiple GDPR investigations for systematic data breaches because it broadcasts people's personal data to countless companies. Google's RTB shares data with hundreds of companies, without any assurance of who that data is then shared with or how it will be used.

As already mentioned, technology corporations, such as Alphabet, Meta and Amazon, have been collecting very large amounts of data from their users (e.g., those who conduct searches on Google and those who communicate with their friends on Facebook), and doing so without their knowledge, let alone consent—or, at the very least, without their consent until the recent enactment of the GDPR, which only covers the EU and those who interact with the EU. Accordingly, this bulk data—or, at least a good deal of it, depending on which particular kind(s) and extent of data is in question—has been collected in violation of the privacy/data-control rights of users of Google and Facebook services. (There are complex ethical questions concerning privacy/data-control rights in relation to the data collected and onsold by the technology corporations; see below for discussion.) Moreover, data analytics (e.g., ML techniques) have been deployed to structure this data in a manner suitable for commercial purposes—notably, advertising purposes. For instance, profiles of customers are developed to enable better targeted, and therefore more efficient and effective advertisements. The corporations using this data for commercial purposes include not only the corporations who originally collected the data, but also the myriad of other corporations to whom Google and Facebook onsell the data (until recently without the knowledge or consent, of those from whom it was extracted).

According to Zuboff, these commercial activities are not simply to be understood as violations of privacy/data control rights, [26] or, as she puts it, the extraction of "behavioural surplus." For the quantum of data in question, and the power of the data analytics used, is such as to enable the creation of 'predictive products'. For instance, a bank might use such data to create profiles of their customers that are much more accurate than the profiles they could create based on their customers' use of the bank's existing products alone. They could then go on to use these more detailed profiles to create

---

[25] https://www.zdnet.com/article/google-accused-of-leaking-personal-data-to-thousands-of-advertisers/. Accessed 6/7/2020.
[26] Zuboff, *Age of Surveillance Capitalism*.

new financial products that target their customers much more precisely. Thus, "one recent study used the mobility data generated by 100,000 bank customers' cell phones over a one-year period to predict with very high accuracy their likely demand for a given loan product."[27] Given this predictive ability and the ability to use manipulative techniques (e.g., subliminal advertising and so-called nudges), the possibility of behavioural modification emerges, although Zuboff herself emphasizes the predictive ability, as opposed to what we take to be its conceptually distinct manipulative techniques.[28] Of course, the power of manipulative techniques is enormously enhanced by predictive ability.

Arguably, Zuboff has, firstly, overstated the extent of the privacy/data-control problem in market-based institutions (e.g., the technology corporations are only one industry, albeit a very important one). Secondly, Zuboff has understated the benefits of bulk databases and associated data analytics (e.g., health benefits and the benefits to consumers from better alignments of products with consumer desires). Thirdly, Zuboff has somewhat misdiagnosed the problem (e.g., the power conferred on the state by bulk data and data analytics, including the power of foreign states to intervene in democratic processes, such as occurred with the involvement of Cambridge Analytica in the 2016 US presidential election).

That said, there are, nevertheless, various ethical problems in this area that need to be identified. Here it is helpful to invoke the distinction between the micro-, or individual, level and the macro-, or institutional, level.

At the microlevel, the supposedly freely given data is not in fact autonomously given if, firstly, one is unaware that this data is being extracted, or, secondly, one does not have a reasonable opportunity not to freely give it (e.g., if one is dependent on Google or another oligopolist for access to a search engine, which is a basic necessity and moral right in today's digital world). Nor can the principle of caveat emptor ('buyer beware') be invoked at this point to justify these practices of the technology corporations. For the principle of caveat emptor can only reasonably be used to justify these actions of sellers, if, firstly, the buyer is aware that s/he is a buyer in a market, and, secondly, the

[27] Mariano-Florentino Cuéllar and Aziz Z. Huq, "Review of Zuboff's *Age of Surveillance Capitalism*," in *Harvard Law Review* 133 (2020): 1291n51, who reference in turn Cagan Urkup et al., "Customer Mobility Signatures and Financial Indicators as Predictors in Product Recommendation," *PLOS ONE* 13 (2018): 1, 2–5.

[28] See Richard Thaler and Cass Sunstein, *Nudge* (London: Penguin, 2009) for an account of nudging in the service of good ends but nudging (e.g., making one product more salient than another by placing it in a more prominent position, is a means and, as such, can be used for bad ends).

market is not a market in an essential good to which buyers do not have access other than via the monopolistic or oligopolistic seller in question.

Further, the data in question once analysed can, at least in principle, be used to develop detailed profiles of individuals and/or track their movements and activities to an extent that is inconsistent with the individual right to privacy. Thus, cell phones transmit the location of their users to service providers, yielding a detailed map of users' movements. Arguably, any such map should be under the control of the individual whose movements it maps. It would be one thing for law enforcement to have access to such a map if the individual in question was suspected of a serious crime, but it is quite another for this map to be aggregated with others and analysed for advertising or other commercial purposes without the consent of the individuals concerned.

At the macrolevel, one important concern pertains to the institution of what purport to be free-markets and the imbalances of power between technology corporations and the users of their services. The technology corporations are de facto monopolies or oligopolies in the provision of what are in many cases essential services (e.g., search engines and access to social media). As for the users of these services, their moral rights to privacy/data control and their economic interests (e.g., the economic benefits of the bulk data extracted from their activities) are without adequate institutional including legal protection—notwithstanding recent regulation, such as GDPR. However, under additional proposed EU legislation—specifically, the Digital Services Act (DSA)—new obligations on Facebook, YouTube, and Twitter would be created—backed by substantial fines for noncompliance—to reveal information and data to regulators about how their algorithms work, how decisions are made to remove content, and how adverts are targeted at users. Moreover, under another EU legislative proposal, the Digital Market Act (DMA), monopolistic technology companies and anticompetitive practices would be targeted. Sanctions would include the possibility of breaking up companies, such as the parent companies of Google (Alphabet) and Facebook (Meta).

2.2.3 Bulk Data and Surveillance: Some Principles

The spectre of the surveillance society is rightly a matter of profound concern and calls for ethical analysis at several levels. At one level, there is a need for piecemeal analyses of specific categories of data and associated technology (e.g., bulk metadata collection, DNA databases, and facial-recognition technology). At another level, there is a need to have a comprehensive view of the emerging—or, at least, potentially emergent—surveillance society, with a view to establishing some general principles to curtail and constrain processes and practices that are inimical to individual privacy, and which do not have countervailing benefits. In subsequent chapters, we provide many of these piecemeal analyses. Here we articulate some of the more important general principles that are constitutive of liberal democracy, such as democratic accountability, the preservation of fundamental individual freedoms (including a substantial degree of individual privacy), and the integrity of market-based institutions (including the removal of monopolistic, oligopolistic, and other power imbalances between producers and between producers and consumers).

First, privacy in relation to sensitive personal information consists in large part in the right to control the access to and use of this data. Accordingly, there is a presumption against the collection, analysis, and use of bulk sensitive personal data without the consent of the person concerned. Moreover, the consent in question needs to be genuine consent. One does not freely consent to hand over one's wallet if there is a loaded gun pointed at one's head, since one has been coerced into doing so. Likewise, in modern society citizens have rights of access to the internet and to search engines, and arguably also to some forms of social media as de facto powerful channels of public communication. Accordingly, if Google or Facebook require one to consent to their collection and use of one's personal information so as to have access to their services, then there must be alternative providers of these services who do not require consent to the collection and use of one's personal information—and the service provided must be of reasonable quality and at a reasonable price.[29] If this is not the case, then the consent given to Google and Facebook is not genuine consent.

Second, while the presumption against the collection, analysis, and use of bulk sensitive personal data can be overridden for law enforcement purposes, the specific purposes in question need to be justified in terms of their moral weight (e.g., saving lives), and in terms of the efficiency and effectiveness of the bulk databases in question. Justification cannot simply consist in a general appeal to community security or safety. Nor can justification consist in a general appeal to the economic and social benefits that follow from the

---

[29] Miller, *Moral Foundations of Social Institutions*, Chapter 10.

exploitation of this data for advertising and other commercial purposes by technology corporations. After all, the technology could be utilized to provide these benefits under somewhat different institutional arrangements (e.g., ones in which the technology corporations were public utilities).

Third, bulk database cross-linkages also need to be justified. Accordingly, it is unacceptable for data, including surveillance data, that was originally and justifiably gathered for one purpose (e.g., taxation or combating a pandemic) to be interlinked with data gathered for another purpose (e.g., CT) without appropriate justification. The way metadata use has expanded from initially being shared with only a few agencies engaged in CT, to now being used quite widely by many governments in the Western World, is an example of function creep. Likewise, the utilization by law enforcement in their investigations of, for instance, the genomic data of citizens held by private companies and created for other, quite specific purposes, infringes privacy rights and, at the very least, needs to be justified on a case-by-case basis.

Fourth, bulk databases give rise to new security concerns, such as the possibility of identity theft, and future untoward uses and implications that are not currently understood may come to light with greater scientific and technological advancement. In recent years, there has been an ongoing series of major breaches of data security,[30] including Yahoo 2013–2016 (three billion user accounts exposed), First American Financial Corporation 2019 (885 file records leaked), Microsoft 2021 (sixty thousand businesses worldwide hacked, providing access to their emails), and Facebook 2021 (names, phone numbers, account numbers, and passwords of 530 million customers leaked). Again, the problems in this area cannot be framed in terms of a simple weighing of, let alone trade-off between, individual privacy rights and the community's interest in public safety.

Fifth, insofar as the use of bulk data collected for law enforcement, health, or other purposes can be justified for the investigation of serious crimes, and privacy and other concerns mitigated, it is imperative that their use be subject to accountability mechanisms to guard against misuse. Moreover, the citizenry should be well informed about these systems and have consented to the use of these systems for the specific, justified purposes in question. They should be publicly debated, backed by legislation, and their operation subject to judicial review.

---

[30] K. Chin, "Biggest Data Breaches in US History," *Upguard*, 13 July 2023. https://www.upguard.com/blog/biggest-data-breaches-us.

Sixth, integration of the bulk databases of the personal and public information of citizens and the application of face recognition, phone metadata, and the like to track the movements and activities of citizens has the potential to create a power imbalance between governments and citizens that favours governments, and between corporations and consumers that favours corporations. Accordingly, these developments risk undermining important principles hitherto taken to be constitutive of the liberal democratic state. For instance, in a liberal democratic state, it is generally accepted that the state has no right to seek evidence of wrongdoing on the part of a particular citizen, or to engage in selective monitoring of that citizen, if the actions of the citizen in question have not otherwise reasonably raised suspicion of unlawful behaviour and if the citizen has not had a pattern of unlawful past behaviour that justifies monitoring.

Seventh, changes of organizational ownership may cause individuals to lose control of data in a way that they would not approve of. One such way this occurs is through demutualization. Health funds by their very nature collect sensitive data. When they demutualize, the individual may not have an easy path to retaining control of their data. Access to the fund online, made increasingly necessary by reduced telephone response services, may require agreeing to terms and conditions that include the onselling of personal data. Since health funds have waiting periods and exclusions, churn may be a difficult option, leaving the individual little choice but to comply with the funds demands.

More generally, the power imbalance between governments and citizens needs to be in favour of citizens.[31] It is a cornerstone of liberal democracy that government of the people needs to be not only for the people but *by* the people. Moreover, the power imbalance between corporations and consumers needs to be in favour of consumers. As argued by Miller (and noted in Chapter 1, Section 1.7), the raison d'être or defining moral purpose (the "collective good," in Miller's parlance) of a market-based industry is to ensure an adequate and sustainable supply of a good or service at a reasonable price and of reasonable quality.[32] This is only likely in circumstances in which there is a power imbalance in favour of the consumers of the good or

---

[31] For an individualistic analysis of the notion of the power of collectives, see Miller, *Institutional Corruption*, Chapter 2, on social power. Suffice it to say here that the notion of joint action, and of multi-layered structures of joint, have a prominent role. See Chapter 7, Section 7.1.

[32] Miller, *Moral Foundations of Social Institutions*, Chapter 10 on business corporations.

service in question, since it is their aggregate needs that ultimately ought to guide transactions.

2.3 Encryption

Encryption is obviously a good thing since it protects privacy. This is especially so in the context of the above-mentioned tsunami of cybercrime. Worldwide, as described in Chapter 1, there has been an explosion in cybercrime. Moreover, in some instances, cybercriminals work hand in glove with state actors. Thus, many Russian ransomware gangs operate under the direction of the Russian state. It goes without saying that citizens, businesses, and government agencies need to be able to protect themselves against cybercrime. Encryption has a crucial role to play in the cyberdefences of citizens, businesses, and government agencies. Encryption is too, of course, a cornerstone of the commercial web. Therefore, encryption, including high-level encryption, is a good thing and should be supported.

However, encryption is potentially problematic if it unreasonably impedes legitimate law enforcement investigations, such as CT operations. The ethical dilemma in this area is illustrated by the following two relatively recent events.[33]

Firstly, there was the conflict between Apple and the FBI.[34] In December 2015, Syed Farook killed fourteen people in San Bernardino.[35] The FBI suspected that his phone may have contained information that could implicate others involved in the planning of the attack, or in possible future attacks. However, an Apple iPhone allows by default only ten attempts to unlock the phone by its six-digit passcode before the phone is wiped. Apple refused the FBI request to remove the ten attempts limit. Ultimately, Apple did not have to back down, since a third party succeeded in cracking the phone—including, conceivably, bypassing or shutting down the autoerase feature by some means).

---

[33] Seumas Miller and Terry Bossomaier, "Privacy, Encryption and Counter-Terrorism," in *Counter-Terrorism, Ethics, and Technology: Emerging Challenges at The Frontiers Of Counter-Terrorism*, ed. A. Henschke, A. Reed, S. Robbins, and S. Miller (Dordrecht: Springer, 2021).

[34] L. Grossman "Inside Apple CEO Tim Cook's fight with the FBI," *Time*. https://time.com/4262480/tim-cook-apple-fbi-2/. Accessed 28/10/23.

[35] M. Schmidt and R. Perez-Pena, "FBI: Treating San Bernadino Attack as Terrorism," *New York Times*. https://www.nytimes.com/2015/12/05/us/tashfeen-malik-islamic-state.html. Accessed: 28/10/23.

Secondly, in mid-2020, Operation Venetic in the UK and coordinated operations in Europe, made news when very large criminal networks in the UK and in Europe were destroyed as a result of access to their supposedly secure EncroChat mobile phones. Joseph Cox, in a thorough article on Vice Motherboard, reported that in the Netherlands alone, "the investigation has so far led to the arrest of more than 100 suspects, the seizure of drugs (more than 8,000 kilo cocaine and 1,200 kilo crystal meth), the dismantling of 19 synthetic drugs labs, the seizure of dozens of (automatic) fire weapons, expensive watches and 25 cars, including vehicles with hidden compartments, and almost EUR 20 million in cash."[36] In the UK, over seven hundred arrests—including of crime bosses—have been made, and two tons of drugs (worth over £100 million) have been seized. While Operation Venetic concerned criminal organizations primarily engaged in drug dealing, money laundering, weapons distribution, and murder of rival criminals, phones with end-to-end encryption are known to be widely used by terrorists, thus this law enforcement achievement is highly germane to CT operations as well.

The phone type accessed in Operation Venetic, which was basically a customized Android phone, provided end-to-end encryption (i.e., email, text messages, and voice calls are encrypted on the phone and not decrypted until they reach the destination phone). It is thought the phone was not decrypted but rather hacked, since malware was apparently found on the EncroChat device itself, meaning that it could potentially read the messages written and stored on the device before they were encrypted and sent over the internet (see 2.3.1 below). This was an early example of *client-side scanning* (CSS) in which documents and data are scanned before encryption or after decryption.[37] Evidently, CSS could potentially be required to be installed on all devices and, if so, it would enable not simply the interception of encrypted communication prior to encryption or after decryption, but also use algorithms to scan the information stored on the devices to identify,

---

[36] J. Cox, "How Police Secretly Took over a Global Phone Network for Organised Crime," *Vice Motherboard* 2020. https://www.vice.com/en/article/3aza95/how-police-took-over-encrochat-hac ked. Accessed 28/10/2023.

[37] CCS has come to the fore in the UK Online Safety Bill which is under consideration in the House of Lords at the time of writing. Some of its draft provisions are very contentious. For instance, according to Electronic Frontiers Foundation: "Clause 110 of the bill requires websites and apps to proactively prevent harmful content from appearing on messaging services. This will mandate the screening of all user content, all the time." https://www.eff.org/deeplinks/2023/05/uk- online- safety- bill- must- not- violate- our- rights- free- speech- and- private. Accessed 28/10/2023.

for instance, child sexual abuse material (CSAM).[38] Since algorithms could be developed for illegitimate reasons, such as to scan stored information in order to identify political dissidents, this level of access to personal information if afforded to law enforcement authorities is potentially morally problematic. Moreover, there is also the potential problem of malevolent actors, rather than legitimate law enforcement agencies, accessing devices on which CSS has been installed.

These two law enforcement operations graphically illustrate the importance of encryption in law enforcement. On the one hand, encryption provides privacy protection to ordinary citizens, confidentiality protection to legitimate businesses, and confidentiality to police and other security agencies fighting crime and terrorism. On the other hand, encryption also affords protection to drug cartels, human traffickers, and terrorist organizations.

To address this ethical question, three main tasks need to be performed. Firstly, there is the task already undertaken above—namely, to provide an analysis of the nature and moral significance of privacy, including its relationship to confidentiality, autonomy, and security. Secondly, we need provide a description of relevant cryptographic technologies. One focus here will be on WhatsApp.[39] We explain how cryptographic keys work and the challenges they present to security agencies. By describing the technical issues in some detail, we show how it is that high level end-to-end encryption is, in effect, invulnerable to decryption, but also how devices that use such encryption are, nevertheless, vulnerable without proper cyber hygiene by virtue of their use of passwords and the possibility of being hacked and the insertion of malware. This section is of particular importance, given the central role this technology has come to play in organized crime and terrorism, and in law enforcement's efforts to combat them, but also because the highly technical nature of this technology means that there is a general lack of understanding of its actual powers and limitations. Our third main task is to provide a discussion of the privacy rights and security needs in relation to encryption in the overall context of CT policies of liberal democratic states.

---

[38] H. Abelson et al., "Bugs in Our Pockets: The Risks of Client Side Scanning," New York, 2021. https://www.schneier.com/academic/archives/2021/10/bugs-in-our-pockets-the-risks-of-client-side-scanning.html https://www.cs.columbia.edu/~smb/papers/bugs21.pdf. Accessed 28/10/2023.
[39] "WhatsApp Security," WhatsApp. https://www.whatsapp.com/security. Accessed 28/10/2023.

2.3.1 Encryption: Description

Modern computer-based cryptography involves several methodologies (e.g., public/private key [PPK] cryptography). To avoid confusion, we distinguish between passwords and keys. For our purposes a password is a combination of symbols or a phrase that is matched against a stored version. If the match it successful, the user is allowed access to whatever the password guards. In the early days of computing, passwords were stored as is. If the password was *donkey*, then *donkey* was stored—meaning that the password file on the computer, containing all the passwords, was extremely powerful—too powerful—endangering the entire computer system if it was stolen. Today, passwords are always stored as hashes. The password the user types in is hashed, converting *donkey* to, for example, *se75)($@P*. The result is compared with the stored hash, and since hash values are effectively irreversible, the hash file is of no use to a would-be intruder.

Passwords are usually quite short. Being short, passwords are susceptible to brute force attack—that is, an attack in which every possible combination is tried in succession, until the solution is found. Thus, protection from unauthorized access is often afforded by a mechanism that wipes all content on the device after, for example, ten attempts to find the password, as with the terrorist's iPhone, mentioned above.

A key is input to an algorithm that encrypts or decrypts a document. They are similar, in a sense, to passwords. However, by contrast to passwords, keys are much longer. The longer the better. Whereas we think of passwords in terms of the number of characters, the length of a key is usually given in bits. A bit is the information in a binary (i.e., two option) choice, a logical yes or no. Thus, a bit can be represented as a zero or one, and we could write the key as a series of zeroes or ones. Since a character is normally eight bits, we could think of a 2048 bit key as equivalent to 256 characters (i.e., 8 x 256). Symmetric encryption uses one key to encrypt and decrypt. If one encrypts a zip file, and then decrypts it with the same key, then symmetric encryption is being used.

It is important to distinguish encryption of documents and data on a device (e.g., a phone) from encryption in transmission. The first involves some sort of encryption control, of which a password is the most well-known, but there are other options, like fingerprint, retinal scan, and so on. Despite huge efforts, people persist in using easy-to-guess passwords, like the name of the dog, house address, or favourite fruit. A brute-force attack on a password requires time proportional to m^n, where m is the number of options for a character and n is the number of characters. Thus an eight-character password, using alphanumeric characters (i.e., the ten digits and the twenty-six letters of the alphabet in both lower- and uppercase) gives rise to 62^8 possibilities (i.e., 200 trillion—which a desktop computer could run through in a relatively short time). If we use the most widely used mapping of letters, numbers, and symbols to bit patterns (i.e., the whole extended ASCII character set of 256 characters) we get 256^8 possibilities (i.e., millions of trillions).[40] So, the number of possibilities is a function not only of the length of the password, but also of the number of available characters. However, there needs to be very large numbers of possibilities to defeat even a standard desktop computer. On the other hand, there can be very, very large numbers of possibilities that a standard computer would take decades to run through.

If we want to send the document over a public channel, we can use an industrial-strength encryption, such as AES, along with a password known to both sender and receiver. The alternative to encrypting the document and sending it over a public channel is to use an encrypted channel, such as WhatsApp, in which all messages are encrypted. Any useful channel must be end-to-end encrypted, meaning that is encrypted on the source device and not decrypted until it gets to the destination device. To avoid compromising of the key—perhaps by blackmail, torture, truth serum, and so on—systems such as WhatsApp use ephemeral keys, of which we will go into more detail below.

It is important to distinguish between the interception of communications in real time and the accessing of stored material (e.g., documents). Stored material, even if encrypted, is susceptible to accessing if the device is retrieved by investigators and its password determined. Real-time interception of, and access to, the content, as opposed to the metadata (e.g., time, date, location, and sender and receiver of call), of communications protected by end-to-end encryption will be extraordinarily difficult unless the communication is intercepted prior to encryption or after decryption. This is because the required decryption is extraordinarily difficult, absent access to encryption keys. Crucially, the encryption keys used for communications in

---

[40] ASCII stands for American Standard Code for Information Interchange. Computers can only understand numbers, so an ASCII code is the numerical representation of a character such as a or @ or an action of some sort. ASCII was developed a long time ago and now the non-printing characters are rarely used for their original purpose.

devices using end-to-end encryption are typically ephemeral. They are only used for a single message transmission and then discarded. Accordingly, since WhatsApp, for instance, uses end-to-end encryption, security agencies cannot usefully wiretap phones using it because anything they acquired would not be decryptable.

Typically, encryption keys resist brute-force attacks by virtue of the vast number of possibilities that would have to be tried in the time available (e.g., a number that would require decades for even a high-powered computer to find). Thus, the RSA[41] algorithm used in PPK requires two very large prime numbers (p and q), which are multiplied together to produce an even bigger number $N = pq$. Take a number such as 1333. This factorizes into 31 x 43, which are both prime numbers. The important thing to know is that as the numbers, such as 1333, get bigger, it becomes very difficult to find the constituent primes (31 and 43). The idea is to make N so big, that finding the two prime factors would take an inordinate amount of time[42]. Hence there has been pressure on governments from law enforcement and security agencies to enforce access to encryption keys. For example, WhatsApp provides encrypted phone calls and messages.

To allow security agencies to eavesdrop on conversions with WhatsApp and its kin is rather complicated, owing to the hierarchy of keys of different lifetimes used in the encryption. Thus, let us consider the simpler case of giving security agencies access to private keys, assuming that there are suitable judicial processes to allow access only in case of real need, along the lines already discussed. Storing all these private keys is itself a security risk: they may be leaked, stolen by hackers, or just left in unsecured places by defective software due to careless programmers. An alternative is a sort of skeleton private key, sometimes referred to as a backdoor key. The same issue of keeping the skeleton key safe applies of course, but there is an additional problem. There is consensus amongst cryptographers that creating the structure for such backdoor access weakens the encryption, thus making it easier for hackers to break.[43]

---

[41] RSA is a type of asymmetric encryption using different keys (such as PPK).

[42] As this goes to press, new algorithms are in the process of standardization which will make such cryptography robust to quantum computers. Current RSA is not.

[43] "If We Build It They Will Break In," *Lawfare*, https://www.lawfareblog.com/if-we-build-it-they-will-break, accessed 29/9/2020; "Exceptional Access: The Devil Is in the Details," *Lawfare*, https://www.lawfareblog.com/exceptional-access-devil-details-0, accessed 29/9/2020; "What if Responsible Encryption Back-Doors Were Possible?" *Lawfare*, https://www.lawfareblog.com/what-if-responsible-encryption-back-doors-were-possible, accessed 29/9/2020.

In the face of this resistance to providing encryption keys to governments, law enforcement's focus has been on finding passwords or on means of attack that do not rely on decryption by virtue of knowing the keys, but rather on bypassing the keys (e.g., by inserting malware into devices, as happened in the EncroChat case). There is also the possibility of legislation, such as exists already in the UK, where a warrant can be obtained to force a suspect to decrypt a document with prison terms for noncompliance.

Of course, we will not know for some time exactly how EncroChat was compromised, since the security agencies are unlikely to divulge this information. The consensus seems to be that this was not a defeat of the encryption, but the capturing of messages through spyware[44] before they were encrypted and sent. The spyware was most likely downloaded from EncroChat servers, which had themselves been infected. The phones were likely infected with something quite ordinary, such as a news release or a software update. One common spyware technique is key logging. Every key pressed by the user is recorded in a place hidden to the user and sent across the internet to the spyware's owner. Here, as elsewhere, knowledge of spyware techniques can afford protection to those being spied upon.

The principal encrypted voice call and message systems at the moment are Signal, Telegram, WhatsApp (owned by Facebook), and Facetime (owned by Apple). Let us consider WhatsApp as illustrative. WhatsApp was very popular, even before it was taken over and became part of Facebook's infrastructure. It is end-to-end encrypted, the gold standard, which means that it is encrypted by the sender, decrypted by the receiver, and not decrypted anywhere along the way. Of course, the provider could have a system in which they keep the encryption keys and save the messages, which means that the message could be decrypted by a third party at a later date. As discussed above, law enforcement has supported this, since it would be to their advantage. At any rate, to give users confidence in their communication being forever secret, the app uses ephemeral keys, which are created for a particular message transmission and then discarded. The user's private keys are never sent anywhere and are not known to the provider.

There are basically two approaches to encrypting a document: block ciphers, such as AES, which break the document up into chunks (blocks) and

---

[44] Pegasus, NSO Group's smarphone spyware, is a case in point. This Israeli company's product is feared to have been procured not only by liberal democracies but by repressive regimes. M. Srivastava and T. Bradshaw, "Spyware Offers Keys to Tap into Big Tech Cloud," *Financial Times*, 20 July 2019, p. 1.

encrypt each individually; and stream ciphers, such as RC4, which operate one character at a time.

Today's block ciphers are both very complicated and very secure. The data is broken up into blocks. Each subblock is individually encrypted using algorithms, then combined with other blocks, and the process repeated for a dozen or so iterations. The current, more secure version is AES256.

Stream ciphers date back to the 16th century, with the invention of the one-time pad, beloved of espionage stories ever since. The pad is some document, say Tolstoy's book *War and Peace*. Starting at some agreed place in the book—the spies must agree on the book and where to start—the message is compared letter by letter with the book, and a reversible algorithm is used to go from one to the other. Thus, if the message has a k and the book at the same point has a q, then the algorithm would output, say, a z. Going backwards taking the z in the encrypted document, comparing it with the q in the book, which the algorithm translates to a k. The algorithm most used is XOR. The computational equivalent is the Vernam cipher, which combines the characters of a document one by one with a random character from the keystream (the letters, one by one, from the book in our Tolstoy example). The one-time pad and, consequently, the Vernam Cipher were shown by Claude Shannon to be unbreakable, provided that the one-time pad is perfectly random.[45] In the Vernam cipher, we use a keystream, which is just a random series of characters. Computer random generators are now very good at producing very long strings of integers/characters with no relationships between them and no recurring patterns of any kind. But they are only ever pseudorandom. The generator will have control parameters and a starting state, and if these are replicated, then the replica will enable the production of exactly the same sequence. As is obvious, in the predigital computing days of cryptography, keeping the code book secure was vitally important. Of course, with the advent of keystream (Vernam) ciphers, the code book has been replaced by a random number generator. However, it is now vitally important to keep the details of its parameters and starting state (not necessarily its algorithm) secure.

An essential point to note here is that cryptographic systems may fail for three reasons: computer power increases, allowing a brute-force attack; the invention of new attack algorithms; and simple flaws in implementation.

---

[45] Claude Shannon, *A Mathematical Theory of Cryptography*—Case 20878, Alcatel-Lucent MM-45–110-92 (1945).

The most effective attacks are not brute-force but exploit some loophole in the cryptography design. Mostly, the problems are in software, but occasional a bug appears at the hardware level. In 2020 The Verge reported on a nasty vulnerability in Intel chips, which could enable the construction of key loggers:

> Security firm Positive Technologies discovered the flaw and is warning that it could break apart a chain of trust for important technology like silicon-based encryption, hardware authentication, and modern DRM protections. This vulnerability jeopardizes everything Intel has done to build the root of trust and lay a solid security foundation on the company's platforms, explains security researcher Mark Ermolov.[46]

Of course, programmers can make errors in implementing cryptographic algorithms. Cryptography is not immune to software bugs.

A fundamental problem in cryptography is agreeing on passwords or encryption keys, using a public channel, where everybody can read the transmissions but cannot infer the password. This is the idea behind a Diffie-Hellman *key exchange*[47] used in PPK and in EEC (elliptical curve cryptography), which is relied upon by WhatsApp.

Let us conclude this section by considering the level of security on Apple devices. Apple has two backup options.[48]

1. Via Finder/iTunes, you can turn on encrypted backup (it is off by default). If you do so you need to create a password. But there is no way of using the backup if you lose the password. Thus, you must create a

---

[46] "A Major New Intel Processor Flaw Could Defeat Encryption and DRM Protections," The Verge. https://www.theverge.com/2020/3/6/21167782/intel-processor-flaw-root-of-trust-csmesecurity-vulnerability. Accessed 24/9/2020.

[47] A simple account of the Diffie-Helman key exchange in lay person's terms is by way of the following analogous scenario. Charlie and Denise have a birthday tradition, where they make a cake to which each contributes a secret mix of spices. However, because of a pandemic lockdown they are on opposite sides of the world. They still want to keep their recipe secret, though. First, they agree on the main ingredients of the cake, which anybody can know. Now comes the clever trick. Charlie adds her spice mix to the ingredients and sends the mixture to Denise. It's very hard to work out exactly what was in the spice mix when mixed with the other ingredients. Denise does the same with her secret ingredient and sends the mixture to Charlie. Now each just adds their own secret ingredient to the mixture received from their partner. Both now have the same cake mix and the cake should turn out the same. This simple version is vulnerable to a man-in-the-middle attack. Evil Edith intercepts the spice mixtures and substitutes her own recipes.

[48] "How to Back Up to iCloud," WhatsApp. https://faq.whatsapp.com/iphone/chats/how-to-back-up-to-icloud/?lang=en. Accessed 28/9/2020.

password that you'll remember, or you must write it down and store it safely, because there's no way to use your backup without this password.

2. Via iCloud (the default and Apple preferred option). Now Apple has the encryption keys. It would argue that this is good for users since if they lose the password, Apple can recover it.

However, note this: although Chinese iPhones will retain the security features that can make it all but impossible for anyone, even Apple, to get access to the phone itself, that will not apply to the iCloud accounts.[49] Any information in the iCloud account could be accessible to Chinese authorities who can present Apple with a legal order. Elsewhere the keys are stored by Apple in the US, which means, under a suitable court order, Apple could be forced to give up the keys and hence the data on the phone. Now it seems that WhatsApp messages are backed up to the cloud unencrypted. Media and messages you back up are not protected by WhatsApp end-to-end encryption while in iCloud.

In a strange twist, Google, which depends heavily on targeted advertising revenue, and obtains this through massive surveillance of how its users employ its services, nevertheless offers similar personal security to Apple. Data backed up to Google is encrypted by a key, accessed by the phone's pin number or fingerprint, and so on, and this key is controlled on Googles' servers by a custom chip, referred to as Titan. Now, since a pin number is a very weak password, the Titan uses the old maximum number of tries principle (although we do not know how many tries).[50] The limited number of incorrect attempts is strictly enforced by custom Titan firmware that cannot be updated without erasing the contents of the chip. By design, this means that no one (including Google) can access a user's backed-up application data without specifically knowing their passcode.

2.3.2 Encryption: Ethical Analysis

In the light of our conceptual analysis of privacy/autonomy, confidentiality and security, and our descriptive account of encryption, we can now

[49] "Apple Moves to Store iCloud Keys in China, Raising Human Rights Fears," *Reuters*. https://www.reuters.com/article/us-china-apple-icloud-insight/apple-moves-to-storeicloud- keys-in-china-raising-human-rights-fears-idUSKCN1G8060. Accessed 28/9/2020.

[50] "Apple May Have Ditched Encrypted Backups, but Google Has Not Done So," *AndroidCentral*. https://www.androidcentral.com/apple-may-have-ditched-encrypted-backups-google-hasnt. Accessed 28/9/2020.

offer an ethical analysis of privacy rights and security needs in relation to encryption.

We have argued that privacy rights, including in respect of smartphone content and metadata, are important in part because of their close relation to autonomy. However, we also noted that privacy rights are not absolute; they can justifiably be overridden, for instance, in relation to an imminent terrorist attack. Therefore, the strong claim that some privacy advocates are inclined to make—namely, that there is, in effect, an absolute moral right to very strong (i.e., uncrackable) encryption, since there are no circumstances in which very strong encryption should be impermissible—is not sustainable. This is, of course, not to demonstrate that very strong encryption is morally impermissible under all circumstances. Perhaps, for instance, citizens who live in an authoritarian state are morally justified in possessing devices equipped with very strong encryption. Or perhaps the threat posed by cyber criminals to citizens and businesses in liberal democracies and elsewhere is so severe because of, for instance, their ability to engage in sophisticated cyber attacks, that they cannot be protected other than by means of devices equipped with very strong encryption. Moreover, even if the threat posed by cyber criminals in liberal democracies did not require devices equipped with very strong encryption, as is evidently the case, nevertheless, very strong encryption might be morally permissible if there were other means by which law enforcement agencies could efficiently and effectively investigate and, if justified, charge terrorists and other criminal suspects.

Such methods might include recourse to lawful coercion such as exist in the UK, where a warrant can be obtained to enforce a suspect to decrypt a document, with prison terms for noncompliance. However, this might be unlikely to work in cases in which the suspect in question has committed a serious crime, carrying a prison term well in excess of that attached to noncompliance with the warrant, and either the evidence that he has committed the crime consists of the encrypted material on the device in question or he is a terrorist committed to a cause—and, therefore, willing to accept the prison term for noncompliance (in addition to that for his terrorist offences).

Again, very strong encryption might be morally permissible if law enforcement could rely on access to bulk metadata and the associated use of ML techniques as well as methods such as hacking and insertion of malware, or the use of CSS, in order to efficiently and effectively investigate and, if justified, charge terrorists and other criminal suspects (as presumably occurred in the EncroChat scenario). On the other hand, CSS, bulk metadata collection, and integrated databases are themselves problematic from a privacy perspective.

Although privacy rights can be overridden under some circumstances—notably by law enforcement investigations of serious crimes—there is obviously a point where infringements of privacy rights are excessive and unwarranted. Security agencies' ongoing, ready access to the personal data of the entire population, as is currently the case in Tibet and Xinjiang, is unacceptable and certainly inconsistent with liberal democratic principles. Moreover, regulation, and associated accountability mechanisms need to be in place to ensure that, for instance, personal information obtained for a legitimate purpose, such as combating organized crime or CT, can be accessed by law enforcement officers to enable them to detect suspects and protect citizens from, for instance, being murdered but not used to identify protesters at a political rally.

We have also argued that the sharp contrast between privacy and security cannot be maintained, since security includes informational or data security (i.e., security of personal data and confidentiality in relation to data held by security agencies). Moreover, it is primarily goods that are not essentially informational that ultimately need to be weighed to achieve an acceptable moral equilibrium—notably, individual autonomy, personal security, and institutional integrity.

Further, we have seen that high-level end-to-end encryption is, in effect, invulnerable to decryption. However, as we have also seen, devices that use such encryption are, nevertheless, vulnerable by virtue of their use of passwords, which carry the possibility of being hacked. They are also vulnerable to the possible insertion of malware or government-sanctioned CSS.

In the light of the above, several interconnected ethical issues have come into view. Some of these arise from the expanding use of bulk data collection and surveillance operations, especially in the context of the interlinkage of databases, data analytics, and surveillance technologies. These developments are relevant to debates surrounding encryption insofar as they provide an advantage to security agencies that might to some extent mitigate the problem of not having access to encrypted communications and documents.

This is not to say that there ought not to be constraints on bulk data collection and analysis. For instance, it is unacceptable for data originally and justifiably gathered for one purpose (e.g., taxation), to be interlinked with data gathered for another purpose (e.g., CT), without appropriate justification.

Another important development that needs to be kept in mind when adjudicating privacy and encryption issues in CT contexts is the blurring of the distinction between the application of the domestic law enforcement and the military combat framework in CT operations, given that terrorist organizations, such as Al-Qaeda and Islamic State, operate in war zones as well as in well-ordered jurisdictions. What are the privacy rights of, for instance, those suspected of travelling abroad with the *intention* of becoming foreign terrorist fighters, but who are yet to fulfil this intention? Should they be treated as ordinary citizens possessed of the full array of privacy and other rights, who are only potential, and not actual, criminals?[51] Again, what are the privacy rights of those suspected of being foreign terrorist fighters who have returned to their home country? Should they be treated as ordinary citizens possessed of the full array of privacy and other rights albeit, if they are returnees, then they are citizens suspected of criminality? Or should they be regarded, in effect, as suspected terrorist-combatants, and therefore suffer a curtailment of their privacy and other rights, even in the absence of sufficient evidence to convict them of terrorist offences (e.g., the ongoing monitoring of their private communications and the retention of their personal data by domestic security agencies, and the disclosure of this data to third parties such as foreign governments and their security agencies).

Finally, it should be noted that there is a danger in relation to the technological developments discussed here (e.g. bypassing encryption and the use of integrated bulk databases), as there is in relation to technological developments discussed elsewhere (e.g., the use of facial-recognition technology,[52] Chapter 4, Section 4.4), that the general principles that constitute liberal democracy are gradually undermined, such as the principle mentioned in Section 2 that the state has no right to seek evidence of wrongdoing on the part of a particular citizen or to engage in selective monitoring of that citizen, if the actions of the citizen in question have not otherwise reasonably raised suspicion of unlawful behaviour and if the citizen has not had a pattern of unlawful past behaviour that justifies monitoring. However, this principle is potentially undermined by certain kinds of offender profiling, and specifically profiling in which there is no specific (actual or reasonably suspected) past, imminent, or planned crime being investigated. We note that not simply communicative content, but also metadata could be used for

---

[51] Although in some jurisdictions, such as Australia, travelling to Syria and other zones of armed conflict is in and of itself a crime. See Section 119.2 of the Criminal Code of Australia.

[52] Miller and Smith, "The Ethical Application of Biometric Facial Recognition Technology.".

profiling, risk assessment, and monitoring of people who are considered at risk of committing crimes. Moreover, in a liberal democratic state, there is a general presumption against the state monitoring the citizenry. This presumption can be overridden for specific purposes, but only if the monitoring in question is not disproportionate, is necessary, or otherwise adequately justified and is kept to a minimum and subject to appropriate accountability mechanisms.

3

Freedom of Political Communication and Computational Propaganda

Rights, Responsibilities, and Truth Aiming by Reasoning with Others

Social media platforms, such as Facebook, YouTube and Twitter, are used by billions of communicators worldwide, as are search engines, such as Google. The advent of these tech giants—or at least of the technology upon which they rely—has enabled the moral right to communication, and especially to political communication, to be exercised on a scale hitherto undreamt of. Consequently, there are unprecedented flows of information and opinion, globally and locally. In doing so, it has enabled ordinary citizens and thought leaders to directly communicate with large numbers of their fellow citizens, and thereby enhanced public communication. It has also enabled dissident political leaders in authoritarian states to directly communicate with their fellow citizens, and thereby very effectively mobilize their supporters, as occurred in the case of the Arab Spring.

However, as with other vehicles of public communication, in authoritarian states, such as China and Russia, social media platforms and, therefore the mass communication of political views and politically relevant factual information, which these platforms enable, have been subject to ongoing censorship by removing specific content, blacklisting particular individual communicators, or the wholesale blocking of a social media platform. Accordingly, it would be a mistake to think that communications on these platforms by and to the citizens of authoritarian governments are beyond the direct or indirect control of these governments. Social media networks, including major ones like Facebook and Twitter, can be blocked wholesale.

Indeed, in the context of the 2022 Russian invasion of the Ukraine, Russia has blocked Facebook.[1] As for China, it has established the Great

---

[1] Dan Milmo, "Russia Blocks Access to Facebook and Twitter," *The Guardian*, 5 March 2022. https://www.theguardian.com/world/2022/mar/04/russia-completely-blocks-access-to-facebook-and-twitter.

Cybersecurity, Ethics, and Collective Responsibility. Seumas Miller and Terry Bossomaier, Oxford University Press.
© Oxford University Press 2024. DOI: 10.1093/oso/9780190058135.003.0004

Firewall: a combination of regulations and technical devices (e.g., IP address blocks, analyzing and filtering URLs, packet inspection) designed to censor content on the internet. China's position is that national governments have the ultimate right to control the internet within their borders, and that this covers foreign companies, citizens, and anyone who attempts to interfere by, for example, creating software to undermine its control. China has also a declared intent to be the world leader in artificial intelligence, and this is an important factor in maintaining internet control, since messages of all kinds can be screened for content considered undesirable. Moreover, even encrypting content through VPNs is difficult in China since their servers are blocked en masse.

Further, the advent of social media platforms and the associated cybertechnologies, such as algorithms and automated software (e.g. bots that mimic real people), has not only, somewhat predictably, brought with it a desire and a capacity on the part of authoritarian governments to censor legitimate political communications, but it has also, less predictably, gone hand in hand with an exponential increase in the spread of disinformation, misinformation, conspiracy theories, hate speech, and propaganda on the part of a wide array of actors, and a concomitant increase in use of hate speech in particular to publicly shame, indeed often 'cancel', those individuals whose communications and views disagree with those of the members of the group doing the shaming or cancelling.[2] These actors include individual citizens, single-issue pressure groups, right-wing and left-wing extremist groups, terrorist groups, criminal organizations, and in some cases governments (e.g., Russia). Following Woolley and Howard,[3] we will refer to this latter phenomenon, insofar as it is undertaken in the service of political agendas, as computational propaganda. A particular feature of computational propaganda is its contribution to the generation of echo chambers in which users are exposed to information that reinforces their own point of view. Thus, social media algorithms adjust the content that users are exposed to, thereby creating filter bubbles. As a result, the individual user is isolated from a wide spectrum of

---

[2] Cocking and van den Hoven, *Evil On-line*. Jeff Horwitz, *Broken Code: Inside Facebook and the Fight to Expose Its Toxic Secrets* (New York: Penguin Doubleday, 2023). A notable instance of hate speech and cancelling was that directed at the internationally famous philosopher, Sir Roger Scuton, who held conservative views intolerable to many on the left wing and, therefore, according to some of those on the left, ought to suffer, and in fact did suffer, hate speech, including death threats, on an ongoing basis. Dominic Green, "Roger Scruton: A Conservative for Modern Times," *Wall Street Journal: Opinion*, January 3, 2020. https://www.wsj.com/articles/roger-scruton-a-conservative-for-modern-times-11578955867. Accessed 12/01/2023.

[3] Woolley and Howard, *Computational Propaganda*, 4–5.

views and is exposed principally to users with similar views to their own. This strengthens the user's views at the expense of competing views and of information that might challenge the users' view, thereby leading to an increase in entrenched 'hard shelled' perspectives that are not open to revision. The result is a weakening of evidence-based discussions and a polarization of political discourse that facilitates unevidenced extremist views.[4]

Of particular relevance to this chapter on cybersecurity and its relation to freedoms and responsibilities in respect of political communication is the advent of the tech giants, such as Meta (parent company of Facebook), Alphabet (parent company of Google), and Twitter (now known as X and owned by Elon Musk), the widespread use of AI, powerful cyber technologies, such as social media bots, and an unregulated cyberspace— or, at least, a cyberspace in which regulations are not effectively enforced.

Social media bots are used inter alia to automatically generate disinformation (as well as information), propagate ideologies (as well as nonideologically based opinions), and function as fake accounts to inflate the followings of other accounts and to gain followers. Bots are diverse and may range from the very simple, to Turing-test competitive AIs. Consider a *chatbot*. It might be triggered when a web page is opened, saying "Welcome to Super Velocipedes, the one-stop-shop for electric bikes." It might be a bit more sophisticated and check if the website has already stored a cookie, and if so, say "Welcome Back." This is a one rule AI. But it could be more complicated. If the user types a message, it could scan it for keywords and display one of a set of canned responses, which best match the keywords. "Do you have a cheap electric bike?" "Yes, our entry-level model is $300 and comes with a two-year warranty." It could go even further and use natural language AI to parse the user message and formulate an original reply. The user says he is 5ft 4 and weighs 200kg. What sort of electric bike should he get? The response is that he should get a mountain bike and lose some weight. A concern in this chapter is with bots that masquerade as human on social media, distributing propaganda or falsehoods. Since creating bots is relatively easy, it is possible for a social media platform to be dominated by them.

Moreover, as this book goes to press generative AI is making waves. Thus, the problem of fake material is going up exponentially along with the rise of generative AI such as chatGPT. ChatGPT can produce very convincing text

---

[4] F. A. D'Alessio, "Computational Propaganda: Challenges and Responses," *Academia Letters*, Article 3468 (2021). https://doi.org/10.20935/AL3468. Accessed 28/10/2023.

with little direction from users. As a result, educational institutions and other epistemic organizations feel threatened by its outputs. Importantly, for our concerns in this chapter, there is the capacity to produce content that is convincing in itself (and perhaps substantially true) but that appears (falsely) to emanate from trustworthy sources (e.g., fake *Guardian* newspaper articles).[5] Moreover, there is the further possibility of producing large amounts of disinformation and propaganda that is convincing in itself and that appears to emanate from trustworthy sources.

The upshot of these developments is that the moral right of freedom to communicate has frequently not been exercised responsibly. Moral obligations to seek and communicate truths rather than falsehoods have not been discharged, resulting in large-scale social, political, and ultimately physical harm. Thus, the advent of the tech giants, social media bots and an unregulated cyberspace has enabled extremist political groups, such as Islamic State to flourish (at least initially), facilitated interference in the democratic process in various countries by foreign powers, and accelerated virulent politically motivated hate speech, leading in some instances to murder and mayhem, as in the case of the attacks on Rohingya Muslims after hate speech on Facebook emanated from the Myanmar military.

Nor have liberal democracies, such as the US, UK, and Australia, avoided these problems. It seems that whether Brexit was a good or a bad decision (objectively considered), a decisive factor in generating the result in favour of Brexit may well have been driven by a media/social media political campaign, largely based on disinformation, xenophobic conspiracy theories, and propaganda. In Australia, social media has exacerbated the prior problem of trial by media (e.g., the media/social media role in the unsafe conviction of the well-known social conservative Cardinal George Pell on child sexual abuse charges, later overturned by the High Court of Australia).[6] As for the US, it has witnessed the rise of home-grown extremist political groups fed on a diet of disinformation, conspiracy theories, hate speech, and propaganda via social media. Nor have the mainstream political parties been immune to these problems; indeed, at times they have been key contributors. For instance, former US President Donald Trump consistently claimed, and

---

[5] Chris Moran, "ChatGPT is Making Up Fake *Guardian* Articles." *The Guardian*. https://www.theguardian.com/commentisfree/2023/apr/06/ai-chatgpt-guardian-technology-risks-fake-article. Accessed 28/10/2023.

[6] Virginia Miller, *Child Sexual Abuse Inquiries and the Catholic Church: Reassessing the Evidence* (Firenze: Firenze University Press, 2021), 112–17; Gerard Henderson, *George Pell: The Media Pile-On and Collective Guilt* (Melbourne: Connor Court, 2021)

continues to claim, that the 2020 US presidential election, which he demonstrably lost, involved massive voter fraud. The culmination of these processes in the US was the violent attack in January 2021 on the Capitol building, which houses the US Congress.

In short, liberal democracies have now had more than a taste of the destabilizing effects on their institutions and, indeed, on national security of an unregulated cyberspace.

Nor are the sources of these security problems entirely internal to the liberal democracies themselves. Revelations concerning the data firm Cambridge Analytica's illegitimate use of the data of millions of Facebook users highlighted the ethical issues arising from the use of machine-learning (ML) techniques in relation to social media for political purposes by malevolent foreign actors. As we show in Chapters 1 and 4, ML is a process in which a computer system is trained to identify patterns on the basis of a vast amount of data fed into it, and thereby enabled to make accurate descriptions or predictions (Chapter 1, Section 1.6.2; Chapter 4, Sections 4.1 and 4.2). For instance, ML might enable an antispam filter to identify spam or a medical device to distinguish malignant melanoma lesions from benign moles. Importantly, ML does not rely on a human-written algorithm that instructs the system how to tell the difference between spam and nonspam or malignant melanoma from benign moles.

Cambridge Analytica is or was—the revelations brought about its demise—a firm that used ML processes to try to influence elections in the US and elsewhere by, for instance, targeting vulnerable voters (micro-targeting) in marginal seats with political advertising.[7] Of course, there is nothing new about political candidates and parties employing firms to engage in political advertising on their behalf. However, if a data firm has access to the personal information of millions of voters and is skilled in the use of ML techniques, then it can not only identify patterns, and thereby develop detailed, fine-grained voter profiles, it can also match individual voters with these profiles (e.g., as advertisers might tailor their ads to individual consumers rather than merely relying on hit-and-miss advertising). This ML-based process enables political actors to appropriately target those voters with the profiles in question, and thereby in their political communications to not simply echo the preexisting views and concerns of these voters but use this

---

[7] Von Hannes Grassegger and Mikael Krogerus, "Ich Habe Nur Gezeigt, Dass Es Die Bombe Gibt," *Das Magazin*, 3 December 2016. https://www.dasmagazin.ch/2016/12/03/ich-habe-nur-gezeigt-dass-es-die-bombe-gibt/.

detailed knowledge of their views and concerns to influence their behaviour. However, this influence is frequently manipulative and worryingly so.

In the first place, unknown to the voters there is a knowledge gap or epistemic disparity between the political actors and the voters in question. For instance, the political actor is in possession of these detailed voter profiles and the voters are unaware of this. In the second place, the political actors rely on manipulative techniques (e.g., appeals to emotion, false or misleading claims) to influence the voters. However, the use of these manipulative techniques is even more effective in the context of this epistemic disparity. In short, the political actors have reached a whole new level of manipulative influence over the voters in question, and, if these voters are in marginal electorates, then there is the distinct possibility of decisively influencing the electoral outcome.

The ethical consequences of the situation are potentially far-reaching. One set of ethical issues pertains to privacy and confidentiality: illegitimate access on the part of Cambridge Analytica to private information, and in the case of Russian hackers accessing the democratic party's emails, to confidentiality. The latter is a data security concern. Another set of ethical issues pertains to institutional corruption—corruption of the democratic process. The problem here is compounded by home-grown corruption of liberal democratic institutions by, for instance, populist leaders who wilfully undermine electoral and other institutional processes in the service of their own political and personal goals. Arguably, Trump is such a leader given this insistence that the presidential election he lost was unfairly won by Biden because of electoral fraud. While some forms of corruption are not necessarily crimes (and some crimes are not necessarily forms of corruption), nevertheless, institutional corruption in general, and corruption of democratic institutions in particular, is ultimately a national security issue.[8] A further set of ethical issues pertains more obviously to national security. The use of ML techniques by foreign powers, such as Russia, to favour one candidate over another in the service of their own political agenda (e.g., to sow discord in liberal democratic polities that they have no right to participate in, let alone to undermine). Such manipulative political influence over users of social media utilizing a combination of new technological tools, such as ML, and psychologically based, manipulative marketing techniques, raises more directly the emerging ethical issue of the tension in cyberspace between

---

[8] Miller, *Institutional Corruption*.

freedom of communication, on the one hand, and the need to restrict computational propaganda (i.e., certain forms of disinformation, political propaganda, and hate speech), on the other.

The ongoing and widespread dissemination of disinformation, hate speech, and extremist propaganda on the internet and on social media in particular is not only an abnegation of the moral responsibilities that come with freedom to communicate, it corrodes epistemic and moral norms, such as truth telling and trust in the truth telling of others, polarizes attitudes, promotes race, ethnicity, gender, religion, economic class and single-issue based forms of factionalism (inter alia), and sows discord in a polity. As such, it is inherently socially, morally, and politically destabilizing and, therefore, needs to be curtailed, prohibited, or otherwise combatted. In doing so, it needs to be kept in mind, firstly, that freedom of political communication is a very fundamental moral right which, arguably, ought never to be entirely extinguished and, therefore, needs to be respected even in many circumstances in which it is not responsibly exercised.[9] Secondly, notwithstanding what has just been said, freedom of communication is not an absolute right; it can justifiably be overridden under limited circumstances (e.g., in wartime), and, if not exercised responsibly by (especially) those who communicate to the public at large (e.g., if used by political leaders to tell lies), potentially curtailed, again, under limited circumstances. Accordingly, while combating ongoing and widespread dissemination of disinformation, hate speech, and extremist propaganda on the internet and on social media is a necessary undertaking, it gives rise to a host of difficult ethical problems.

There are, of course, criminal laws against disinformation that incites violence (e.g., propagated by terrorists), perverts the course of justice (e.g., trial by media), or directly causes serious harm (e.g., shouting "fire" in a crowded theatre), as well as civil laws regarding defamation that enable noncriminal lawsuits. There are also, more problematically, laws against seriously offensive language and (in some jurisdictions) hate speech. However, liberal democracies jealously guard the moral right of freedom of speech. Freedom of speech is not simply a moral right but a fundamental legal right. Crucially, the right to freedom of speech entails that at least some disinformation and political propaganda will be legally permissible (e.g., the false claims that 9/11 or climate change is a hoax). Moreover, the right to freedom of speech entails that some speech that is offensive to some will be legally permissible (e.g.,

---

[9] Unless, of course, indirectly by virtue of one's right to life being justifiably overridden or forfeited.

claiming that Jesus Christ or the prophet Muhammad did not exist is permissible in many jurisdictions, even if it offends at least some of the faithful). Accordingly, a problem arises or, rather, a set of problems. Where are the lines to be drawn between (respectively) what ought to be legally permissible and what ought to be legally impermissible disinformation, propaganda, and offensive speech? Further, in relation to legally permissible disinformation, propaganda, and offensive speech, what measures other than reliance on underlying sociomoral norms and, therefore, social disapproval can be taken to reduce their prevalence in cyberspace? More generally, what measures can be taken to strengthen the relevant sociomoral and epistemic norms in cyberspace?

In addition to these issues, there is the problem of enforcement. It is one thing to legislate against certain forms of disinformation, propaganda, and offensive speech, but it is quite another to interpret this legislation in particular instances, and especially to enforce this legislation in cyberspace. Is it, for example, simply a matter of relying on the tech giants themselves to remove unlawful communications on their platforms? Perhaps. However, thus far their record is very poor. Indeed, in the case of Facebook, for instance, it has been argued[10] that their policies and, in particular, their use of algorithms to expand their user base and average user time per day on its platform in the context of their avowed principle of neutrality and the widespread appetite for sensationalist content, have greatly exacerbated the problem. But ought they be allowed to remove some legally permissible disinformation, propaganda, or offensive speech on the grounds that it violates moral or other principles that *they* happen to believe should be complied with or that *they* judge to be the principles favoured by their various online communities (i.e., whom they ultimately rely on for their revenue), even if some of these communities largely consist of adolescents not yet old enough to vote?

At this point, an argument from private ownership is sometimes invoked. Roughly, the idea is that if someone wants to use their own money to set up a social media platform—to set up, let us assume, a mechanism of public communication (we set aside 'private' communications on social media)—in order to enable individuals and organizations to communicate messages, opinions, and so on to other individuals and organizations, then they ought to be allowed to decide who gets to communicate what on their platform.

---

[10] Horwitz *Broken Code.*

After all, the person(s) *own* the platform. There are several problems with this line of argument.[11]

Firstly, it does not follow from the fact that a person has a moral right to set up a mechanism of public communication that the person has a moral right to determine who gets to communicate what on that platform (including by recourse to procedures of cancellation and amplification). The claim that the one follows from the other is as it stands simply an unargued assertion. What is called for at this point is a descriptive and normative account of the nature, function and audience reach of the mechanism of public communication in question. For instance, is the content that is publicly communicated political in its nature and function, and does it reach a large percentage of voters? If so, what are the moral and epistemic standards, if any, that are required or, at least, complied with, bearing in mind the serious harms that can result from the widespread dissemination of disinformation, propaganda and hate speech? Secondly, given the importance of social media platforms as mechanisms of public communication in contemporary polities, arguably *everyone,* whether they are owners of social media platforms or not, has a moral right to communicate and to be communicated to via a dominant social media platform. This is, or ought to be regarded as, the modern equivalent of the moral right to speak to fellow citizens at public forums, such as the Agora in ancient Athens or the Forum in ancient Rome. Moreover, this right includes the moral right to freedom to communicate whatever content they wish to their fellow citizens, albeit constrained by justified moral principles, including ones enshrined in the law (e.g., laws against incitement). Further, this basic right does not derive from, and is not enlarged or extended by, property rights in general, and not by the right to set up a mechanism of public communication, in particular. The right of a citizen to engage in public communication with fellow citizens on matters of importance to all, including in modern societies via social media mechanisms, is not somehow increased by the fact that citizen has the moral right and the financial means to establish and own a mechanism of public communication. Indeed, any such extension of the right of such owners would constitute an infringement of the pro tanto *equal* right of citizens to utilize the mechanism of public communication.

Part of the answer to some of the above questions of combating disinformation, hate speech, and extremist propaganda (and some related ones) lies,

---

[11] Miller, *The Moral Foundations of Social Institutions,* Chapter 10.

we suggest, in strengthening epistemic institutions as well as the sociomoral norms, such as evidence-based truth telling, which support and are supported by epistemic institutions. (In making this suggestion we concede that these institutions have doubtless in many cases themselves been weakened by the exponential rise in computational propaganda, given their occupants are members of the broader society and, therefore, not immune to the influence of computational propaganda). What do we mean by epistemic institutions? The term *episteme* refers to knowledge. Therefore, epistemic institutions are those institutions that have as a principal institutional purpose the acquisition and/or dissemination of knowledge (understood broadly to include factual knowledge; reasoning processes, like induction and deduction; evidence-backed economic, political, and ethical perspectives; and understanding). Accordingly, epistemic institutions include schools, universities, and media organizations responsible for news/comment. They also include private or government research laboratories, think tanks and, for that matter, national security intelligence agencies.[12]

These practical ethical questions mentioned above presuppose answers to some more fundamental theoretical questions. For instance, can the distinction between politically motivated fake news,[13] hate speech, and propaganda, on the one hand, and factual and other objective claims and perspectives, on the other hand, be sustained? What is the nature and extent of the moral right of freedom of communication?

In this chapter on security and freedom of communication in cyberspace, our concern is with countering computational propaganda (comprised of politically motivated disinformation, hate speech and propaganda/ideology), while respecting the moral right to freedom of communication and in particular the moral right to freedom of political communication.[14] In the first section, we offer definitions of disinformation, hate speech, and propaganda/ideology (and, relatedly, what we refer to as quasi propaganda/ideology and single-issue groupthink), respectively. As we shall see, these phenomena have at least one important feature in common: they are not truth aiming

---

[12] Miller, *Moral Foundations of Social Institutions*; Miller, "Epistemic Institutions: A Joint Epistemic Action-based Account," *Nous-Supplement: Philosophical Issues* 32 (2022): 398–416.

[13] Fake news is a problem for citizens and their political leaders worldwide. See, for instance, the struggles of Maria Ressa, Nobel Peace Prize winner and CEO of Rappler in the Philippines in relation to fake news and freedom of the press in that country. Maria Ressa, *How to Stand up to a Dictator* (New York: Harper Collins, 2023). See also Horwitz *Broken Code*.

[14] Miller, "Freedom of Political Communication, Propaganda and the Role of Epistemic Institutions in Cyberspace."

(in a certain sense). In the second section, we elaborate the right to freedom of communication and its relation to epistemic institutions. In the third and final section, we discuss the general problem of countering computational propaganda. That is, we discuss how to counter politically motivated disinformation, hate speech, and propaganda/ideology (and quasi propaganda/ideology and single-issue groupthink) in cyberspace and the role of epistemic institutions and social media platforms in this enterprise. In doing so, we offer some specific regulatory recommendations.

3.1 Disinformation, Hate Speech, and Propaganda/Ideology

Computational propaganda comprises politically motivated disinformation and hate speech, and propaganda/ideology in cyberspace, and utilizing various forms of cybertechnology. However, an important initial, essentially philosophical, task is to characterize disinformation, hate speech, and propaganda/ideology. The communicative technology and the scale of these forms of communication may have changed, but the underlying concepts are unchanged (i.e., disinformation is disinformation irrespective of whether it is delivered face to face or by a bot). Interestingly, while there is a large and growing literature on computational propaganda, its extent, modes of delivery, and so on, the underlying concepts of disinformation, hate speech and propaganda/ideology are typically taken for granted. Yet, unless the distinction between disinformation and accurate information can be made out, the enterprise of successfully combating computational propaganda is unlikely to get off the ground.

3.1.1 Disinformation

The definition of disinformation is contested and, therefore, the following definition is necessarily somewhat stipulative.[15] News, by definition, purports to be true, or in the case of visual images and the like, purports to be an accurate representation of reality, even if it is in fact false. However,

---

[15] Michael Lynch, "Fake News and the Internet Shell Game," *New York Times* (2016). https://www.nytimes.com/2016/11/28/opinion/fake-news-and-the-internet-shell-game.html. Accessed 29/10/2023.

news items are frequently disseminated on the internet by persons who do not endorse them. Indeed, on occasion, by persons who explicitly state that they are false.

We use the terms, *disinformation* and *fake news* to refer to news that is in fact false and not believed by its originator—as opposed to subsequent disseminators—to be true. The originator is an individual person or, perhaps, a group of persons acting jointly (e.g., a joint communication—a species of joint epistemic action).[16] Naturally, such an originator might be acting *qua* member of an organization (e.g., a news reader acting *qua* employee of Fox News, in which case this individual's moral responsibility for his or her communications, including disinformation, might be substantially diminished). We note that, whereas organizations and other collectives might have legal responsibility per se, they do not possess moral responsibility per se. The latter always rests on individuals, individually or jointly (e.g., on employees or their managers).[17]

Notice that on the above definition, news that is false and believed to be false by its originator is disinformation and fake news, but so is news that is false and neither believed nor disbelieved by its originator to be true. This is because news by definition—and whether it is in fact true or false—*purports to be true*. Thus, whatever its originator believes or does not believe, he or she presents the news item as being true. Notice on this definition of disinformation, there is a distinction between disinformation and misinformation—but it is not the one that is sometimes drawn. On this definition, misinformation is false; however, it is believed by its disseminator to be true.

Disinformation is morally problematic for at least three reasons. Firstly, it is false and yet, given the communicative reach of the internet, and of social media—and the use of automated dissemination techniques (e.g., bots)—it is likely to be believed by many, even if disbelieved by many others (or, at least, they suspend belief). We note that somewhat paradoxically the credibility of fake news on social media platforms—notably Facebook—is likely to be enhanced by the copresence on these platforms of objective news emanating from relatively high-quality news outlets, such as the BBC which is not to say that such news outlets are not themselves at times lacking in

[16] Seumas Miller, "Joint Epistemic Action: Some Applications," *Journal of Applied Philosophy* 35, no. 2 (2018): 300–18; "Assertions, Joint Epistemic Actions and Social Practices," *Synthese* 193, no. 1 (2016): 71–94.
[17] Miller, "Collective Moral Responsibility: An Individualist Account."

objectivity and prone to bias, in the case of the BBC, so-called woke bias.[18] Secondly, especially in the case of an ongoing series of mutually supportive, politically motivated, fake news items, there are likely to be untoward political consequences arising from large numbers of people believing such news items, including potentially the undermining of democratic processes that rely on voters making judgements based on facts rather than falsehoods. Thirdly, the originators of disinformation, insofar as they engage in fake news or other forms of disinformation as a matter of habit, or otherwise as an ongoing practice, are likely to be corrupt and can be corruptors of others. How so?

Widespread disinformation, especially if it is emotionally appealing (e.g., sensationalistic) or otherwise utilizes manipulative techniques, undermines trust that others comply with the sociomoral and epistemic norms of evidence-based truth telling and may weaken one's own commitment to comply with these norms.[19] Commitment to norms of evidence-based truth telling is weakened by the costs associated with compliance when others do not comply (e.g., others reciprocate one's truths with lies), and the pull of emotionally appealing messages, or by messages that otherwise utilize manipulative methods. Moreover, widespread emotionally appealing disinformation weakens the role of social disapproval in maintaining the norms of evidence-based truth telling. The anonymity of communicators also weakens the role of disapproval since the identity of the persons to be disapproved of is not known, which means they can communicate disinformation with impunity. This weakening of the role of social disapproval undermines the norms of evidence-based truth telling because such norms are in part constituted by social approval of compliance and social disapproval of noncompliance.[20] Further, habitual liars are morally, and not simply causally, responsible for their negative impact on the norms of evidence-based truth telling, and if these habitual liars are frequent and somewhat effective communicators (e.g., by virtue of the emotional appeal of their messages in the light of the felt grievances of the members of their audience), and large in number, then the extent of disinformation and the corresponding corrosive impact on norms of evidence-based truth telling is likely to be very significant. The problem

---

[18] E. Haigh, "Biased BBC," *Daily Mail*, 30 December 2022. https://www.dailymail.co.uk/news/article-11581105/Biased-BBC-rewriting-British-history-promote-woke-agenda-leading-academics-warn.html. Accessed 29/10/2023.

[19] Seumas Miller, "Truth-Telling and the Actual Language Relation," *Philosophical Studies* 49, no. 2 March (1986): 281–94; Miller, *Social Action*.

[20] Miller, *Social Action: A Teleological Account*, Chapter 4, on social norms.

is compounded by even a small number of highly influential, frequent, and very effective communicators who reach large audiences, directly and/or indirectly via message forwarding—perhaps in large part by virtue of their powerful and/or high-profile institutional position—and habitually lie, as has been argued in the case of Trump.[21]

It is sometimes suggested, especially in communicative contexts in which the norms of evidence-based truth telling are weak, that ultimately there is no important distinction between fake news (i.e., disinformation) and factual news—and, as a corollary, politicians, academics, news media, and other disseminators cannot provide objective communicative content of high quality since the notion of objective truth is itself believed to be meaningless or hopelessly naïve. Accordingly, one media or other report cannot be of higher quality than another by virtue of being correct, more accurate, or more balanced. It is suggested that the reasons for this are manifold, and they include: the fact that communicative content is a representation, and as such always reflects a standpoint, means that the mechanisms of media communication necessarily mediate, and therefore distort. Quality is believed to be simply in the eye of the beholder. There is not the space to deal with all these kinds of arguments in detail, though it is easy to show that they do not justify the strong position they are intended to demonstrate.[22] Suffice it to say here that the notion that one cannot aim at truth, and on occasion approximate to it, and the notion that every piece of analysis and comment is as good as every other, are self-defeating. If accepted, they would render communication pointless. It is a presupposition of communication in general, including both linguistic communication and audiovisual representation, that there is a truth to be communicated or some fact of the matter to be represented, and that on many occasions this is achieved. If this were not so, communication of news would be rendered pointless and cease taking place. Thus, if communicators and audiences alike thought that there really were no truths to be communicated, let alone believed, then it would be pointless to report that on 9/11 two planes were flown by terrorists into the Twin Towers building in New York City, killing some three thousand people. Likewise, it would be pointless to show footage of the planes flying into the towers, and their subsequent collapse, pointless because

[21] Rattner, "Trump's Election Lies Were among His Most Popular Tweets"; Kessler et al., "Trump's False or Misleading Claims Total 30,573 over Four Years."

[22] Sissela Bok, *Lying: Moral Choice in Public and Private Life* (New York: Pantheon Books, 1978).

media communicators and their audiences alike would be operating on the assumption that there was no fact of the matter. Moreover, it is a presupposition of comment and analysis that not every piece of analysis and comment is as good as every other one, since there is always as least one that is regarded by the communicator as inferior—namely, that which is the negation of the one put forward (e.g., that terrorists did *not* fly planes into Twin Towers and did *not* kill anyone). Accordingly, when QAnon-supporting US congresswoman Marjorie Taylor Greene claimed that 9/11 did not happen, she was not denying that there was *some* fact of the matter; rather, she was denying the particular fact that on 9/11 two planes flew into the Twin Towers building and killed three thousand as a result. And, of course, she was affirming the contrary fact—namely, that this event did not take place. Greene's claim was false and inconsistent with the evidence; indeed, her claim was preposterous, and she later retracted it. But the nonsense that she and others propound does not undermine the reality of objective truth. Rather, it merely serves, firstly, to remind us of the tenuous grip that many people have on objective truth, and, secondly, to underline the importance of seeking the truth in the context of widespread disinformation. This is perhaps especially the case for those who occupy roles in epistemic institutions, such as universities and the news media.

Naturally, academics, journalists, criminal investigators, intelligence officers, and judges typically regard truth-seeking as of great importance. However, while most academics believe that intellectual work in universities is an end-in-itself, members of other epistemic institutions are perhaps less clear on this point, since for them the truth is a means to some further end (e.g., justice, in the case of criminal investigators and judges, and the public interest, in the case of journalists). In one sense, this claim is true. In the criminal justice system, truth-seeking is a means to justice, and truth-seeking by investigative journalists serves democracy by providing knowledge that citizens need to possess in order to make their collective decisions, and that politicians need to possess in order to make their decisions on behalf of their citizens. Again, the intelligence acquired by national security intelligence officers does need, as they say, to be actionable—that is, ultimately it needs to serve national security needs.

However, the claim that truth-seeking is *merely* a means is false. For the acquisition of the truth—or, at least, of probable truth—is (or ought to be) an end-in-itself for journalists, criminal investigators, judges, intelligence officers, and so on. Let us explain.

The activities of the occupants of these epistemic roles are not related to knowledge merely as means to end, but also conceptually. Truth is not an external, contingently connected end that some investigatory activities might be directed towards, if the journalists, detectives, judges, and so on happen to have an interest in it, rather than, say, in falsity, 'playfulness' (as with postmodernists), or self-interest (as with demagogues who say whatever might be useful to them and without regard for the truth). Rather, truth is internally connected to epistemic activity, whether it be a journalist investigating corruption, a detective investigating a crime, a judge or members of a jury determining the guilt or innocence of a defendant, or a scientist seeking the cure for COVID-19. Thus, aiming at truth is aiming at truth as an end-in-itself. This is, of course, consistent with *also* aiming at truth as a means to some further end, such as apprehending an offender or saving lives. In other words, supposed epistemic activity that *only* aimed at truth as a means to some other end would not be genuine epistemic activity, or would be defective *qua* epistemic activity, since for such a pseudo-truth-seeker, truth would not be internal to his or her activity. Such a pseudo-truth-seeker would abandon truth aiming if, for example, it turns out that the best means to the journalist's, detective's, judge's (and so on) ultimate end is not after all truth, but rather falsity. Obviously, such pseudo-truth-seekers would be extremely dangerous since their reports, judgements, findings, and so on would be very unreliable. For they are not simply persons who aim at (and often acquire) the truth, but who, nevertheless, often provide false reports, judgements, or findings they know to be false—or, more likely, to be somewhat misleading because unpalatable truths are omitted or downplayed. Rather, these pseudo-truth-seekers do not aim at truth in the first place. That is, having little interest in the truth, they do not seek the truth, and as a result do not themselves acquire knowledge. Therefore, they do not have knowledge to provide to others. Of course, in the real world such pseudo-truth-seekers are unlikely to exist in a pure form; indeed, they probably could not exist in a pure form, given the impossibility of aiming at falsity when one makes a judgement.[23] However, the commitment to the truth might well weaken in a newspaper office, police service, intelligence agency, or science laboratory that lacks independence and in which the desire to please or not to antagonize one's superiors (i.e., newspaper owners, senior police, politicians or funders, respectively) is overwhelming. This is especially the

---

[23] Miller, "Joint Epistemic Action and Collective Responsibility".

case when one considers the inherent difficulties in acquiring accurate, significant knowledge in many of these areas. As a consequence, such epistemic
professionals might initially merely report what they know to be false or misleading on some occasions when it is politically or otherwise expedient to do
so, but end up over time largely abandoning the practice of evidence-based
truth-seeking in favour of selective data collection and skewed analyses that
serve personal, political, or other nonepistemic agendas—that is, they end up
becoming something akin to pseudo-truth-seekers.

There is an important institutional implication of this. As we have just seen,
whereas the primary institutional ends of journalists, criminal investigators,
intelligence officers, members of juries, and academics are essentially epistemic, in each case the realization of their epistemic end serves a further
important purpose, which is only realizable by the activity of the occupants
of other institutional roles: the military, in the case of national security intelligence officers; and doctors and nurses, in the case of medical researchers.
Accordingly, there is an institutional division of labour. Hence in the context of the criminal justice system, the criminal investigators provide a brief
of evidence to the prosecuting agency; in the context of the news media,
the journalist investigates and communicates a news item to members of
the public. Prosecutors, members of the public, and others provided with
knowledge by epistemic institutional actors, in turn act or refrain from
acting on that knowledge. For instance, prosecutors prosecute offenders on
the basis of evidence provided by criminal investigators, and citizens vote
against governments on the basis of news provided by journalists. In order
for these various institutional divisions of labour to function successfully, it
is critical that the reports, briefs of evidence, scientific findings, and so on
that are provided are reliable, and therefore that the epistemic activity of
the journalists, detectives, academics, and so on is not unduly influenced
or otherwise undermined by those to whom they provide these epistemic
products (e.g., by their bosses, political masters, funders—or, indeed, in the
case of journalists, by consumers of news seeking a diet of sensationalistic
stories). Accordingly, consistent with an appropriate level of responsiveness
to their various superiors, it is necessary that the commitment of epistemic
professionals to the truth ultimately override personal, economic, or political
considerations, other than perhaps in extreme circumstances. Accordingly,
they may need to speak unpalatable truths to power, including to the masses
of people who might have a dominant voice on social media. Perhaps less obviously, epistemic professionals may need to speak the truth in circumstances

in which by doing so the ultimate ends of their epistemic activities might not in their view be well served. For instance, criminal investigators might need to provide evidence that is exculpatory of an accused person they know to be guilty, or journalists might need to resist their natural desire to ignore or downplay inconvenient truths the dissemination of which would harm some morally worthy cause they advocate. It is important that journalists, criminal investigators, members of juries, intelligence officers, academics, and so on not engage in institutional overreach by second-guessing the actions of those who are the institutionally authorized, or otherwise morally appropriate, recipients of the knowledge that these epistemic professionals have unearthed.

While the distinction between disinformation and factual claims is relatively clear—notwithstanding claims to the contrary—distinctions, firstly, between politically motivated hate speech and strident pejorative criticism, and, secondly, between, political propaganda/ideology and political comment/opinion, are more problematic. Moreover, there is a threefold distinction, again not clear-cut, between political propaganda/ideology, quasi propaganda/ideology, and what might be termed single-issue groupthink (of which more below).

3.1.2 Hate Speech

The definition of hate speech is contested. Let us, assume, however, that it is speech that incites hatred against some group—or, at least, is intended to do so and has some reasonable chance of doing so.[24] Accordingly, hate speech is to be distinguished from strident pejorative criticism insofar as the latter is truth aiming (i.e., has truth as an end in itself). By contrast, hate speech is not truth aiming in this sense. The truth is only of interest insofar as it can serve to incite hatred.

Hate speech does not necessarily incite violence or other serious crimes, though these may well be longer term, indirect consequences of hate speech. The hate speech of interest to us here is politically motivated hate speech, which is speech that incites hatred against a target group and is performed to serve some political purpose (e.g., ISIS hate speech directed at minority Christian groups with a view to inciting violence against them

---

[24] Jeremy Waldron, *The Harm in Hate Speech* (Cambridge: Harvard University Press, 2012).

and, ultimately, eliminating them from the region).[25] Evidently, such hate speech should be criminalized if only on the grounds that it incites violence. Naturally, politically motivated hate speech, often features abusive language, and manifestly incites hatred against the target group. However, sometimes it is couched in moderate language and consists in advocating policies that are ostensibly based on facts (i.e., factual claims which turn out to be false or highly misleading). In the latter cases, context is all important, if the speech in question is properly to be regarded as hate speech. Consider a right-wing politician's speech advocating that immigrants from a certain racial group should be sent back to their homeland and that there should be a ban on any further immigrants from that group, based on his false claims that the immigrants are mostly criminals and/or welfare recipients. Suppose this speech is disseminated on social media and, on a targeted basis, to members of an audience likely to be receptive to these views because of their preexisting prejudice or unfounded fears.[26] The speech is racially discriminatory, and given its pattern of dissemination and ultimate intention (i.e., to incite racist sentiment in the service of a political agenda), it arguably constitutes hate speech. However, in being couched in moderate language, and in the light of the moral importance of free speech, it nevertheless might not reach a threshold that warrants its criminalization (e.g., it does not clearly constitute incitement to violence). Here, as elsewhere, morally repugnant speech is not necessarily justifiably criminalized, especially if it can be countered by other means.

In the light of this account of hate speech, it is clear that it is potentially harmful not only to individuals and groups who are the object of its attack—and not only because it is likely to be false—but because it is likely to sow discord in a liberal democratic polity, and even, for that matter, in authoritarian states, though the latter have been known to exploit racial, ethnic, or class divisions to maintain their own power (e.g., Hitler's amplification and exploitation of anti-Jewish sentiment). As is the case with fake news, hate speech in cyberspace is especially problematic because of the unprecedented communicative reach of the internet and of social media platforms.

---

[25] H. Kanso, "Symbol of ISIS Hate becomes Rallying Cry," 20 October 2014. https://www.cbsnews.com/news/for-christians-symbol-of-mideast-oppression-becomes-source-of-solidarity/. Accessed 29/10/2023.

[26] Of course, some such fears might not be unfounded. For instance, the immigrants in question might take their jobs by virtue of being prepared to work for lower wages. More generally, from the fact that a politician is right wing (or, for that matter, left wing) it does not follow that they are engaging in hate speech.

3.1.3 Propaganda, Ideology, and Single-Issue Groupthink

Political propaganda is frequently, but not invariably, communication in the service of a political ideology.[27] Accordingly, political propaganda seems to presuppose political ideology. Therefore, we need a serviceable account of political ideology, and one that enables a distinction to be maintained, firstly, between ideology and the more generic notion of systems of political ideas, which might or might not be instances of ideology and, secondly, between politically motivated communications that are elements of an ideology and those that are not, although they may be untruthful, manipulative, or otherwise epistemically untoward.[28]

We also need to keep in mind that political ideologies are often framed within a wider historical narrative of social and economic life (e.g., a history of class, nation, ethnicity, race, religion, or gender-based exploitation). Moreover, with respect to ideologies, we need to distinguish between those that are relatively comprehensive, systematic, and underpinned by a pseudo-theoretical body of ideas (e.g., National Socialism, Maoism, various extremist Islamist ideologies such as those of Islamic State and Al Qaeda)), and those that are not. Let us refer to these two categories as ideologies and quasi ideologies, respectively. While the distinction is often not precise in practice or even conceptually, the latter (quasi ideologies) tend to be more narrowly focused on a single issue (e.g., the neo-Luddites), or a jumbled set of issues that have not been systematically connected (e.g., the Proud Boys' ideology comprising the use of violence to promote pro-'Western values', the antivaxxer stance, or that of Antifa comprising the use of violence to promote antifascist, antiracist, pro-LBGT, proworker, environmentalist, anarchist stances). Moreover, quasi ideologies have not developed even a pseudo-theoretical body of ideas to underpin them and often lack an organizational structure that might enable the development of a consistent theory, or at least pseudo-theory. Rather than being underpinned by a theoretical body of ideas, quasi ideologies are often based on far-fetched conspiracy theories, or beliefs (e.g. the belief that the members of certain race groups have a biologically-based very low intelligence or that whether a person is a man or a woman is essentially a matter of individual subjective preference rather than

---

[27] Jacques Ellul, *Propaganda: The Formation of Men's Attitudes,* trans. Konrad Kellen and Jean Lerner (New York: Random House/Vintage, 1973).

[28] Although, the distinction between an ideology and a system of political ideas is often not clear in practice.

of objective underlying biological features).[29] As Sutton and Douglas point out, "research indicates that conspiracy theories may play a powerful role in ideological processes." They are associated with ideological extremism, distrust of rival ideological camps, populist distrust of mainstream politics, and ideological grievances.[30]

Accordingly, quasi ideologies (unlike many political ideologies, such as Marxist-Leninism) are deeply irrational and, for this reason, while often promoted by rational political actors to serve their own political ends, these rational political actors often do not actually believe in these quasi ideologies (unlike Lenin, in the case of Marxist-Leninism). Indeed, in some cases, rational political actors promote these quasi ideologies without believing in them and do so by clandestine means (e.g., the Russian government's promotion during the US presidential campaigns of both right-wing and left-wing extremist quasi ideologies). Quasi ideologies and conspiracy beliefs have flourished in cyberspace as a result, in large part, of the unprecedented communicative reach of the internet and of social media, both in terms of the extent of this reach, which includes potential audiences of hundreds of millions, and the fact that it is instantaneous.

There is a further distinction between ideologies and quasi ideologies, on the one hand, and what we referred to above as single-issue groupthink.[31] The beliefs of antivaxxers are a paradigm example of this. These beliefs are focused on a single issue and not only largely false (e.g., that vaccines do not generally protect against COVID-19, but rather kill large numbers of people), but often far-fetched and based on unevidenced conspiracy theories or crackpot ideas (e.g., that COVID-19 is a hoax). The prominent, indeed influential, conspiracy theorist in the US, Alex Jones, is a case in point. Jones was recently forced by a court to pay $1.5 billion as a result of his false claim that the 2012 Sandy Hook shootings were a hoax.[32] These beliefs are generated and maintained in large part not by evidence, but typically by a

---

[29] Uwe Steinhoff, "The Transgender Craze," *Philosoph.* https://uwesteinhoff.com/2022/06/11/the-transgender-craze-and-the-babble-about-self-identifying-as-a-woman-why-men-who-think-theyre-women-are-psychotic-and-the-politicians-humoring-them-are-opportunistic/. Accessed 6/11/2023.

[30] R. Sutton and K. Douglas, "Conspiracy Theories and the Conspiracy Mindset: Implications for Political Ideology," *Current Opinion in Behavioral Sciences* 34 (2020): 118.

[31] The term *groupthink* is sometimes used to refer to a belief among a small group that is essentially reliant on group pressure. Our use of the term is related but adjusted for application to internet and social media users (i.e., potentially very large groups).

[32] Sam Cabral, "Alex Jones Files for Bankruptcy after Sandy Hook Verdict," BBC News, 2 December 2022. https://www.bbc.com/news/world-us-canada-63837309.

combination of felt self-interest (e.g., the belief that one is personally healthy and therefore will be unaffected by COVID-19, even if others are at great risk if they become infected), emotion (e.g., anxiety in the midst of a global pandemic, fear of the effects of vaccines), and group influence and pressure (e.g., in the context of antivaxxer propaganda and protests). As is the case with quasi ideologies, single issue groupthink is irrational, at the very least in the sense that it does not substantially consist of evidence-based truths.

Let us now elaborate some of the key properties of a political ideology, keeping in mind the distinctions between ideologies, quasi ideologies, and single-issue groupthink.[33] Firstly, it is important to note that in order for something to be a political ideology (as opposed to a quasi ideology, or instance of single-issue groupthink) it must comprise a set of *systematically* connected beliefs, assumptions, or claims underpinned by a (pseudo-) theory. Moreover, this systematically connected set of beliefs or claims must, if it is an ideology at least, be susceptible of instantiation in the minds of a group of people. After all, the whole point of an ideology is to powerfully influence people's thinking in a manner that shapes their behaviour. Such a group must constitute a community of sorts, and not simply a set of unrelated individuals. Moreover, if the beliefs and claims are to be systemically connected, and if their instantiation is to be sustained, then the adherents to the ideology will not only need to rely on an underpinning (pseudo-) theory but also on an organization. Inter alia, this organization will curate the ideology, including its underpinning (pseudo-) theory, and ensure that the ideology is instantiated—that is, ensure that the constitutive beliefs and claims continue to be accepted in the correct form by the adherents. Naturally, instantiated quasi ideologies consist of adherents with shared beliefs and claims. However, quasi ideologies are inherently unstable, given their lack of systematization and of underpinning (pseudo-) theory, as well as their lack of organizational structures to sustain them. The same point holds even more so for instances of single-issue groupthink, such as the anti-vaxxer movement, and related forms of groupthink and quasi ideology. Indeed, groupthink and quasi-ideology as social media phenomena are inherently unpredictable, unstable and uncontrollable although, as is the case with an infectious disease for which there is no vaccine, their spread can be amplified by irrational and, in some cases, malevolent actors or, alternatively, contained by the cooperative action of rational actors motivated by the common good (see below).

[33] Seumas Miller, "Ideology, Language and Thought," *Theoria* 74 (1989): 97–105.

Consider in this connection the QAnon conspiracy according to which the world is controlled by the so-called "Deep State."

Further, it must be emphasized that the key constitutive elements of ideologies and quasi ideologies (and, for that matter, instances of group-think) are beliefs and claims. It is sometimes supposed that their key constitutive elements are actions, at other times appearances, and at still other times that these elements are words or concepts (as opposed to beliefs and claims that are, of course, expressed in some language). But an ideology (or instance of groupthink) cannot consist of actions, social practices, and the like per se since, unlike beliefs or claims, actions are not *about* the world and are not true or false. It is a constitutive feature of an ideology and of group-think that they are about the world, and that they are true or (more likely) false. Naturally, while an ideology consists in beliefs and claims, it has a defining purpose to change behaviour (i.e., actions and practices), and, once changed, these actions and practices will reinforce the ideology since they have become permeated by the ideology—see below).

Moreover, an uninstantiated ideology or instance of groupthink cannot consist of appearances per se, even though the way the world appears to be may bring about false beliefs and indeed ideological beliefs. Here a perceptual analogy may be useful. A stick placed in water has the appearance of being bent and may cause the perceiver to believe that it is in fact bent. Yet, from the fact that the world appears to a subject to be a certain way, it does not follow that the subject believes that the world is the way it appears to be. We do not, for example, believe that the stick is bent, although it certainly appears to us to be bent. But if appearances are not necessarily accepted as true, then they cannot be constitutive of instantiated ideologies or of group-think. For if someone adopts an ideology or participates in groupthink, then the person accepts its content as true. Again, it is surely clear that it is only beliefs and claims, as opposed to unitary items, such as words or concepts, that constitute commitments to this or that view of the world—and as such can be true or false. By contrast words and concepts as such do not constitute such commitments and make no truth claims. Thus, the word *unicorn* is consistent with there being or not being unicorns; however, the belief that unicorns exist is a *commitment* to the world being a certain way and is true if the world is that way and false if it is not.

Secondly, we suggest that for any set of shared beliefs to count as a political ideology, whether it be an ideology or a quasi ideology, it must have a certain kind of origin. In particular, the existence of the ideology cannot ultimately be caused by reality being as the ideology says that it is. This is also the case for instances of groupthink. Thus, a particular systematically connected set of beliefs (e.g., liberalism) would potentially qualify as an ideology according to our definition, if it were brought into existence not by the world being as liberalism says it is, but rather because it was fashioned as an expedient account of things by the economically ascendant classes.[34] Moreover, the beliefs of, say, members of the Proud Boys would potentially qualify as a quasi ideology on our definition, if it were brought into existence not by the world being as the ideology says it is, but rather as an expression of the felt grievances of various disaffected groups of white males in the US, which rationalizes their violent responses to these felt grievances. Again, the beliefs of antivaxxers would potentially qualify as an instance of single-issue groupthink on our definition if it were brought into existence not by the world being as the antivaxxers' groupthink says it is, but rather as an expression of the felt grievances, anxieties, and so on of the antivaxxers which rationalizes their responses to these grievances and anxieties.

Thirdly, we suggest that to count as a political ideology or quasi ideology, a set of beliefs must serve some kind of political purpose; indeed, that is its raison d'être. It might, for example, have the purpose of undermining, or alternatively preserving the *political status quo*. Similarly, some, but not all, forms of single-issue groupthink have the purpose of changing some public policy (e.g., mandatory vaccination). Here we note that the political purpose served by a quasi ideology or groupthink might not in fact be the ostensible purpose enshrined in the quasi ideology or content of the groupthink. Accordingly, to return to a point we made above and its illustrative example, if Russian-based propagandists are seeking to polarize, and thereby destabilize the US or elsewhere by spreading competing quasi ideologies, they nevertheless may not themselves believe the content of the quasi ideologies that they are spreading. Indeed, those quasi ideologies might well be contradictory, and therefore, if a person believed one, then he or she would not believe the other.

---

[34] It is logically possible (at least) that a set of ideas might be (substantially) true, but nevertheless that they are held true by some group might be accidental rather than as a consequence of the world being as the ideas represent it as being. Moreover, it may even be that the members of the group in question were aiming at the truth in coming to accept these ideas, albeit they came to accept them as a result of a faulty methodology. However, this would not necessarily be an instance of an ideology since those in the grip of an ideology will abandon truth if and when it does not serve their political motives or nontruth aiming motives. Thus, while arguably the members of this group do not have knowledge, nevertheless, they are not necessarily in the grip of an ideology.

Finally, it should be noted that there is a high probability that an ideology or quasi ideology will be substantially false, given that its causal origin cannot be the world being the way the ideology says it is, and given that it must serve some or other political purpose. That said, it is important to keep in mind that political ideologies and quasi ideologies typically consist in part in truths, as well as falsehoods and half-truths. Moreover, political ideologies rely in part on legitimate grievances. If not, they are likely to have little or no credibility. By contrast, quasi ideologies do not necessarily rely on legitimate grievances, and as a result typically lack credibility. Accordingly, they are likely to be relatively ephemeral or mutable phenomena, although the erosion of epistemic norms in cyberspace and the communicative reach of the internet and of social media platforms provides fertile ground for their incubation, sustaining them to a degree that would probably not otherwise have been the case.

For their part, the instances of single-issue groupthink of interest to us here—namely, those with some (possibly inchoate) political purpose—are also highly likely to be substantially false and for the same reasons. However, they are less likely to rely on truths than political ideologies or even quasi ideologies. Moreover, as is the case with quasi political ideologies, and for the same reasons, they are likely to be ephemeral or mutable phenomena, only more so.

Although, political ideologies and quasi ideologies typically consist in part in truth, the core or constitutive elements of a political ideology, and even more so of a quasi political ideology are likely to be false or fanciful (e.g., the classless society, the Third Reich, the Volkstaat, the Caliphate, lesbian separatism, and the Queer Nation). Moreover, the propagation of an ideology relies on falsehoods, half-truths, and hate speech. Modern propaganda is likely to rely on a suite of psychologically based, manipulative marketing techniques and fake news disseminated on the internet. Aside from the constitutively ideological components of an ideology (i.e., its content), ideology impacts itself causally on communication and thought, by way of permeation, implication, and presupposition. Accordingly, and notwithstanding what was claimed above, actions, practices and so on can be used to convey ideological content.

Sometimes processes of permeation, implication, and presupposition enable the ideology to be influential while going undetected. Consider an advertisement consisting of a video clip of a well-dressed, handsome man ostentatiously smoking an identifiable brand of cigarette, while getting into the driver's seat of an expensive car and making the statement "This is a fine car."[35] Here there may be a non-political-ideological core belief—that is, the conviction that in virtue of its being mechanically sound and fuel-efficient, the car is fine. However, in addition to this nonideological core belief, and overlaying it, may be ideological beliefs, such as that the car is fine, not simply in virtue of being mechanically sound, but also in virtue of being socially prestigious and expensive. Here a core visual image and accompanying statement of nonideological meaning is *permeated* by ideological meaning: in effect, a consumerist ideology is being sold. And, of course, there is the *implication* that smoking this brand of cigarette will make one attractive and bring prestige, when *in reality*, of course, it is more likely to lead to an addiction to nicotine and, ultimately, to health problems, such as cancer or heart disease.

A further kind of example entails the notion of a presupposition as well as implication, although there is no attempt to conceal the ideological message. Consider for instance the ISIS video entitled "Flames of War."[36] Inter alia it shows what is purportedly a captured Syrian soldier speaking submissively as he digs his own grave. The implication of this scene in the context of the video, which frequently refers to those whose actions are "favoured by God" and those not favoured ("unbelievers"), is not only that the enemies of ISIS can be killed, but that, as required by the ISIS ideology, they morally ought to be killed and that they will be killed, since the extremist group is far more powerful than its enemies and will prevail in its enterprise of extending its Caliphate (the video was released in late 2014 after ISIS successes in Iraq and Syria). It is a presupposition of the video that there is a God, Allah, to whose will all should bend, and that the world is divided into loathsome unbelievers, deserving of death and true Muslims, who are doing Allah's will and will receive Paradise as their reward. *In reality*, of course, the upshot of the acceptance and implementation of this extremist ideology, and its associated murderous actions, will be—and indeed, in the case of the Islamic State, has been—the establishment of short-lived, dysfunctional communities characterized by ignorance, poverty, injustice, and extraordinary brutality,

---

[35] The example was provided by Richard Freadman. See Richard Freadman and Seumas Miller, *Rethinking Theory: A Critique of Contemporary Literary Theory and an Alternative Account* (Cambridge: Cambridge University Press, 1992).

[36] See W. H. Allendorfer and S. C. Herring, "ISIS vs US Government: A War of On-line Video Propaganda," *First Monday* 20 (2015): 12, https://firstmonday.org/ojs/index.php/fm/article/downl oad/6336/5165. Accessed 29/10/2023.

which are exemplars of extreme forms of social pathology rather than of paradise on earth.[37]

A final important point needs to be kept in mind. Propaganda on its own has little political effect. If it is to undermine, for instance, a liberal democracy, it needs to be a component of an integrated package comprising the existence of a felt grievance against some group, such as injustice suffered at the hands of the political elite, a technological means for wide dissemination (e.g. printed matter and social media), and, at least in conflict situations, some form of kinetic capacity (e.g. armaments) and strategy (e.g. terrorism).[38] Needless to say, as is the case with fake news and hate speech, the unprecedented communicative reach afforded by the internet, social media platforms, and associated AI and cybertechnology to propagandists, when taken in conjunction with the availability of an arsenal of manipulative techniques—including not only well-known marketing and mass communication techniques, but ones based on recent research in psychology and neuroscience, such as forms of cognitive bias—has greatly increased the potential impact of political ideology, notably quasi ideology,[39] In short, computational propaganda is, in many respects, more formidable and harder to combat than previous forms of propaganda—or, more precisely, than propaganda using previous technological and other means of dissemination.

3.2 Freedom of Communication, Truth, and Liberal Democracy

As we saw above, notwithstanding the individual, collective, and institutional harms caused by politically motivated disinformation, hate speech, and propaganda/ideology and associated quasi ideology and groupthink in cyberspace (hereafter propaganda/ideology/groupthink) (i.e., computational propaganda)—not to mention their inherent epistemic and moral undesirability—there are good reasons not to enact laws to prohibit them

---

[37] Paul Burke, Doaa El Nakhala, and Seumas Miller, ed., *Global Jihadist Terrorism: Terrorist Groups, Zones of Armed Conflict and National Counter-Terrorism Strategies* (Cheltenham: Edward Elgar, 2021).

[38] Haroro J. Ingram, "A Brief History of Propaganda During Conflict," International Centre for Counter-Terrorism—The Hague (2016). https://icct.nl/publication/a-brief-history-of-propaganda-during-conflict-a-lesson-for-counter-terrorism-strategic-communications/. Accessed: 29/10/2023.

[39] Tzu-Chieh Hung and Tzu-Wei Hung, "How China's Cognitive Warfare Works," *Journal of Global Security Studies* 7, no. 4 (2020): 1–18. See also Chapter 6, Section 6.2.

entirely, although these reasons are consistent with placing some legal restrictions on them. For instance, there should be laws against incitements to violence and there is a need for legal redress if one is defamed by, for instance, being maliciously and falsely accused of child sexual abuse. Naturally, it does not follow from the need to limit the reach of the laws restricting some forms of communication that there should not be individual, collective, and indeed institutionally based opposition to disinformation, hate speech, and propaganda/ideology/groupthink. Such opposition would necessarily rely heavily on social disapproval in the context of resilient social norms of evidence-based truth telling.

As already mentioned, the historically most important reason for not enacting laws to prohibit disinformation, hate speech, and propaganda/ideology/groupthink is the moral right to freedom of communication.[40] The moral right to freedom of communication is a very fundamental right. Moreover, it is a moral right the expression of which has been greatly enhanced by the internet and perhaps especially by social media, such as Facebook and Twitter. More specifically, given our concerns here, social media has enabled speakers, as individuals and as members of groups, to communicate *directly* through channels of public communication, and therefore potentially to very large audiences. So, these speakers can communicate publicly to large audiences without the mediating role of print and electronic media organizations. So far, so good. However, in the absence of these mediating institutions, these speakers are engaging in public communication without the quality control that a regulated news media provides by means of the editorial independence of owners, the professionalization of journalists, the de facto (as well as de jure) liability of news organizations for defamatory material (i.e., since news organizations and journalists do not enjoy anonymity), and so on. Thus, as already described, there is a downside to this extension of the right to freedom of communication. Public communication of disinformation, hate speech, and propaganda/ideology/groupthink has been accelerated (i.e., there has been an exponential growth in computational propaganda). Accordingly, there is a need to ensure much higher levels of compliance with the relevant epistemic and moral norms, whether by the promulgation and enforcement of laws and formal regulations and/or by other means. Certainly, reliance on self-regulation has

---

[40] Frederick Schauer, *Free Speech: A Philosophical Inquiry* (Cambridge: Cambridge University Press, 1981).

failed miserably. We return to this issue in the next section. In this section, the focus is on achieving greater clarity on the moral rights to freedom of communication, and relatedly freedom of rational inquiry, and on the relationship of these rights to the roles of epistemic professionals, such as academics and journalists. A key notion here is that of truth aiming by reasoning with others, which is a form of joint epistemic action. Joint epistemic action, we suggest, is the basic building block of epistemic institutions.[41] Epistemic institutions are critical in combating computational propaganda. Hence, the importance of the notion of truth aiming by reasoning with others. This essentially cooperative model of human reasoning—and of epistemic activity, more generally—stands in some tension with atomistic individualist models.

There are two especially salient arguments for freedom of communication, and relatedly freedom of intellectual inquiry. The first is associated with the English philosopher John Stuart Mill, while the second is loosely associated with the German philosopher Immanuel Kant.[42] (We do not mean to imply that these arguments are the only ones advanced by these philosophers, much less that the versions of them we propound below are precise renderings of the work of these philosophers.) We note that, in the first instance, freedom of communication and freedom of rational inquiry are fundamental human rights, and specifically natural moral rights. Here we refer to natural rights by way of contrast with institutional rights. The former are rights individual human beings possess by virtue of being human beings, while the latter are rights that attach to occupants of institutional roles at least in part by virtue of the existence of the institutions in question. It is a further question how these natural moral rights are transformed into institutional rights, as when the moral rights to academic freedom and freedom of the press emerge because of the existence of universities and news media organizations. However, there are several preliminary points that can be made.

Firstly, these latter institutional rights are derived in part from the more basic natural moral rights, and in part from the constitutive purposes of the institutions in question (e.g., the epistemic purposes of universities, the news media and social media platforms concerned with public political communication). Secondly, these derived moral rights—which are also institutional rights—are typically, other things being equal, weaker than the

---

[41] Miller, "Epistemic Institutions: A Joint Epistemic Action-based Account." Miller, "Joint Epistemic Action: Some Applications."
[42] John Stuart Mill, *On Liberty* (London: Longman, Roberts and Green, 1869); Immanuel Kant, *Groundwork of the Metaphysics of Morals*, trans. H. J. Paton (New York: Harper Collins, 1956).

fundamental natural moral rights from which they are (in part) derived. Thirdly, insofar as these institutional rights are also moral rights (as opposed to, say, merely legal rights), then they attach to the individual human beings who occupy the institutional roles in question (e.g., they are moral and institutional rights of individual academics or journalists or social media professionals). These moral rights do not attach to collective entities, such as universities or media organizations. Collective entities per se are not possessed of mental states and are not, therefore, moral agents or possessed of moral rights or obligations, although they are possessed of legal and other institutional rights and duties.[43] It is worth noting in this context that the First Amendment to the US Constitution lumps together the individual right of free speech and freedom of the press, and the right of free speech has been interpreted as applying with equal weight to individual human beings and to organizations.[44] Given the initial power imbalances between large organizations such as media organizations on the one hand, and individual human beings on the other, the net result of this conflation of these very different kinds of bearers of rights of free speech has been, we suggest, to privilege the 'voices' of (especially large) organizations over those of individuals, further entrenching the power of the former at the expense of the latter. Given the ability of individual human beings to use social media to directly communicate to large audiences, the advent of social media has the potential to somewhat redress this power imbalance. Unfortunately, for various reasons we discuss in the next section this has not happened, at least to any great extent. For instance, according to Horwitz,[45] "whales" such as Trump, Rihanna and other "influencers" are given special treatment and shielded from the normal requirements to comply with moral and epistemic norms. Let us now turn to the analysis and justification of the rights to freedom of communication and freedom of rational inquiry.

According to John Stuart Mill, new knowledge will only emerge in a free marketplace of ideas. If certain ideas are prevented from being investigated or communicated, then the truth is not likely to emerge, since those suppressed ideas may in fact be the true ones. We note that the notion of a marketplace

---

[43] We are aware that this common-sense perspective is disputed by some philosophers. However, Miller has addressed their arguments elsewhere. Miller, *Moral Foundations of Social Institutions*; Miller, "Joint Epistemic Action"; Miller, "Joint Rights: Human Beings, Corporations and Animals."

[44] There are various other important distinctions with respect to forms of speech that have normative implications. For instance, commercial speech is not accorded the same weight as other forms of free speech.

[45] Horwitz, *Broken Code*, 180–81.

in play here might need to be somewhat loosely construed so that, for instance, Wikipedia might be understood as a marketplace insofar as there are no barriers for anyone to participate by adding or correcting information, although there are no buyers and sellers in the conventional sense. We take it that Wikipedia involves a form of collective epistemic action, or joint epistemic action. It relies on the epistemic (i.e., knowledge) contribution of multiple actors.

Let us look more closely at this argument. We will restrict ourselves to political ideas in the sense of politically relevant factual claims, hypotheses, unsubstantiated claims, interpretations, and theories, the epistemic resolution of which often requires relatively complex processes of reasoning and justification undertaken in a public forum. In recent times, this has been done on to a considerable extent on the internet and social media platforms. The internet and social media platforms, as well as more traditional mass media outlets, constitute the de facto modern public forum. Importantly, political communication in the public forum must be conducted in accordance with moral and epistemic norms of evidence-based truth telling and trust if it is to serve its fundamental institutional purposes (or collective ends in our parlance) rather than become corrupted and undermine those purposes. Since the purposes of the enterprise are, in the first instance, epistemic (i.e., understanding of public policy problems and potential solutions), the content of these purposes is necessarily underspecified; the content is, by definition, unknown to, or at least in the case of the potential solutions, undecided in the minds of, the participants in advance of the collaborative communicative and reasoning process undertaken in the forum.[46] Accordingly, the communicative and epistemic process can be derailed not only by fake news, disinformation, ideology, and the like, but also by the vulnerability of the citizenry to manipulative techniques used in the communication of fake news, disinformation, ideology, and so on. In short, there is a great need for the exercise of rational capacities on the part of leaders and citizens, communicators and audiences. Here Mill appears to rely on a distinction between rational inquiry and justification, on the one hand—a possibly solitary activity—and freedom of communication, on the other.

---

[46] Naturally, one of the aims and effects of ideology is to frame and infect such epistemic decision-making or judgement in a manner that settles, at least to a high degree, the outcome of these judgements in advance of them being made. Moreover, as already mentioned, ideologues, including both extreme right wing and extreme left wing ideologues, frequently seek to cancel their political opponents by recourse, inter alia, to disinformation and hate speech. Such acts of cancellation are, of course, paradigm instances of violating the moral right to freedom of communication.

This argument needs to be unpacked.[47] We suggest the following rendering of it:

(1) Freedom of communication is necessary for rational inquiry.
(2) Rational inquiry is necessary for knowledge. Therefore:
(3) Freedom of communication is necessary for knowledge.

The argument is valid, and premise (2) is plausible in relation to the sort of knowledge at issue here. What of premise is (1)?

The justification for (1) is evidently that rational inquiry requires: (i) a number of diverse views or perspectives, possessed by different persons and different interest groups, and (ii) a substantial amount of diverse evidence for/against these views, which is available from different sources. Moreover, (iii) regarding (i) and (ii), there is no single (a) infallible and (b) reliable authority.

Notice, firstly, that Mill's argument distinguishes between freedom of communication and rational inquiry and relies on the proposition that freedom of communication is a necessary condition for rational inquiry. However, freedom of communication may not be a sufficient condition for rational inquiry—that is, one can have freedom of communication without rational inquiry. Indeed, to a considerable extent this is what we currently have in an unregulated, or at least ineffectively regulated cyberspace; that is, freedom to communicate disinformation, hate speech, and propaganda/ideology/groupthink rather than freedom to communicate knowledge based on rational inquiry. Notice, secondly, that Mill's argument for rational inquiry is instrumentalist or means/end in its form. (The argument for freedom to communicate is also instrumentalist; however, we have already assumed,[48] that the freedom of communication is a basic right and an intrinsic good.) The claim is not that rational inquiry is good in itself, but rather that it is a means to another good—namely, knowledge, and it should be added that the knowledge of interest to us here and to Mill is collective knowledge, generated by joint epistemic action. So the notion of collective knowledge in play here is, roughly speaking, that of knowledge shared among members of a population (e.g., a polity, global audience, or an academic community

---

[47] Seumas Miller, "Academic Autonomy," in *Why Universities Matter*, ed. Tony Coady (Sydney: Allen and Unwin, 2000).

[48] As does Mill, in effect, in his *On Liberty*, although this is not uncontroversial given his commitment to utilitarianism.

(which is frequently a global community of sorts).[49] It is then an open question—as far as Mill's argument is concerned—whether or not knowledge is an intrinsic good, or merely a means to some other good. By contrast, we assume that knowledge is an intrinsic good. To this extent the moral weight to be attached to freedom of rational inquiry is weaker than it would be in an argument that accorded freedom of rational inquiry the status of fundamental moral right, as we have taken the right to freedom of communication to be. This is important given our claim that the problem of disinformation, hate speech, and propaganda/ideology/groupthink in cyberspace is in large part the problem of a weak commitment to norms of evidence-based truth telling, and therefore to rational inquiry. In short, while there is freedom of communication in cyberspace it has become decoupled from rational inquiry.

The second argument for freedom of rational inquiry is not inconsistent with the first, but is nevertheless quite different. Notice that freedom of rational inquiry presupposes that rational inquiry is the result of the exercise of that right. Freedom to engage in irrational thought is not freedom to engage in rational inquiry. So, the right to freely engage in rational inquiry, if it is exercised, necessarily results in rational inquiry.

The second argument for freedom of rational inquiry accords freedom of rational inquiry greater moral weight by treating it as having the status of a fundamental moral right. This second argument—or at least our neo-Kantian rendering of it—relies on a wider sense of freedom of rational inquiry, one embracing not only freedom of thought and reasoning, but also freedom of communication and discussion. The argument begins with the premise that freedom of rational inquiry thus understood is a basic, as opposed to derived, moral right. Here the term *rational inquiry* is intended to be taken to refer to reason-based inquires directed to the achievement of understanding. As such, is does not refer exclusively to those matters that can only be understood by experts or intellectuals. Rational inquiry is a human practice that should not be the preserve only of academics and other experts. This is not to say that academics and others with specialist knowledge or more developed levels of understanding ought not to be accorded due respect as epistemic authorities. Climate scientists are a case in point. Moreover, it is not to say that epistemic professionals more generally—notably journalists, teachers,

---

[49] There are a host of further issues here concerning the definition or rather definitions of collective knowledge which we cannot pursue here. See the journal, *Social Epistemology*.

and academics—do not have a critical institutional role in maintaining and strengthening the practice of rational inquiry. Clearly, they do have this role and an important question is how to ensure that they can continue to perform this role in cyberspace.

As we have rendered it, freedom of rational inquiry (in our wide sense) is not an individual right of the ordinary kind. Although it is a right that attaches to individuals, as opposed to groups per se, it is not a right which an individual could exercise by him/herself. Communication, discussion, and intersubjective methods of testing are social, or at least interpersonal, activities. However, it is important to stress that they are not activities that are necessarily relativized to certain designated social groups. In principle, rational inquiry can and ought to be allowed to take place between individuals in interpersonal and communal settings, including online, irrespective of whether they belong to the same social, ethnic, or political group. In short, freedom of rational inquiry, or at least its constituent elements, is a basic natural moral right.[50] Note that being a basic moral right, it can in principle override collective interests and goals, including national economic interests and goals. Hence, the dilemmas that can arise between security and freedom of communication, or between security and freedom of rational inquiry, in relation to political action. However, in the context of our discussion of computational propaganda, freedom of rational inquiry, far from being at odds with security, is a necessary condition for it. Currently in liberal democracies, security in a broad sense is, as we have seen, being undermined by a de facto unregulated cyberspace that is characterized by high levels of disinformation, hate speech, and propaganda/ideology/groupthink. Moreover, even security in the narrow sense of respect for the right to personal security, including the right to life, is now threatened by the currently unregulated (or, at the very least, under regulated) cyberspace, as the January 2021 attack on the US Congress building has made clear. Ultimately, the antidote to disinformation, hate speech and propaganda/ideology/groupthink is, we suggest, rational inquiry in our favoured sense. However, rational inquiry requires institutional promotion and some (albeit constrained) degree of direct and indirect regulatory protection and enforcement thereof, at both the national and international levels, due to the global nature of cyberspace. However, given the importance of the right to freedom of communication, the precise

---

[50] This is not to say that freedom of intellectual inquiry in respect of certain topics is not in practice restricted to those with expert knowledge and training.

form that such regulation should take poses a significant challenge (to which we return below).

If freedom of rational inquiry is a basic moral right, then like other basic moral rights, such as the right to life and to freedom of the person, it is a right that all humans possess, and it should be promoted and protected in liberal democracies. Here we need to get clearer on the relationship between the basic moral right to freely engage in rational inquiry, on the one hand, and knowledge or truth, on the other.

To reiterate: the term, *knowledge*, as used in this context, embraces not only factual information, but also understanding. Note also, that in order to come to have knowledge in this sense, one must possess rational capacities (i.e., capacities that enable not only the acquisition of certain kinds of information, such as via a Google search, but especially the development of understanding). Here the term, *rational* is broadly construed. It is not, for example, restricted to deductive and inductive reasoning. This point holds irrespective of whether the communicative context is offline or online, the coffeehouse or Twitter, and notwithstanding the advantages and disadvantages—and ultimate intellectual upsides and downsides—of some of these modes of communication over others (e.g., lengthy single speeches to a small audience versus brief tweets to thousands).[51]

Freedom of rational inquiry and knowledge, in this extended sense of knowledge, are not simply related as means to end, but also conceptually. To freely inquire is to seek the truth by reasoning. As we saw above, truth is not an external contingently connected end that some inquiries might be directed towards, if the inquirer happened to have an interest in truth, rather than, say, in falsity or playfulness. Rather, truth is internally connected to rational inquiry. A rational inquiry, which did not aim at the truth, would not be a rational inquiry, or at least would be defective *qua* rational inquiry. Moreover, here aiming at truth is aiming at truth as an end in itself. (This is not inconsistent with also aiming at truth as a means to some other end.) In other words, an alleged rational inquiry that only aimed at truth as a means to some other end would not be a rational inquiry or would be defective *qua* rational inquiry, since for such a pseudo-inquirer truth would not be internal to his/her activity. Such a pseudo-inquirer is prepared to abandon—and indeed would have in fact abandoned—truth aiming if, for example, it turns out, or if it had turned out, that the means to their end was not after all truth, but rather falsity.

[51] Lynch, *The Internet of Us.*

Further, to engage in free rational inquiry in our extended sense involving communication with, and testing by, others, is to *freely seek the truth by reasoning with others*. Rational inquiry in this sense is not exclusively the activity of a solitary individual. Moreover, here reasoning is broadly construed to embrace highly abstract formal deductive reasoning, at one end of the spectrum, and informal (including literary) interpretation and speculation, at the other. Further, it embraces ordinary political discourse among nonspecialists, as well as technical discourse among experts, and discourse attempting to bridge these divides (e.g., between scientists and ordinary citizens on climate change).

There are, of course, methods of acquiring knowledge that do not necessarily, or even in fact, involve free inquiry (e.g., A's knowledge that he has a toothache, or B's knowledge that the object currently in the foreground of her visual field is a table), but these taken in themselves are relatively unimportant items of knowledge as far as public discourse is concerned, and certainly as far as epistemic institutions, such as the press and universities, are concerned. (Obviously other items of knowledge of the same species can be very important in the context of some rational inquiry, e.g., an inquiry into whether a recently developed drug eases pain or an inquiry into ordinary perception.)

Given that freedom of rational inquiry is a basic moral right and given the above described relationship between rational inquiry and truth (or knowledge), we can now present our second argument in relation to freedom of rational inquiry.[52] This argument in effect seeks to recast the notion of freedom of rational inquiry in order to bring out the potential significance for liberal democratic polities in particular, of the Kantian claim that freedom of rational inquiry is a basic moral right:

(1) Freedom of rational inquiry is a basic moral right.
(2) Freedom of rational inquiry is (principally) freedom to seek the truth by reasoning with others.
(3) Freedom to seek the truth by reasoning with others is a basic moral right.

Our discussion has yielded the following plausible propositions. First, the kind of knowledge in question is typically attained by reasoning with others

---

[52] Miller, "Academic Autonomy."

(whether conducted offline or online, whether in the coffee house or via Twitter and so on). Second, to freely engage in rational inquiry is to seek truth (or knowledge) for its own sake.[53] Third, freely seeking the truth (or knowledge) for its own sake, and by reasoning with others, is a basic moral right.

Let us grant the existence of a basic moral right to freely pursue the truth by reasoning with others. The political implications of this are fourfold. Firstly, liberal democracies in particular need to ensure that this moral right of members of the citizenry is respected, indeed institutionally cultivated. As Mill stressed, the ability to exercise this right, and the habit of exercising it, are preconditions of liberal democracy.

Secondly, liberal democracies in particular need to ensure that there is a sphere of public political communication in the sense of a *public forum*, even if it is a privately *owned* forum, such as Twitter, which most of the members of the citizenry access frequently and in which all the main political actors communicate their divergent perspectives and policies in a manner that enables at least the policies (if not the theoretical perspectives) to be reconciled by a process of public deliberation and discussions in the context of an overall commitment to the underlying framework of liberal democratic principles and a shared epistemic background.[54] The contrast here is with multiple, highly differentiated spheres of political discourse. Political participation in highly differentiated spheres consists in the participants of any given sphere only communicating with other members of that sphere (i.e., political communication takes place in 'echo chambers'). For the millions who receive their news and political commentary via social media platforms, such as Facebook, the problem of echo chambers is increased by these platforms' use of algorithms and ML techniques to increase user numbers, average user time on-line, and advertising revenue as overriding priorities. The tendency is to offer sensationalistic news/commentary and news/commentary of a political stripe that the past online behaviour of the type of 'consumers' in question has indicated their preference for.[55]

Thirdly, liberal democracies need to ensure that this right is institutionally embedded in epistemic institutions. Since the basic building block of

---

[53] This is, of course and as mentioned above, consistent with pursuing the truth for the sake of other additional ends (e.g., to relieve poverty or arrest an offender).

[54] A shared epistemic background entails substantial overlap between the belief structures of members of the citizenry in respect of relevant factual matters (in addition to their beliefs with respect to liberal democratic principles).

[55] Cass Sunstein, *Republic: Divided Democracy in the Age of Social Media* (Cambridge: Harvard University Press, 2017). Horwitz, *Broken Code.*

epistemic institutions is joint epistemic action and in particular seeking the truth by reasoning with others, the process of embedding ought to be, at least in principle, relatively unproblematic. Indeed, in the case of functional epistemic institutions it is, so to speak, in their DNA. For instance, the exercise of the moral right to freely pursue the truth by reasoning with others is a central feature of universities.[56] Naturally, the truths in question are sometimes difficult to acquire without intellectual training of various kinds (e.g., empirical methods). Again, the moral right to pursue the truth by reasoning with others is a central feature of media organizations functioning as the Fourth Estate,[57] or, at the least, ought to be a central feature of these organizations, even if it is often not.[58] Naturally, the truths in question pertain to matters of public interest and are often subject to political contestation.

In the light of the above, one problem with some epistemic institutions is that they might have been diverted or distracted from their core epistemic tasks and methods of rational inquiry to the point, in extreme cases, where they have become dysfunctional epistemic institutions. A news media outlet focused only on sensationalist news coverage is a case in point. Another problem is the negative impact on these institutions of an unregulated cyberspace that is characterized by computational propaganda. This brings us to our fourth point.

Fourthly, liberal democracies need to ensure that public discourse, including in cyberspace, is conducted in accordance with the norms of evidence-based truth telling (i.e., in accordance with the norms in part constitutive of the exercise of the moral right to freely pursue the truth by reasoning with others, e.g., the social norm to aim at the truth and norms governing evidence collection and analysis). Here there is a need for qualifications when the communication in question is understood to be of an informal or casual kind (e.g., between Facebook friends, or when the communicators are, say, children). We note that disinformation, hate speech, and propaganda flout these norms—although they are parasitic on them (see below)—and are antithetical to the proper exercise of the right itself (the right to freely pursue the truth with others).[59] Accordingly, the question that now arises is how politically motivated disinformation, hate speech, and propaganda/ideology/

---

[56] Miller, *Moral Foundations of Social Institutions*.
[57] Ibid.
[58] Al Gore, *The Assault on Reason* (New York: Penguin, 2007).
[59] We note that this right is a *joint* right. See Miller, *Moral Foundations of Social Institutions*, Chapter 2; Miller, "Joint Epistemic Action."

groupthink is to be countered in cyberspace—that is, how is computational propaganda to be successfully combated.

3.3 Epistemic Institutions, Market-Based Social Media Platforms, and Combating Disinformation, Hate Speech, and Propaganda/Ideology/Groupthink

3.3.1 Countering Computational Propaganda: Diagnosis

Effectively countering computational propaganda (i.e., political propaganda/ideology/groupthink)—including political propaganda/ideology/groupthink impregnated with disinformation and hate speech—is a complex undertaking. For one thing determining what is propaganda/ideology/groupthink and what is not is problematic, especially given that, as shown above, nonideological content often only implies or is permeated by ideology. For another thing, it is inconsistent with the liberal democratic value of freedom of communication to prohibit all propaganda, all fake news, or even all hate speech. Moreover, as is to be expected, different liberal democracies take a different view on where to draw the line here. The US does not prohibit hate speech, unless it directly incites serious crimes such as violence (the ISIS video clearly does), whereas many EU jurisdictions do.[60] This is, of course, not to say that propaganda might not be curtailed, without necessarily being prohibited, as is the case with advertising. Cigarette advertising is curtailed without being prohibited in many jurisdictions (e.g., no cigarette ads on TV or on sites accessed by children).

However, even if the legal issues could be sorted out, on the basis of cogent ethical analysis, and agreed to nationally and perhaps globally—since international regulations might be required for certain platforms and content—there remains the enforcement problem. Consider extremist jihadist propaganda that incites violence, and as such is prohibited. According to J. M. Berger, extremist jihadist propaganda has three dimensions: content, dissemination methods, and identity.[61] Accordingly,

---

[60] Waldron, *Hate Speech*.

[61] J. M. Berger, "Defeating IS Propaganda. Sounds Good, But What Does It Really Mean?," International Centre for Counter-Terrorism—The Hague (2017), https://icct.nl/publication/defeating-is-ideology-sounds-good-but-what-does-it-really-mean/. Accessed 29/10/2023.

in the case of extremist jihadist propaganda, social media sites can be quickly taken down, undermining that dissemination method. On the other hand, terrorist attacks themselves continue to be widely reported in the local and global media, thereby giving oxygen to terrorists' ideologies and causes.

Moreover, there are more sophisticated, and potentially more effective methods of dissemination of propaganda—for example, the profile-based microtargeting of vulnerable groups by state actors, such as Russian state agencies or their proxies. As mentioned above, these can make use of large databanks and ML techniques to build profiles and target the vulnerable. Such methods are not so easy to counter, although providing adequate protection of personal information held by social media companies, such as Facebook, would be a good start, one that has been made in the EU by way of the GDPR (see Chapter 1, Section 1.5, and Chapter 2, Section 2.2.2). Moreover, there are various other countermeasures that can be, and to some extent, have been put in place. For instance, there is the NATO Strategic Communications Center of Excellence and the EU East Stratcom Task Force. These organizations combat disinformation inter alia by developing an understanding of disinformation techniques, building public awareness of these techniques, fact checking, and training election personnel in relation to disinformation techniques. More generally, media companies can, and to some extent have, established fact checking units. Arguably, well-resourced, independent fact-checking institutions using ML and other sophisticated techniques, should be established with a public dissemination function, and used by other institutions, such as the media, and also by members of the public.

Directly countering content with countermessaging may have limited effect on those susceptible to propaganda, whether extremist jihadist Muslims or those with extreme left-wing or right-wing views. After all, it is these groups' felt alienation from liberal democracy that is in part the source of the problem. Successful propaganda, as was suggested above, is always anchored in reality, but is also vulnerable to the communication of reality (i.e., facts inconsistent with its content—inconvenient truths).[62] Accordingly, there is likely to be a need to address felt grievances, at least to the extent that they are justified (e.g., if in part based on economic injustice).

---

[62] Gore, *Assault on Reason.*

Naturally, propaganda could be countered by counterpropaganda, disinformation campaigns, and the like, as frequently happens in wartime, and as is now being advocated by some individuals in the name of protecting liberal democracy (e.g., some members of the US Democratic Party in an Alabama Senate election).[63] However, it is a violation of its underlying principles and, therefore, inherently morally problematic, for liberal democracies to eschew a commitment to truth (notably facts), evidence-based rational inquiry, and open discussion, in favour of propaganda (i.e., disinformation, half-truths, manipulation, hate speech, and so on). Moreover, this strategy is ultimately likely to be counterproductive and end up simply devaluing the liberal democratic currency. In short, ultimately it is likely to weaken liberal democracies, and thereby play into the hands of its authoritarian political enemies, although matters are somewhat more complicated in the circumstances of war (see Chapter 6, Section 6.2, on cognitive warfare).

What of identity? Certainly, an appeal to class, national, religious, ethnic, racial, gender, or other identity, and an attempt to drive a wedge between 'them' and 'us' is an important feature of computational propaganda and political propaganda/ideology/groupthink, more generally. The propaganda in question might or might not be unlawful, depending on the nature of it and the jurisdiction in which it is disseminated. Given legal limitations or enforcement problems, what is the way forward here? Naturally, if a polity has processes and pursues policies that are just, both procedurally and substantively, inclusive (e.g., of marginalized groups), and effective (i.e., have beneficial outcomes), then this will mitigate the harms of identity-focused propaganda. However, as is the case with other strategies, this strategy, while necessary, is not sufficient. It is not a silver bullet. Moreover, when the identities in question are national identities and the 'us/them' wedge is being driven by their own governments (e.g., the United States under the former Trump administration's "America First" policy and China and Russia under President Xi Jinping and President Vladimir Putin respectively), then this strategy is unlikely to succeed in the global sphere, even if it can be implemented to some extent.

---

[63] Editorial Board, "Democrats Used Russian Tactics in Alabama. Now They Must Swear Them Off," *Washington Post*, 27 December 2018. https://www.washingtonpost.com/opinions/democrats-used-russian-tactics-in-alabama-now-they-must-swear-them-off/2018/12/27/5b97c332-0941-11e9-a3f0-71c95106d96a_story.html.

On the other hand, within any given liberal democratic polity, developing and implementing an explicit set of policies to rebuild, maintain, and strengthen what we referred to above as the *public forum* is something that governments can and should address. These policies should include rethinking the role of epistemic institutions in cyberspace.

In the context of the legal limitations and/or enforcement problems confronting the enterprise of countering computational propaganda, and assuming that counterpropaganda, disinformation, and the like are not a morally acceptable option, we want to suggest a different strategy, a strategy which should be seen as complementary to the other strategies already mentioned. In doing so, we draw attention to three somewhat neglected, related, underlying conditions that facilitate political propaganda, namely: (1) the strength of epistemic norms in a population targeted by propaganda; (2) the intellectual health of the epistemic institutions in that population; and (3) their degree of embeddedness in, and influence on, the population that hosts them.

We note at the outset the importance of maintaining not only the distinction insisted upon between propaganda/ideology/groupthink and knowledge acquisition/dissemination—typically involving truth aiming by reasoning with others, a species of joint epistemic action—but also between knowledge acquisition/dissemination and entertainment (e.g., soap operas, cartoons).[64] The latter does not generally purport to be true. However, the emergence in recent decades of infotainment, including in cyberspace, is corrosive of this distinction, which is a point we cannot pursue further here. While insisting on the distinction between propaganda/ideology/groupthink and knowledge acquisition/dissemination, it is also important to draw attention to a central aspect of their relationship: propaganda/ideology/groupthink is parasitic on knowledge acquisition/dissemination and the epistemic norms that underpin it. Fake news, for instance, purports to be true, otherwise, it would have little effect. However, while pretending to comply with the epistemic norm of aiming at the truth, it flouts it; it is not required by its originator to be true and, indeed, its originator often knows it is false. It is a lie.

As with many parasites, propaganda/ideology/groupthink undermines the health of its host while simultaneously relying on the continued existence

---

[64] Advertorials are a form of communication comprising fictional, factual, and manipulative content. We do not have space to consider this problematic form here.

of its host. Accordingly, computational propaganda is a species of corruption: institutional corruption.[65] If unchecked, computational propaganda corrupts epistemic norms within a population and may also corrupt epistemic institutions—notably, media organizations responsible for news/comment that lack independence from an authoritarian government or which are subject to powerful and pervasive financial pressures that tend to cause them to espouse, for instance, a virulent form of capitalist ideology. On the other hand, as is the case with other species of propaganda/ideology/groupthink, because computational propaganda is parasitic on epistemic norms, it is susceptible to criticism for failing to live up to the epistemic and *moral* standards with which it purports to comply. It purports to be true, and hence is discredited, when shown to be false. Propagandist/ideologues/participants in groupthink fail to meet moral standards not simply because they fail to comply with epistemic standards by being incorrect or insufficiently attentive to the evidence, but because they are dishonest. They pretend to be aiming at the truth while actually telling lies, or, at the very least, they purport to be aiming at the truth and complying with epistemic norms, such as evidence-based reasoning, while in reality indulging in spurious reasoning and self-deception in the service of a collective fantasy. Accordingly, computational propagandists can be criticized by independent, credible, and reliable public communicators, not only for being incorrect, but also for being dishonest—indeed, for being corrupt. The charge of corruption is more likely to generate moral disapproval and, ultimately, rejection among members of a population than are purely epistemic offences. Moreover, the considerable reach of social media and other forms of mass media can be mobilized by these public communicators to amplify this message in cases where it is demonstrably true.

In a liberal democratic polity, epistemic institutions, notably the free and independent press, and schools and universities, have a key role in combating computational propaganda, or so we suggest. Epistemic institutions, such as schools and universities, have a key role in building resilience to disinformation, hate speech, and propaganda/ideology/groupthink, whether it be online or offline propaganda, by cultivating the skills and habits of rational inquiry and, relatedly, the development of well-informed, rationally defensible political perspectives among children and adults. Moreover, epistemic institutions, such as a free and independent press and universities have a key

[65] Miller, *Institutional Corruption*.

role in not only ensuring that the citizenry is reflective and well informed, but also in helping to ensure that public discourse, whether online or offline, is conducted in accordance with the epistemic norms constitutive of free and open rational inquiry, consistent with the proper exercise of the right to freely pursue the truth by reasoning with others. For instance, experienced investigative journalists based in well-resourced news media organizations, such as the BBC, are the source of much of the important news necessary to enable informed opinions on the part of voters, which, to reiterate, is not to say that the same journalists are not, at times or in relation to certain issues, themselves prone to ideological influence, even groupthink. Moreover, those responsible for politically motivated fake news, hate speech, and, more generally, propaganda/ideology/groupthink can be held to account by a free and independent press. Consider the BBC. It is both independent of government and, as a public broadcaster, independent of private sector companies. Moreover, its news division is a well-resourced, epistemically competent, genuinely public communicator, as opposed to an epistemically incompetent or narrowcast communicator—or platform facilitating the dissemination of epistemically deficient, narrowcast content. It is a genuinely public communicator by virtue of having a UK national and a global audience, composed in part of most of the key national and international opinion makers and most of the other influential public communicators. As such, it is well positioned to hold governments and powerful private sector actors alike to account, particularly in the context of a range of other well-resourced, epistemically competent, independent news/comment providers.

Here we note that the widely held view that the advent of the internet and of social media platforms, such as Facebook, Twitter, YouTube, TikTok and the like, has made redundant traditional epistemic institutions, such as a free and independent press, has proven to be spectacularly false. On the contrary, the advent of global social media platforms, such as Facebook, Twitter, YouTube and TikTok, has gone together with an exponential increase in the spread of fake news, hate speech, and propaganda/ideology/groupthink, and as a consequence has undermined the practice of rational inquiry and the existence of well-informed political perspectives among the citizenry. It has done so, in part, by undermining epistemic norms and undermining the strength and influence of epistemic institutions (e.g., by enabling the dissemination of fake news, hate speech and propaganda/ideology/groupthink on a vast scale). Moreover, these global tech companies have failed to adequately self-regulate in a manner that ensures that the content on their platforms complies with epistemic norms.[66] Indeed, the tech giants often dis-avow responsibility for these untoward developments by arguing that they are merely platforms and not publishers of the noxious content in question. More generally, the commercial interests of the tech giants tend in practice to override their stated commitments to the public good and to upholding epistemic norms regarding the content their platforms support (see discussion in Chapter 2, Section 2.2.2).

There are at least four salient features of the developments just described. First, the social media platforms are in fact more like platforms than publishers and certainly unlike news organizations, although they curate the content on their platforms in various ways, such as using algorithms that bring content to the attention of users of their platforms. For instance, as Balkin says: "an end user's Facebook feed does not offer every possible posting from the user's Facebook friends in the order they were posted; instead, Facebook decides which posts are most relevant and in what order to display them."[67]

However, unlike publishers, and especially news and media organizations, their central function is the provision of communication infrastructure, and, to this extent, they are more akin to telephone companies. We note that telephone companies, whether they be publicly or privately owned, do not or, at least, ought not, have any decision-making role in relation to communicative content, other than perhaps to ensure, or assist law enforcement to ensure, that it is not unlawful. The same point holds for internet payment systems (e.g., PayPal), Domain Name System registrars, cyberdefence services, and so on.[68] At any rate, an important consequence of this conception that social media platforms are not publishers is that they have been able to escape legal liability for illegal content supported by their platforms, even though they have been conscripted by governments to take down illegal content on their platforms, and additionally to take down offensive but, nevertheless, legal content that is (supposedly) inconsistent with their terms of use. However, in the US there exists a somewhat incoherent and institutionally damaging legal arrangement under which the social media companies can, on the one hand, escape liability under Section 230 of the Communications Decency

---

[66] Adam Henschke, "On Free Public Communication and Terrorism Online," in *Counter-Terrorism: The Ethical Issues*, ed. Seumas Miller, Adam Henschke, and Jonas Feltes (Cheltenham: Edward Elgar, 2021) and Alastair Reed and Adam Henschke, "Who Should Regulate Extremist Content On-line?," in *Counter-Terrorism, Ethics, and Technology: Emerging Challenges at the Frontiers of Counter-Terrorism*, ed. A. Henschke, A. Reed, S. Robbins, and S. Miller (Cham: Springer, 2021).

[67] J. M. Balkin, "Free Speech is a Triangle," *Columbia Law Review* 118 (2018): 2040–41.

[68] Ibid.

Act 47 U. S. C. for content posted on their platforms, while on the other hand, simultaneously, enjoying the right to decide whether and how to post users' content under the free speech rights of the First Amendment (i.e., they have publication rights).

The second salient feature of the development is the extraordinary communicative reach of the technology, and more specifically the ability of multiple fake accounts and bots to massively and anonymously amplify messages, including messages that attack and seek to discredit those communicating the truth about computational propagandists. Moreover, as already mentioned, the social media companies themselves have exacerbated the problem by amplifying disinformation, hate speech and ideology/groupthink, and shielding influential communicators who propagate it. Here we need to stress that while there is a moral right to freedom of political communication, there is no moral right to *amplify* one's political communications by recourse to automation (e.g., using bots). Nor is there a moral right to amplify political communications by using multiple individuals operating under the direction of a single individual or authority (e.g., using troll farms in which an organization hires multiple people to engage in its computational propaganda).

Relatedly, there is no moral right to tell lies or otherwise engage in deception (e.g., by using fake accounts), and manipulation (e.g., by using bulk data and algorithms to target unknowing vulnerable individuals). Here we need to distinguish deception from anonymity and, as we did above, anonymity from privacy. There is no moral right to lie or otherwise deceive or to engage in manipulation per se, although the moral right to communication implies that, speaking generally, deceptive, and manipulative communications ought not be criminalized or otherwise subject to government regulation. Indeed, telling lies, deception, and manipulation might be morally justified under some circumstances, such as in response to unavoidable yet morally unacceptable requests for personal information. However, it would not follow from this that there is a moral right, whether on the part of individuals or organisations (including social media companies), to set up one or more fake accounts (in order to deceive) or a moral right to use bulk data and algorithms (in order to manipulate). There are no such moral rights and, therefore, it is in principle, at least, morally justifiable for governments to ban fake accounts and to ban the use of bulk data and algorithms to manipulate. What of telling lies?

Lying or otherwise deceiving in some contexts is rightly criminalized (e.g., perjury or fraud, which is theft by deception). Moreover, telling lies about another person or organization, if it is harmful, may well constitute defamation, which is a tort, making the offender liable to civil action, and which in some circumstances is a crime. Thus, Dominion (an election technology company) recently won $800 million from Fox News in an out of court settlement on the grounds that the claim made by Fox that Dominion had rigged the 2020 US presidential election was defamatory. Of course, Fox has not been and perhaps cannot and should not be convicted of a crime. However, Fox News is a private sector media company and Rupert Murdoch is not a public official. Accordingly, it may well be that a further appropriately restricted category of lies (i.e., false explicit statements, deliberately made)[69] may well be justifiably criminalized under the following conditions. These lies are told by public officials during their time in office or for a period thereafter (to be determined), including by elected officials (if speaking outside of the legislature and, therefore, not protected by a legislative privilege), and the lies have been massively amplified and are demonstrably very seriously institutionally harmful. What of the restriction on the content of the lies in question? A case in point would be a public official who repeatedly, deliberately, and falsely, explicitly claims on the channels of public communication to an audience of millions that the US election was rigged—and in doing so convinces millions of voters that this is so.[70] Moreover, in light of the Fox News scandal and related recent trends, a more nuanced, and in fact traditional, understanding of the freedoms and responsibilities of the press is called for, one which insists on role differentiation in news/comment media organizations, and if necessary enforces it by regulation. Accordingly, the following institutional roles need to be distinguished and defined in terms of rights and duties.[71] Owners of news/comment media organizations (e.g., Rupert Murdoch), have a commercial function that, although necessary, ought to be subservient to the fundamental institutional purpose of news/comment media organizations, and is potentially at odds with the principles governing a free and responsible press. Occupants of this role ought not to be able to influence news/comment content. Editors must have independence from owners. Well-resourced investigative journalists, who operate in the context of a distinction between news and comment, even if they do provide some comment, fulfil a function that is central to a free press. Commentators who are paid

---

[69] Stuart Green, *Lying, Cheating and Stealing* (Oxford: Oxford University Press, 2006) Chapter 5 Section 2.

[70] Naturally, the electoral office in question would not only have to be appropriately independent of the legislature but also itself to be subject to stringent accountability measures to ensure the integrity of the electoral process.

[71] Miller, *Foundations of Social Institutions*, Chapter 10.

employees of the news media firm (e.g., Sean Hannity on Fox News), and therefore whose visibility and influence in respect of news/comment content should be circumscribed, do not have a function that is central to a free press. Expert commentators who are independent of the news/comment media organization, possessed of relevant expertise and representative of a wide spectrum of viewpoints, have a function that is central to a free press. Nonexpert interviewees/communicators—notably, members of the public—who are participants in newsworthy events and/or representative of relevant organizations and groups in relation to important matters of public interest—their function is central to a free press.

A further related point pertains to anonymity. As discussed above (see Chapter 2, Section 2.1.1.) privacy is not the same thing as anonymity, and whereas there is a basic moral right to privacy, there is no such basic right to anonymity. Naturally, the right to privacy implies the derived right to a degree of anonymity, and specifically the right not to have one's identity publicly disclosed in settings in which one is not a willing participant (e.g., the right of a person who is not a public figure not to have her name and photo displayed in a national TV program that she has declined to participate in). However, if a person is a willing participant in a highly publicized activity, then the question of anonymity become more complex. Does one have a moral right to use an anonymous Twitter account to engage in public discourse? In some circumstance one does, and in other circumstances one does not. We suggest that if one is a highly influential public communicator on political matters, whether via Twitter or some other channel of public communication, then one does *not* have a moral right to anonymity. On the contrary, one's fellow citizens have a moral right to know who you are.

The third feature is the global institutional character of the big tech companies. Here it is important to stress that the moral right of freedom of communication, and more specifically freedom of speech, attaches to individual *human beings* and pertains to their interpersonal communicative interaction. From this it does not follow that collective entities, like governments, corporations, and other organizations, have a *moral* right to freedom of speech—though they may have a legal right to it. Or rather, since organizations per se do not literally have minds and, therefore, do not literally speak, it does not necessarily follow that the occupants of organizational and other institutional roles have the same moral right to freedom of speech *qua members of those institutions*, as they do qua ordinary human beings. Indeed, they do not. Rather, the nature and extent of the right to freedom of speech of institutional role occupants qua role occupants is determined in large part by the nature and normative purposes of the institutions to which they belong, since these purposes are collective goods, according to our normative teleological theory of institutions (see Chapter 1, Section 1.7; Chapter 7, Section 7.1; and Chapter 5, Section 5.2).[72] So, the institutional and moral right of members of corporations to engage in speech pertaining to their business activities (e.g., to advertise their goods and services), may be very wide indeed, whereas their moral right qua members of these corporations to speak on matters unrelated to their business activities may be quite limited or even nonexistent. For instance, prima facie the CEO of a corporation does not have a free speech moral right qua CEO to publicly pronounce on matters that do not pertain to the business activities of the corporation. On the other hand, the same individual has a free speech moral right qua ordinary citizen to express any view he or she likes (with some limitations mentioned earlier), assuming he does not use his position as CEO to do so (e.g., he is not entitled to amplify the communication of this private view of his by emailing it to all the email addresses held by his corporation). Moreover, the free speech rights and, indeed, obligations of other institutional role occupants, such as journalists, may be much wider that those of the CEO of a corporation, because the institutional purpose of journalists is to communicate content that is in the public interest—an institutionally-based communicative purpose that is much wider than that of the CEO. On the other hand, it may be morally permissible (and potentially obligatory) for the journalist not to publish material if he or she correctly judges that members of the public have no right to know the content of the material in question.

The fourth and final feature is the business model of these market-based technology platforms: to provide free access in return for provision of private data, which can be exploited commercially. Relatedly, as already mentioned, there is the user growth-at-all-costs of some social media companies, such as Facebook. Here we need to invoke our favoured normative theory of market-based industries and do so by contrasting it with the prevailing shareholder value theory (SVT) of corporations in particular.[73] On our normative teleological account of social institutions, including market-based institutions, these social media platforms have as their institutional purpose to provide a collective good, or at least to contribute to the collective good provided by the industry as a whole industry—just as, for instance, the collective good of

[72] Miller, *Moral Foundations of Social Institutions*.
[73] Lynn Stout, *The Shareholder Value Myth* (Oakland, CA: Berrett-Koehler, 2014).

the housing industry is an adequate quantum of affordable, well-built houses for the relevant community.

SVT holds that the ultimate institutional purpose of corporations is to maximize profits, and thereby maximize shareholder value. Since big tech companies are typically corporations, then their ultimate purpose must also be, on this view, to maximize profits and shareholder value. However, SVT is not compelling; it confuses institutional purpose with the reward system that exists to ensure that institutional purpose is achieved, and indeed focusses narrowly on one beneficiary of that reward system—namely, shareholders (as opposed to managers or workers).[74] By contrast with this fixation of SVT on the reward system, our normative teleological theory focuses on the institutional purpose, a collective good, that the reward system exists to serve. In the case of social media platforms, the collective goods in question are (at least) effective channels of public communication that are accessible to all, a public forum for political communication, and (with respect to this public and political communication) general compliance with evidence-based norms of truth telling. These collective goods are not necessarily to be provided by a single media platform, acting alone. Certainly, this was not the case with traditional news media corporations. Rather, it is the industry that is to deliver these collective goods. Moreover, as is the case with traditional news media organizations in some countries, such an industry might consist of several competing social media platforms, some market-based and others publicly funded, though independent of government control, assuming this is practicable. It might not be practicable if it necessarily led to a process of 'balkanization' in which there was little or no possibility of the emergence and sustained existence of a genuinely public forum.

3.3.2 Countering Computational Propaganda: Strategy for Institutional Redesign

What is called for at this point is a strategy for institutional redesign of the global technology companies and of public communication on social media platforms. Here there are several guiding principles. These principles should be understood against a background assumption that the tech companies and the technology they use have provided enormous communicative and

---

[74] Miller, *Moral Foundations of Social Institutions*, Chapter 10.

epistemic benefits and these should not be sacrificed; the baby should not be thrown out with the bathwater.

Insofar as the big tech companies are to remain market-based companies they need to respect the principles of free and fair competition. Accordingly, they might need to be downsized to achieve this, as has been foreshadowed in recent legislative proposals—notably, the EU's Digital Market Act (DMA) (see Chapter 2, Section 2.2.2). The presence of Chinese-based tech giants, such as Tencent and Baidu, complicates matters here. Insofar as they are infrastructure providers of platforms, then each must be redesigned to ensure that it provides the required collective good(s). Commercial considerations cannot be allowed to trump its provision of the collective good, as is allowable in the case of an ordinary commercial enterprise considered on its own (as opposed to a market-based industry considered in its entirety).[75] This may require them, or perhaps some of them, to be transformed into publicly owned enterprises, and at the very least, it would require large monopolist and oligopolist tech companies to operate under a licence held conditionally on their compliance with legally enshrined minimum epistemic and moral standards (e.g., with respect to illegal content on their platforms). Moreover, such a licence should be held conditionally on the provision of greater transparency in relation to the algorithms they use, as would be required under the EU legislative proposal, the Digital Services Act (DSA) (Chapter 2, Section 2.2.2). At any rate, the general principle that needs to determine institutional design policy in this area is, to reiterate, that the public interest in liberal democracies in efficient, effective channels of public communication that are accessible to all, a public forum for political communication that is accessible to all, and compliance with norms of evidence-based truth seeking, which overrides private interests, commercial or otherwise.

Secondly, the regulation of content to ensure compliance with epistemic and moral norms, including but not restricted to legally required norms, is a task that cannot be left to the tech giants—although for reasons of practicability, they may need to continue to be the ones to take down unacceptable content, as opposed to determining the rules governing what content is to be taken down).[76] Certainly, it cannot be left to them in the absence

---

[75] Miller, *Moral Foundations of Social Institutions*.

[76] Admittedly, there is a grey area here. For discussion of some ways to handle this see S. Theil, O. Butler, K. Jones, H. Moynihan, C. O'Regan, and J. Rowbottom, "Response to the Public Consultation on the Online Harms White Paper," *Bonavero Reports Series*, 1 July 2019, https://www.law.ox.ac.uk/sites/files/oxlaw/bonavero_response_online_harms_white_paper_-_3-2019.pdf.

of their legal liability in the circumstance that they fail adequately to ensure compliance on the part of their users with legal requirements, such as the legal requirement not to incite violence. That is, the regulation of content cannot be left to the tech giants in the absence of their having the legal status of publishers. Perhaps the entire compliance task needs to be performed by an external, independent institution, if this is practicable—although, if it is practicable, it is a task that should be paid for by the tech companies themselves and/or their advertisers or others who use their platforms. Here it is important to distinguish between holding someone and/or some organization legally liable for publishing illegal content—and, in the absence of a publisher other than the tech giants providing the platforms, this might need to be the tech giants themselves—and ensuring the compliance of communicative content with epistemic norms by means of a mandatory editorial process. The latter process of epistemic quality assurance includes more than ensuring that legal requirements are met. Perhaps it might involve ensuring that public communicators of political content, which consistently reaches very large audiences (e.g., hundred thousand persons or more), achieve some minimal threshold of compliance with epistemic norms. For instance, these communications could be subjected to a fact check by a competent independent body and the results posted alongside the epistemically offending communication.

This external, independent institution might need to have several roles. As suggested earlier in this section, it might need to have a fact checking role in addition to its role in monitoring content for the purpose of identifying illegal content, perhaps in collaboration with the tech companies. However, it seems that this independent institution could usefully be provided with greater powers. For instance, anonymity is one of the main obstacles to combating illegal content, and more generally to curbing disinformation, hate speech, and propaganda/ideology/groupthink. The identity of these malevolent public communicators is not known. Another obstacle is amplification by automated bots. Let us first consider anonymity. Here it is important to reiterate, firstly, that anonymity is not the same thing as privacy (see Chapter 2, Section 2.1.1). For instance, an anonymous communicator is not per se exercising a moral right to privacy, even though in some instances anonymity can function to protect privacy or confidentiality rights. Secondly, public communicators of politically significant content (e.g., content pertaining to public policy) do not have a moral right to anonymity, at least in liberal democracies. This is not to say that some public communicators might not have a moral right to anonymity in other contexts (e.g., dissidents in authoritarian states), or with respect to other kinds of content.

What of amplification? As argued above, there is no moral right to amplify one's public communications by means of bots or troll farms. Relatedly, there is no right to use fake accounts or to manipulate by means of bulk data and algorithms. Nor do social media companies have a moral right to amplify disinformation, hate speech and ideology/groupthink on their platforms.

In light of the above discussion, the following additional powers for our independent institution suggest themselves.

Firstly, social media account holders (e.g., Twitter and Facebook account holders), including individuals, groups, and organizations, are required to register with the independent statutory authority (e.g., the Office of e-Safety Commissioner in Australia), in the liberal democracy in which they reside. The authority issues a unique identifier, based on the driver's licence, passport, and so on provided by these account holders. Social media providers (e.g., those based in the US) are required by law only to provide accounts to those who have registered with some statutory authority.

Secondly, the authority must provide to law enforcement under warrant the identity of those persons who breach laws. There is no anonymity for lawbreakers and, as a result, many will be deterred from engaging in posting illegal content. If not deterred, they are at risk of being identified, arrested, and charged, including under extradition provisions if they live overseas.

Thirdly, social media communicators of politically significant content (e.g., content pertaining to public policies) on mass media channels of public communication who have large audiences (e.g., hundred thousand followers) are legally required to be publicly identified. There is no right of anonymity for public communicators of politically significant content. Accordingly, there will be a deterrence effect in relation to lawful, as well as unlawful, disinformation, propaganda, and hate speech. Also note that under this arrangement, bots cannot use fake accounts; there will be no fake accounts, or at least their number will be greatly reduced.

Fourthly, content that is otherwise legal, but which, nevertheless, fails to meet minimum epistemic and moral standards (e.g., is demonstrably false, *and* which is significantly artificially *amplified* by one of the above means)— or is otherwise illegitimately amplified—is to be liable to removal by the social media platform or by the independent fact-checking authority, but only in accordance with the adjudication of the independent statutory authority. This adjudication will need to be justified by the authority, and the adjudication and its justification made public. The minimum epistemic and moral standards in question are to be established by the independent statutory authority following on a process of public debate, expert input, and so on. However, to reiterate, amplification is a further necessary condition for liability for removal of content.

A final point concerns the business model that involves the provision of a service in return for handing over one's private information. Recent EU regulation and proposed regulation (such as the GDPR, DSA and DMA) has been enacted or proposed, among other reasons, to ensure informed consent on the part of those who might be asked to provide private information in return for a service. Such legislation might ultimately undermine this business model, and this might not be a bad thing, not least in relation to reducing the quantum of computational propaganda on social media.

4

Criminal Justice, Artificial Intelligence, and Liberal Democracy

Justice is a complex and contested concept that is closely related to various other concepts, including equality and desert. There are also different forms of justice. There is distributive justice, concerned with the allocation of benefits, and there is commutative justice, concerned with punishment for offences. A range of technologies impact on justice. Indeed, the lack of access to technology can itself be an injustice (e.g., the so-called digital divide). However, our concern in this chapter is with criminal justice rather than distributive justice or the somewhat narrower notion of commutative justice, and, more specifically criminal justice in a liberal democratic setting. Here we need to note that injustices of a serious nature are not always perpetrated by criminals or criminal organizations, and may not even involve actual criminality. For instance, there are cases of AI-generated miscarriages of justice emanating from governments and their agencies.[1]

In this work, we have referred to the tsunami of cybercrime, as well as cyberconflict, disinformation, and so on, which has been enabled by the internet and associated cybertechnology, AI, and so on. Of course, criminal activities (see Chapter 1, Section 1.5, for crime statistics) are almost by definition injustices to their victims. Moreover, the boundaries between cybercrime, cyberconflict—notably, as a national security issue (see Chapter 6)—and disinformation, hate speech, propaganda, groupthink, and so on (see Chapter 3) have become increasingly blurred, creating problems

---

[1] An egregious example was the UK Horizon affair, in which software led to the assertion that numerous post office staff had embezzled money. Over seven hundred people were prosecuted, some were jailed, some committed suicide. It was, however, the software which was faulty. The compensation paid out is approaching half a billion dollars. Helen Margetts noted that blind trust in outsourced software has become commonplace in government. https://theconversation.com/post-office-scandal-reveals-a-hidden-world-of-outsourced-it-the-government-trusts-but-does-not-understand-159938. Accessed 6/6/2021. In Australia the Robodebt scheme sent many users of the government Centrelink facility erroneous debt demands. Eventually the errors were discovered and the debts rescinded, albeit at considerable stress to the victims. The affair was of such magnitude it generated a royal commission. https://robodebt.royalcommission.gov.au/. Accessed 25/10/2023.

Cybersecurity, Ethics, and Collective Responsibility. Seumas Miller and Terry Bossomaier, Oxford University Press.
© Oxford University Press 2024. DOI: 10.1093/oso/9780190058135.003.0005

for law enforcement and national security agencies alike. For instance, evidently several Russian ransomware gangs operate under the direction of Russian state intelligence agencies,[2] and the Chinese state is involved in disinformation campaigns against Taiwan.[3] Indeed, according to one influential view, cybercrime is now a national security problem.[4] We discuss some of the ethical issues that arise in this context in Chapter 6.

The challenges confronting law enforcement have been compounded by the advent of the internet, end-to-end encryption (see Chapter 2), and related technologies. One of the most dramatic illustrations of this is the so-called Dark Web which relies heavily on encrypted packets to support onion routing—but, incidentally, is not necessarily used for criminal activities since its structured encryption enables benign activities to remain hidden, as well as malign ones.

Many internet addresses are hidden. There are no names associated with them through Domain Name Servers. But they are still accessible. They constitute the Deep and the Dark Web. The Deep Web is just the large volume of web data behind corporate firewalls and other privileged sites. The Dark Web sits outside, but nevertheless has its own search engine, GRAMS. Since a lot of the activity on the Dark Web is criminal in nature, from illicit drugs to child pornography, it makes use of something known as onion routing which protects access by means of a cryptographic structure. The Dark Web's IP addresses comprise random-looking alphanumeric strings that end in *.onion*. The *.onion* suffix is an indicator of how it works. Routing is done through a long path of intermediate nodes. Finding the ultimate address is like peeling the layers off an onion. Each server, which is a step along the way, rips off one layer of the onion, enabling it to find the next destination of the onion. The cryptographic structure does not allow it to peel off further layers and thus determine what is at the centre. There is a special browser, ToR (The Onion Router), for the Dark Web. If you type, for instance, the name of an illicit drug, into GRAMS you will find sites selling them. To make sure that your purchase does not appear on your credit card statement, an anonymous payment method is needed. Cryptocurrencies, such as bitcoin, protected by

---

[2] According to the UK's National Cyber Security Centre. See UK government press release entitled "UK Cracks Down on Ransomware Actors," 23 February 2023. https://www.gov.uk/government/news/uk-cracks-down-on-ransomware-actors.

[3] Hung and Hung, "How China's Cognitive Warfare Works."

[4] Federal Bureau of Investigations, *Internet Crime Report 2022* (Washington, DC: Internet Crime Report Centre, 2022). https://www.ic3.gov/Media/PDF/AnnualReport/2022_IC3Report.pdf. Accessed 29/10/2023.

blockchain technology (see Chapter 7, Section 7.1.5), have thus been of considerable importance to the Dark Web.

Notwithstanding the challenges confronting law enforcement as a result of what amounts to a technological revolution, the internet, bulk data, AI, and biometric and genomic technology related to cybertechnology (e.g., universal DNA profile databases and facial-recognition technology) also afford opportunities and new technological tools to law enforcement agencies working to combat crime, including but not restricted to cybercrime. This development, while welcome in general terms given the extraordinary challenges law enforcement agencies face, raises a raft of ethical problems, some of the most salient of which we discuss in this chapter.

Machine learning (ML) is a technology impacting criminal justice—notably, in the rise of predictive policing and in relation to legal cases in which there is decision making based on the outcomes of large numbers of past court cases that have been subject to ML analytical techniques. Recent developments in biometrics and the use of DNA databases and facial-recognition technology are also having a major impact on policing. Accordingly, this chapter is divided into four sections: predictive policing; legal adjudication; DNA profiles and databases; and facial-recognition technology. In each case, an important focus is on the implications of the uses of these technologies for the moral rights and principles constitutive of liberal democracy.

4.1 Predictive Policing

Predictive policing (PP) is a term that refers to a range of crime-fighting approaches that utilize crime mapping data and analysis, and more recently social network analysis, big data, and predictive algorithms.[5] Historically, police services have used statistical information to target specific locations in relation to particular types of crime (i.e., they have utilized various methods of crime mapping data and analysis).[6] For instance, police resources have been directed to crime hotspots identified not simply on the basis of past crimes committed at that location, but on the basis of the location in

---

[5] Seumas Miller, "Predictive Policing: The Ethical Issues," in *Future Morality*, ed. David Edmonds (Oxford: Oxford University Press, 2021), 73–82; and Seumas Miller, "Machine Learning, Ethics and Law," *Australasian Journal of Information Systems* 23 (2019): 1–13.

[6] See, for instance, Miller and Gordon, *Investigative Ethics*, Chapters 2, 3, and 10.

question having features that correlated with crimes of the relevant type (e.g., high incidences of theft at locations in which there are a number of tourist attractions and good escape routes for thieves). In doing so, police are, in effect, predicting a crime of theft at the location in question, and acting on that prediction to prevent it. Moreover, police services have also had the practice of targeting known offenders as opposed to merely reactively investigating crimes, as for example, when they target a recently released offender, known to use a particular modus operandi (MO), in response to a recent spate of burglaries involving the same MO. In doing so police are, in effect, predicting that a past offender will continue to offend and taking action to prevent him or her doing so.

Predictive policing is in many ways simply a continuation of these historical practices of police services. However, PP has introduced some new methods.[7] In relation to high volume crimes, such as burglary and car theft, it has utilized big data analytics, including the use of predictive algorithms, to establish a wider set of statistically based correlations that rely on much larger data sets from big data (e.g., to show that a burglary in a given location often generates additional burglaries in that area). That is, burglary is 'self-exciting', to use the jargon, creating a 'crime hotspot' that is analogous to the spread of an infectious disease, such as in a COVID-19 hotspot.[8] In relation to offenders—notably, violent offenders—PP has utilized social network analysis (as well as statistically based correlations). Social network analysis involves initially identifying offenders, then tracing their associates, then the associates of their associates, and so on. In doing so, it utilizes social media (inter alia) as a source of these links in the network.

The rise of PP, especially in many police jurisdictions in large cities in the USA, such as Los Angeles, Chicago, New York, and New Orleans, has raised the spectre of the surveillance society described in the film, *Minority Report*, in which citizens can be arrested by police for crimes they have not yet committed, and have no intention of committing at the time of arrest, on the basis of supposedly reliable evidence that they will commit them. By analogy, police utilizing PP in a city somewhere in the world today might, let us suppose, arrest a citizen, John Smith, for a violent crime, even though he is yet to

---

[7] For detailed discussions see Andrew G. Ferguson, *The Rise of Big Data Policing* (New York: New York University Press, 2017).

[8] The term *hotspot* is used in relation to high crime locations and areas of high COVID infection. This is no accident, given the use by police in PP of techniques used in infectious diseases control and the use in infectious disease control of techniques used in intelligence-based policing (e.g., in counterterrorism).

commit this crime and has no intention of doing so. However, Smith is found loitering in a violent crime hotspot in the neighbourhood where he lives, and he is a known associate of members of the local violent youth gang. But how realistic is this scenario?

Let us consider a pioneering example of predictive policing, the Los Angeles LASER program (Los Angeles Strategic Extraction and Restoration Program). The program commenced in 2011 with the aim of reducing violent, gang-related crime in LA, and especially in certain black neighbourhoods in which most of the gangs operated. It included a location-based hotspot component and an offender-based component.[9] The offender-based component involved identifying chronic offenders based on criteria such as gang membership, past violent crime offences, and prior arrests with a handgun. Chronic offenders are then contacted, but not with a view to arresting them since, for one thing, there is at least at this stage not enough evidence for this. Rather, the purpose of this criminal justice intervention is deterrence, and crime reduction as a result of deterrence. The offenders are contacted and advised of available programs and services designed to reduce their risk of recidivism. However, this serves, in effect, as a deterrence since they are also put on notice that they are being watched and that, if arrested, their failure to avail themselves of these programs and services will count against them in future sentencing (e.g., other things being equal, will result in longer prison terms). The location-based component involved identifying quite specific spatiotemporal locations (e.g., five hundred square metre locations during a six-hour period on certain days of the week) for intervention by police officers. These hotspot locations were selected based on a historical analysis of gun-related crime data. This was followed by an analysis to try to determine what the causes of the high-level of gun-related crime is in each location (e.g., a border-area between rival gangs of a weekend evening), and to develop an appropriate crime prevention strategy for that location (e.g., high police visibility). The effect of the strategy in terms of statistical levels of gun violence in each location was continuously monitored and the strategy adjusted as required. According to the official review by the Inspector General of the Los Angeles Police Commissioner, the LASER program met with considerable success in terms of reducing violent gang-related crime, at least initially and in a number of high-crime neighbourhoods.[10] The LASER

[9] Mark P. Smith, *Review of Selected Los Angeles Police Department Data-Driven Policing Strategies* (Los Angeles: Los Angeles Police Commission, 2019).
[10] Smith, *Review of Selected Los Angeles Police Department Data-Driven Policing Strategies*.

program was discontinued in 2019 partly on the grounds that police were failing to comply with its protocols and unfairly targeting some community members, including some who had not previously been arrested. However, police in the LAPD and in other US cities have continued to use predictive policing models in one form or another.[11]

While collecting and analysing location-based crime data of the kind in question is not morally problematic per se, moral problems might arise with some of the crime prevention strategies utilized in response to that data. For instance, possession of data indicating a crime hotspot would not in and of itself justify stopping and searching or arresting a person merely because s/he happened to be at that location. The data in question would not justify this since it would not reach the threshold of reasonable suspicion—or, therefore, the stronger requirement of probable cause—of a *particular* person. Accordingly, if such a person was to be stopped and searched or arrested there would need to be additional evidential facts about that person based on, presumably, real-time observation of her/him (e.g., visual evidence that s/he was carrying a gun). Of course, this is not to say that police in the US and elsewhere always comply with this understanding of the principle of reasonable suspicion, a principle enshrined in the law in most liberal democracies, including the US. Indeed, unfortunately, there are instances in which they do not. To this extent, the John Smith scenario is not unrealistic.

Nor would collecting and analysing offender-based data of the kind in question be morally problematic. After all, the persons in question are known to be violent offenders and known by virtue of their past convictions for violent crime. However, the level of intrusive attention may well be morally problematic if this was not the case. It would be morally problematic because individuals have a moral right to freedom from state interference absent prior evidence of violation of its laws.[12] Accordingly, the level of intrusive police attention to John Smith in our fictitious scenario is not morally or, in most liberal democracies, legally justified. This, once again, is not to say that it does not happen—notably, it does happen in black neighbourhoods in the US—and, to the extent that it does, the John Smith scenario is realistic.

---

[11] Andrew G. Ferguson, "Predictive Policing Theory," in *Cambridge Handbook of Policing in the United States*, ed. Tamara Rice Lave and Eric J. Miller (Cambridge: Cambridge University Press, 2020), 491–510. Recently the EU has sought to constrain predictive policing methods. *AI Act* European Parliament, 11 May 2023. https://www.europarl.europa.eu/news/en/press-room/20230 609IPR96212/meps- ready- to- negotiate- first- ever- rules- for-safe-and-transparent-ai. Accessed 29/10/2023.

[12] Miller, "Machine Learning, Ethics and Law," *Australasian Journal of Information Systems*, 1–13.

It might be suggested that there are certain categories of legitimately targeted putative offenders who do not have, or ought not to have, this moral right to freedom from state interference (e.g., members of a terrorist organization who have not themselves been convicted or even performed a terrorist act). However, in such cases a morally justifiable law will in fact, at least typically, have been broken—namely, membership of a terrorist organization. This is, of course, not to say that membership of a violent youth gang should be a criminal offence, and certainly, associating with members of such a gang, as John Smith does, should not be a criminal offence (and it is not in liberal democracies)—after all, gang members have parents, brothers, sisters, and so on who are not gang members. Another category of legitimately targeted putative offenders in relation to which this is not the case are, arguably, the police themselves. Consider, for example, the case of police officers who are not known to have committed any crime. Is there a justification for monitoring their behaviour, other than for ordinary work performance purposes? Are police any different from ordinary citizens in this regard? It might be argued that given their position of trust, and the fact that they have extensive powers of arrest and use of lethal force not possessed by ordinary citizens, such monitoring is morally justified.[13] Moreover, the occupational role of a police officer is one freely chosen, so if a police organization has established this practice of monitoring its officers, then presumably its new recruits have found the practice to be acceptable, or at least not intolerable.

Speaking generally, PP faces a number of problems. Some of these problems are problems for predictive policing, even on its own terms of contributing to crime reduction. Others are moral problems, even if they do not directly impact on the effectiveness of PP as a crime reduction approach (e.g., because they involve violations of moral rights or are unjust). Moreover, many of these problems directly or indirectly impact the effectiveness of PP *and* involve some injustice or rights violation. Indeed, they may well reduce the effectiveness of PP precisely because they are held to be unjust or involve rights violations. Aggressive policing tactics, such as so-called saturation policing, have often created more problems than they have solved (e.g., leading to the Brixton riots in this predominantly black neighbourhood in London in the 1980s).

First, there can be problems with the data (e.g., false, or unbalanced data input). To return to our scenario, perhaps the arrest of John Smith is

---

[13] Miller and Gordon, *Investigative Ethics*, 201–23.

an instance of mistaken identity; after all, there are many citizens with that name, and the data might have been wrongly interlinked in the police database to the photo of a John Smith who had no contact with any gang members but only took a wrong turn and found himself in a crime hotspot. At any rate, false data input leads to false positives (i.e., targeting the innocent) and, in addition, false negatives (i.e., failure to target the guilty). Nor is such an error necessarily easily corrected, given the bureaucratic processes involved in changing data in large, confidential police databases.

Second, for some serious crimes, such as murder and terrorism, the databases are comparatively small. For instance, the comparatively small number of known terrorists in databases is an impediment to the use of predictive techniques, such as ML, that are dependent on big data that seek to generate accurate profiles of typical terrorists.

Third, the prediction of future crime, and the use of ML techniques in particular,[14] are based on past reported crimes, arrests, prosecutions, police incident reports, lists of offenders, and so on. Thus, if law enforcement agencies rely on ML, then (other things being equal) offenders and offence types that have escaped detection in the past (e.g., child sexual abuse) are less likely than those who have not (e.g., grievous bodily harm) to be targeted by police. Naturally, other things might not be equal, and police might choose to target offenders and offence types that are now known to have gone undetected in the past, as has recently happened in the case of child sexual abuse. However, if in doing so, police are seeking to predict future crimes by these offenders or of these offence types using ML techniques then problems will arise. Problems will arise because of the inadequacy of the past data sets (i.e., nonexistent, incomplete, and unreliable information about the past offenders and offences in question).

For the same reason (i.e., reliance on past data to predict future crime), communities that have been overpoliced in the past, are likely to continue to be overpoliced relative to other communities. This is especially the case if predictive policing techniques utilize socioeconomic indicators that

---

[14] There are broadly three general categories of errors in ML: inadequate data; bias; and brittleness. Inadequate data is like the requirement for a sufficiently large dataset in statistics. But size is not enough. One might have a very large data set of people from the US, which didn't include any Tibetans. Thus, errors might occur when the ML system encounters a Tibetan. Bias occurs where there is a disproportionate number of cases of a particular category, racial bias being an egregious example. Brittleness is more subtle. If an ML system is used outside its training boundaries, highly anomalous results can ensue.

statistically correlate with crime indicators, as is the case with poor black neighbourhoods in the United States.[15]

Fourth, and related to this last point, one of the more controversial PP techniques is profiling, and profiling can take the form of morally problematic racial profiling. A famous case of profiling in the US is that of Sokolow, who was searched by customs officials at an airport and found to possess drugs. He was searched because he fitted the profile of a drug courier, and this was taken to constitute reasonable grounds for suspicion. Sokolow argued in court that fitting a profile did not constitute reasonable grounds for suspicion. However, he lost the case because it was held that, whereas fitting a profile did not of itself constitute reasonable suspicion, the evidence that constituted the profile in his particular case did.[16] However, there are dangers in the practice of profiling, and what is of interest to us here is profiling that utilizes ML techniques. Specifically, there is the risk of discriminatory algorithms (e.g., an algorithm that relied on a past data set that comprised an unjustifiably disproportionate number of 'stop and frisk' searches of black citizens). Profiling practices that rely on ML techniques that utilize such data sets can end up generating morally unjustified, racially based profiles of offenders, and thereby entrench existing racist attitudes among police officers and others.

Fifth, there is an issue of privacy or confidentiality. Specifically, do some of the databases upon which PP techniques rely consist of personal information or confidential information to which the police do not have a moral or legal right, or are otherwise not morally or legally entitled to access? There are, of course, complications here in relation to what counts as personal information. The content of telephone calls, emails, and so on is typically regarded as personal or confidential information. But what of metadata; data concerning the caller/called, duration of call, and so on? Some have argued that this is not personal or confidential any more than the sender and receiver's name on a parcel sent through the postal. However, a large bank of metadata extracted from a person's phone calls, emails, and so on, can enable the creation of a detailed picture of that person's associates and movements, a picture sufficiently detailed to count as an infringement of their privacy.

Sixth, the existence of large police databases of personal and confidential information pertaining to offenders and other citizens gives rise to data security concerns. Data security paradigmatically consists in ensuring that

---

[15] See Ferguson, *Rise in Big Data Policing*, for an extended discussion of this latter point.
[16] *United States v. Sokolow* (1989). https://caselaw.findlaw.com/us-supreme-court/490/1.html. Accessed 29/10/2023.

such information is protected from unauthorized or otherwise illegitimate accessing. Clearly data security is critical in the face of sustained hacking by criminal organizations inter alia, which can compromise privacy and confidentiality, and thereby put lives at risk (e.g., of informants, undercover operatives, and so on) and the identity security of citizens, in the case of biometric facial images.

Seventh, there are additional moral problems inherent in some of the new techniques used in predictive policing, such as the so-called black box issue in ML. In the case of, for instance, the profiles of offender types generated by ML processes the correlations upon which the profiles are based may not be known or understood by law enforcement, let alone putative offenders or ordinary citizens. Thus, indicators of an offender type generated by ML might include features that do not have any intuitive, or more broadly rational, relationship to the type of offender or offence type in question. Here by 'broadly rational relationship' is meant by the lights of human reasoning. This is especially so if the data used in the ML process is not necessarily restricted to data that is crime-related; for instance, it might include social, economic, medical and consumer data. Accordingly, to use a somewhat far-fetched example, citizen Smith, at least in theory, might be being monitored by police in relation to violent assaults in part on the basis of his identified large shoe size of fourteen and his liking for strawberry ice cream.[17] Here there is no rational relationship between persons with shoe size fourteen who like strawberry ice cream and engaging in violence. Nor has any causal relation been established. Rather, by virtue of some unknown process in the black box a statistical correlation has been established between violent assaults and offender strawberry ice cream consumption and wearing shoe size fourteen. Moreover, some success has been achieved by police who rely on this correlation. Hence, they now regard Smith's wearing size fourteen shoes and liking strawberry ice cream (other things being equal) as grounds for reasonable suspicion that he engages in violent assaults.[18]

Of course, the black box issue arises in contexts other than policing, such as in the provision of health services. For instance, ML techniques enable malignant melanoma lesions to be detected, and to be differentiated from

---

[17] Apparently, in cases of imminent hurricanes the sales figures for strawberry flavoured Pop-Tarts increase significantly. C. L. Hays, "What Wal-Mart Knows about Customers' Habits," *New York Times*, November 14, 2004. http://www.nytimes.com/2004/11/14/business/yourmoney/what-walmart-knowsabout-customers-habits.html. Accessed 29/10/2023.

[18] Other things might not be equal (e.g., if Smith does not have the ability, opportunity, or motive to engage in violent assaults).

benign moles, more accurately than can be done by skin specialists relying on their perceptual abilities. However, the absence of an explanation of the workings of the black box in the case of malignant melanoma lesions is not a moral concern, even if it is a matter of epistemic interest. Evidently, determining whether the moles on a patient's skin are malignant melanoma lesions is a very significant health benefit to the patient, and, in addition, the patient consents to it. Indeed, the patient consents to it knowing that the process involves a black box. For it does not much matter to the patient whether there is an epistemic gap in the process, so long as it works. By contrast, being monitored by police is not a benefit to Smith; indeed, it is an infringement of his right to privacy, although the infringement may well be morally and legally justified and, therefore, not a violation. Moreover, unlike the patient with moles, Smith has not consented to being monitored by the police. Accordingly, there is a presumption against police monitoring of Smith. This presumption can legally and morally be overridden if police have a reasonable suspicion that Smith has committed an offense. Perhaps his past offences reasonably render him an object of suspicion. But let us assume that he does not have any past offences, nor is he known by the police to have any other specific feature that rationally and inductively constitutes sufficient grounds for reasonable suspicion. In that case, as we saw above in relation to Smith's coincidental presence at a crime hotspot, the fact that Smith fits the profile of an offender type generated by the ML process (e.g., wears size fourteen shoes and likes strawberry ice cream) would not justify this monitoring since it would not reach the threshold of reasonable suspicion—or, therefore, the stronger requirement of probable cause—of a *particular* person. (We return to this issue of particularity below.)

However, there might be a probabilistic difference between the two cases. Smith's coincidental presence at a crime hotspot does not reach the threshold of reasonable suspicion for the reason (at least) that there are many innocent persons present at crime hotspots, and Smith might be one of these. Accordingly, the likelihood that Smith is an offender merely because he is present at the relevant crime hotspot is not high. By contrast, in the case of the ML process that generates profiles of offender types, it is far more likely that Smith is in fact an offender, since he meets the relevant profile. Nevertheless, arguably, there would still need to be additional evidential facts about Smith (i.e., additional to his wearing size fourteen shoes and liking strawberry ice cream), if the threshold of reasonable suspicion is to be met, therefore justifying the monitoring of Smith. For wearing size fourteen shoes and liking strawberry ice cream are not causally connected to violent assaults; correlation is not causation. Accordingly, no evidence has been if Smith is ever in a *causal* state relevant to engaging in assault, as he would be if he frequently carried a weapon on his person. Moreover, wearing size fourteen shoes and liking strawberry ice cream are not rationally connected to violent assaults. That is, they do not rationally connect, let alone imply, any motive, ability, or intention to violently assault anyone (i.e., any criminal intention or action on the part of Smith). Accordingly, no evidence has been provided that Smith is ever in a state *rationally* connected to violent assaults as he might be if, for instance, he had a violent temper, had martial arts training, or was carrying a weapon. On the other hand, this argument to the conclusion that the threshold of reasonable suspicion has not been met does not demonstrate that the statistical correlation is epistemically worthless. Rather, it constitutes inductive evidence of a kind that might be sufficient to justify further data collection of a nonprivacy infringing kind in relation to Smith, as opposed to monitoring of Smith. Perhaps it should merely be treated as intelligence, for instance.

Finally, there is the general problem of the effectiveness of PP in criminal justice contexts in which offenders can themselves predict the results of PP predictive processes and, thereby, thwart them. If, for example, criminals become aware of the profiles used by police they can adjust their MOs to escape detection.

Thus far we have focused on PP as it is currently practiced (at least for the most part) in liberal democracies, such as the US, the UK, and Australia. However, PP is constantly evolving and in doing so utilizing new technologies. Fingerprint recognition and DNA identification have a long history of successful use in the criminal justice system to investigate serious crime. However, new biometric identification technologies, such as facial recognition, gait analysis, and voice recognition are rapidly evolving. Moreover, there have been important developments in existing technologies—notably, DNA identification, and the expansion, digitization, and interlinking of existing databases with new ones (see Section 4.3 below). Recently, biometric databases have been established by governments and the private sector; for example, automated facial recognition is now a key part of passports and border security in many countries. Automated facial recognition is a powerful technology with the potential to identify a face in a large

crowd, through integration with CCTV systems, enabling real-time surveillance, identification, and tracking of individuals through public places by police (see 4.4 below). Further notable applications of facial recognition include
the analysis of images taken from the internet, particularly social media.
Images taken from the internet include ones taken from Facebook, Twitter,
LinkedIn, and Google. The significance of this capability is highlighted by
the rapid expansion in the number of images uploaded to the internet. For
example, Facebook alone holds several hundred billion photographs in its
database and uses automated facial recognition software to identify or tag
users in photographs.

The use of these technologies raises a number of pressing ethical concerns.
The capacity to integrate databases of biometric and nonbiometric (e.g.,
smartphone and email metadata, financial, medical, and tax records) information adds to these concerns. For instance, biometric facial image templates
can be used in conjunction with digital images sourced from CCTV, phone
GPS data, and internet history, to provide an increasingly comprehensive
picture of an individual's movements and lifestyle. As is often noted, privacy
and confidentiality in relation to personal data consists in large part in the
right to control the access to, and use of, that data. As such, it is a component
of individual autonomy, a cornerstone of liberal democracy.

China provides an insight into potential developments in the use of integrated databases by government and police. In China, the government
is using biometric facial recognition and gait analysis systems to identify
individuals in public places via CCTV who are suspects of minor crimes,
such as jaywalking. Biometric facial recognition systems now play a role in
China's 'social credit' system, which rewards and punishes citizens based
on their norm-following behaviour, honesty, and courtesy, in concert with
other big data analysis capabilities that facilitate tracking, such as GPS data,
internet use, and financial transaction history. The implications of a low social credit score for Chinese citizens include travel bans, and exclusion from
private schools and higher status professions. In Xinjiang, in particular, the
Uighur population has been subjected to what is now, in effect, a surveillance society. Of recent concern in liberal democracies are companies based
in China (e.g., Tik Tok, owned by Beijing-based tech giant, ByteDance) that
possess vast troves of the personal data of US and other citizens of liberal
democracies. One concern is that China's authoritarian government can
readily gain access to this data, if it chooses to do so, and cannot be trusted
not to do so. Another concern is that China's authoritarian government

could direct Tik Tok to manipulate its algorithms to communicate content in accordance with China's political aims and perspectives (see Chapter 3). As a consequence, bans on TikTok for government use are appearing in western democracies, and the US state of Montana has now banned TikTok on personal devices.[19]

Accordingly, the establishment by governments of comprehensive, integrated biometric and nonbiometric databases of the personal information of citizens, and the utilization of these new technologies in the service of law enforcement under the banner of PP, has the potential to undermine individual autonomy and create a power imbalance between government and citizenry. It is not simply that a single individual, such as John Smith in our scenario, suffers an injustice or rights violation; rather there is widespread rights violations of the citizenry, and as a result a power imbalance is created between government and the citizenry.

The expanding use of biometric facial-recognition databases and other emerging technologies potentially used in law enforcement as part of PP must be clearly and demonstrably justified in terms of efficiency and effectiveness in the service of *specific* law enforcement purposes, rather than by general appeals to community security or safety. Moreover, it has to comply with moral principles constitutive of liberal democracy, such as the principle mentioned in Chapter 3, Section 3.2, that the state has no right to seek evidence of wrongdoing on the part of a particular citizen or to engage in selective monitoring of that citizen, if the actions of the citizen in question have not otherwise reasonably raised suspicion of unlawful behaviour, and if the citizen has not had a pattern of unlawful past behaviour that justify monitoring. Insofar as the use of these technologies and databases, including interlinked databases, can be justified for specific security (and safety) purposes and, therefore, privacy, autonomy, and other concerns mitigated, it is nevertheless imperative that their use be subject to accountability mechanisms to guard against misuse. Further, the citizenry should be well informed about these systems and should have consented to the use of them for the specific, justified purposes in question: they should be publicly debated, backed by legislation, and their operation subject to judicial review.

---

[19] B. Debusmann, "Montana Tik Tok Ban Is First Passed by Any US State," *BBC News*, 15 April 2023. https://www.bbc.com/news/world-us-canada-65281881. Accessed 29/10/2023.

4.2 Legal Adjudication

Another area in which ML techniques are being used is in the legal quagmire of divorce proceedings.[20] Although separation and divorce can be amicable, they can also be monumentally expensive because of legal fees. New ML software now attempts to predict the settlement outcome, based on a huge case history and relatively clear legal criteria for settlement. If the protagonists accept the prediction, a low-cost agreement can be achieved. Accordingly, the utilization of ML techniques in areas of the law involving a high volume of similar types of case and relatively clear-cut legal rules, such as divorce proceedings, may well be hugely beneficial.

As noted above, predicting future legal outcomes of cases based on past outcomes assumes, firstly, a large data set of past cases and, secondly, that new cases have similar features to past ones. Determinations of likelihood of success in divorce proceedings are based on outcomes of prior cases and weighting of criteria used in them. However, prior cases involve judicial errors (e.g., on the part of solicitors, barristers, and magistrates). Accordingly, these errors, especially if made on a systematic basis (e.g., based on prejudice), can now enter the predictive process. However, in doing so, predictions in current cases in which adjudications do not repeat past errors might have turned out to be false predictions and, thereby, mislead those who have acted upon these predictions. Moreover, the possibility of correcting these errors might well be lost if the prediction is simply accepted at face value and acted on without going through a thorough process in which all the various arguments, evidence, and so on are aired prior to a considered adjudication.

Moreover, complex, contested criminal cases are much less amenable to ML techniques than simple, high volume, legal adjudications, given the inherent particularity of many of these cases. Consider the legal adjudication in the case of the serial murderer and rapist, Robert Black. The case warrants detailed description since it is an example of the importance of particularity.[21]

In July 1990, a six-year-old girl, Mandy Wilson, was abducted as she walked in her street in Stow (a town in Scotland, close to the English border). By a stroke of immense good fortune, a neighbour saw her walk towards a

---

[20] John Zeleznikow, "Can Artificial Intelligence and On-line Dispute Resolution Enhance Efficiency and Effectiveness in Courts," *International Journal for Court Administration* 8, no. 2 (May 2017).

[21] This description of the case is taken from Miller and Gordon, *Investigative Ethics*, 127–32.

parked white van, and he could see her feet under the open passenger door beside those of a man. The girl's feet vanished, the van drove off; the witness took the registration number and immediately called the police. The witness was describing the event to the girl's father (a police officer) when the van reappeared and was immediately stopped by the police. The father found his daughter bound, gagged, and stuffed in a sleeping bag behind the driver's seat. She was terrified and had already been sexually assaulted. The driver of the van was Robert Black, a delivery driver who travelled throughout the UK. Black was arrested, charged, pleaded guilty, and was sentenced to life imprisonment.

Black now became the main suspect in other murders—the Maxwell, Hogg, and Harper murders—there being evident similarity between the cases in terms of MO and other factors. Black however declined to speak about the abductions and murders of the three girls. The investigators began a thorough and scrupulous examination of Black's movements and lifestyle between 1982 and 1990, where a key focus was on his job as a delivery driver. Using work records, including wage records and fuel receipts, they built up a picture of his movements and were able to place him in the vicinity of each abduction at the appropriate times, and at the locations of the bodies. The investigators also discovered an attempted abduction of a fifteen-year-old girl, which had failed because she had fought back and a friend came to her assistance. The witnesses' descriptions of the assailant were an exact match to Black.

There was no forensic evidence and no admission by Black. The case was built on the above-described circumstantial evidence, linking the various murders to one another and to Black. But in April 1992, the Crown Prosecution Service elected to prosecute. Importantly, for our purposes here, this was a unique case in UK criminal law. The defence argued there was no direct evidence to establish that Black had committed the offences and argued that each murder should be treated separately. But the court allowed the murders to be presented as a series and allowed evidence from the earlier case, relating to the abduction of Mandy Wilson and the attempted abduction of the fifteen-year-old. Black appeared at Newcastle upon Tyne Crown Court in April 1994, at which point the prosecution detailed the striking similarities between the murder cases, the attempted abduction, and the actual abduction of Mandy Wilson. The series of murders exhibited a common MO followed by Black, and Black was linked to each of by his movements.

Black was convicted of all the charges before the Court and sentenced to life imprisonment on each one. There was a minimum term of thirty-five years recommended for each of the murders.

In October 2011, Black was further convicted of the abduction and murder in 1981, in County Down, Northern Ireland, of the nine-year-old Jennifer Cardy. This offence also relied on painstaking investigation of Black's work records and included testimony from Detective Chief Superintendent (retired) Roger Orr, who was SIO in the Mandy Wilson abduction, and described Black's actions in that offence and in the Maxwell/Hogg/Harper murders. The prosecution alleged the evidence given by Mr. Orr amounted to "a signature for Robert Black and that the case of Jennifer also bears that signature." It is strongly believed that Black may be responsible for a further twelve abduction and murders of young girls, in the UK, France, and Holland. To date, he has declined to assist any police enquiry.

To summarize: Robert Black was a serial rapist and murderer but there was only circumstantial evidence in each of the murder cases. Accordingly, the prediction in each case considered on its own—including predictions using ML techniques—would have been not guilty. However, there was an evidential link between each of these cases: the MO of Black. The prosecution argued that there was a distinctive signature MO in each case and that this MO was used in the abduction case for which he was convicted, as well as the murder cases for which there was insufficient evidence without recourse to the signature MO. The point to be stressed here is not that there was a pattern (i.e., the MO), although the existence of a pattern was a necessary condition for the legal outcome. Nor is it that ML was not necessary to establish this pattern, although obviously ML was not required to discover a pattern in a handful of murder-rape cases. Rather, the point to be stressed is that a discretionary, and inherently particular, legal decision was made, a decision that allowed for the first time an evidential relationship between different cases to be useable in a single discrete case. In short, the Robert Black case was at the time unique, and unique in a manner that made it unable to be predicted based on adjudications in past cases of serial murder and rape. Accordingly, the legal adjudication in the Black case would have been immune to prediction based on ML techniques.

The more general point is that appropriate legal adjudications in complex cases may have an inherent particularity that renders them immune to prediction based on ML techniques. Therefore, there are evidently limitations to the utilization of ML techniques in legal adjudication, and attempts to exceed these limitations may well lead, not simply to error, but to injustice, whether in the form of punishing the innocent or failing to punish the guilty.

4.3 DNA Profiles and Databases

The collection of genomic and biometric data in population-wide or other bulk databases—notably, DNA and facial-image data, is rapidly expanding. Moreover, information obtained from DNA analysis can be integrated with facial-image data, but also with other forms of personal data, such as metadata, health information, and financial information to generate a detailed picture of the individual lives of citizenry. Further, techniques for extracting information from DNA and biometric are constantly evolving. These include ML and other AI techniques, and as such bring with them many of the ethical problems mentioned above, such as bias, although potentially on a much larger scale. However, they include other techniques, such as massively parallel sequencing (MPS) of DNA, which enable not simply the identification of an individual as an offender, or at least as present at a crime scene (as with traditional DNA techniques), but rather the extraction of detailed genetic information associated with a person's externally visible physical traits, ancestry, ethnicity, inherited diseases, and so on.[22] In this section, our focus is on DNA data,[23] and in the following section our focus is on facial-image data. In both cases, a central consideration is the moral significance of the data to be collected, stored, and used, and specifically the potential deprivations and harms that those that have the moral rights to control that data might suffer if they are not able adequately to secure it, including against state authorities engaged in morally illegitimate DNA collection on a mass scale. For instance, in Tibet—the Tibet Autonomous Region (TAR)—Chinese authorities are engaged in mass DNA collection, including the DNA of children and those not reasonably suspected of criminal conduct, in the service allegedly of public security.[24] Accordingly, it is not simply a matter of the cybersecurity issue of data security that arises once the DNA and facial-image data is collected,

---

[22] N. Scudder, D. McNevin, S. Kelty, S. Walsh, and J. Robertson, "Massively Parallel Sequencing and the Emergence of Forensic Genomics: Defining the Policy and Legal Issues for Law Enforcement." *Science and Justice* 58 (2019): 153–58.

[23] Seumas Miller, and Marcus Smith, "Quasi-Universal Forensic DNA Databases," *Criminal Justice Ethics* 41 (2022): 238–56.

[24] "China: New Evidence of Mass DNA Collection," *Human Rights Watch*, 5 September 2022. https://www.hrw.org/news/2022/09/05/china-new-evidence-mass-dna-collection-tibet.

stored, and used by, for instance, law enforcement. For there is the prior question as to whether or not law enforcement (or some other institution or organization) is morally entitled to collect, store, and use the data in the first place, irrespective of how secure the data might be in law enforcement databases once collected—and recent events suggest the level of data security is not in fact all that it ought to be.[25] This prior question is a cybersecurity issue, we suggest, insofar as cybertechnology is a means to collect, store, and/ or use this data. Moreover, it is a cybersecurity issue notwithstanding, firstly, that the cybertechnology in question might be used in combination with other technologies, such as genomic or biometric technologies, and secondly that law enforcement per se is a morally legitimate, indeed necessary, activity. In short, if cybertechnology is a means by which to carry out the morally unacceptable collection, storage, and use of DNA and facial-image data from those who have a moral right to control that data, then such collection, storage, and use is a cybersecurity issue; for it is the use of cybertechnology to violate the security rights and autonomy of those who are morally entitled to control that data.[26]

Population-wide or universal forensic DNA databases include the DNA profiles of all persons in a jurisdiction. To date, there are no such databases other than perhaps in the Tibetan Autonomous Region (see above) and the Xinjiang province in China. However, the technique described as forensic genealogy[27] has the potential to enable the creation of what Marcus Smith refers to as *quasi-universal forensic DNA databases*.[28] These databases already exist in several jurisdictions, including not only China but also the UK and the US.

[25] M. K. McGee, "DNA Test Firm: 2.1 Million Affected by Legacy Database Hack," Government Information Security. https://www.govinfosecurity.com/dna-test-firm-21-million-affected-by-leg acy-database-hack-a-18023#:~:text=An%20Ohio%2Dbased%20DNA%20testing,hacking%20incid ent%20detected%20in%20August; "Home Office Asked to Explain After 150,000 Arrest Records Wiped in Tech Blunder," *Times*, 21 January 2021. https://www.thetimes.co.uk/article/150-000-arrest-records-wiped-in-tech-blunder-krhlf302h?region=global.

[26] Moreover, even if the use of cybertechnology by law enforcement to collect, store and use DNA and facial image data turns out to be morally acceptable there is a question as to the constraints to be placed on law enforcement in this regard (i.e., the constraints to be placed on law enforcement's use of cybertechnology in respect of the security rights and autonomy of citizens). Again, this is a cybersecurity issue.

[27] Nathan Scudder, Dennis McNevin, Sally Kelty, Christine Funk, Simon Walsh, and James Robertson, "Policy and Regulatory Implications of the New Frontier of Forensic Genomics: Direct-to-Consumer Genetic Data and Genealogy Records," *Current Issues in Criminal Justice* 31, no. 2 (2019): 194.

[28] Marcus Smith, "Universal Forensic DNA Databases: Balancing the Costs and Benefits," *Alternative Law Journal* 43, no. 2 (July 2018).

National databases of the DNA profiles of convicted offenders and suspects have been created in many countries. For example, in the United States the National DNA Index System contains approximately thirteen million convicted offender profiles (from a total population of 328 million);[29] the United Kingdom's National DNA Database contains about 6 million offender profiles (from a total population of 66 million);[30] and China has established the world's largest DNA database, believed to include approximately 140 million profiles (from a total population of 1.4 billion).[31]

However, forensic genealogy utilizes DNA profiles to identify suspects who are not included in law enforcement's DNA databases. It does so by relying on the genetic relatedness of an offender—whose DNA is found at a crime scene—to a relative who provided his or her DNA to an organization for genetic health or ancestry testing. The best-known example of forensic genealogy is the 'Golden State Killer' case in California. The offender was identified because a distant relative had been genetically tested to determine their ancestry. The investigators found a partial match between the DNA of the killer derived from crime scenes and the DNA of the distant relative in the ancestry database. They then used standard investigative methods, firstly, to construct a list of suspects from the family tree (derived from birth registry records) and, secondly, to narrow the list of suspects down to the offender.

Using this technique of forensic genealogy and available law enforcement DNA databases together with access to health and ancestry databases, it is estimated by Smith that if five percent of the total population in a jurisdiction were included in a DNA database,[32] then if an offender leaves their DNA at a crime scene, the offender can be identified by their relatives, whose DNA is in an existing database, even if the offender's DNA is not. Accordingly, the need for universal DNA databases is obviated. It is necessary only to have quasi-universal forensic databases, given the availability of the technique of forensic genealogy and law enforcement access to health and ancestry databases.

---

[29] Federal Bureau of Investigation, *CODIS-NDIS Statistics* at http://www.fbi.gov/about-us/lab/biometric-analysis/codis/ndis-statistics, Accessed 2 February 2021.

[30] Marcus Smith, United Kingdom Home Office, *National DNA Database Statistics* at https://www.gov.uk/government/statistics/national-dna-database-statistics. Accessed 2 February 2021.

[31] Emile Dirks and James Leibold, "Genomic Surveillance: Inside China's DNA Dragnet," *ASPI Policy Brief* 34 (2020).

[32] Smith, "Universal Forensic DNA Databases."

4.3.1 DNA and Personal, Familial, and Human Identity

It was mentioned above that a central consideration in respect of data collection, storage and use is the moral significance of the data in question and, specifically, the potential deprivations and harms that those that have the moral rights to control that data might suffer. DNA and other genetic information are unlike, for instance, personal financial data (e.g., funds held in one's bank account) or personal communication information (e.g., whom one contacts), in that it is permanent, unalterable, and indeed in part constitutive of one's identity. That DNA and other genetic information are permanent and unalterable seems clear enough, but what of the claim that DNA is in part constitutive of identity?

At this point, there is a tendency to assimilate moral rights to genetic information to property rights, and specifically intellectual property rights. Indeed, genes and genetic data have in a number of instances been patented. Corporations now hold intellectual property rights, in the legal sense of that term, to some genes and genetic data. However, patents are supposed only to be issued to novel, nonobvious, and useful *inventions*. But, as Koepsell has pointed out, genes are naturally occurring and pre-exist human intention.[33] As Koepsell also points out, genes are, in this respect, akin to one's body or one's body parts (e.g., arms, legs, fingers and toes). Yet, arguably, human beings do not have *property* rights in the moral sense to their body, or more specifically to their brain—nor should anyone else. After all, slavery is agreed by all to be morally unacceptable. This is, of course, not to say that a person ought not to have control over their brain, body, body parts, or DNA. Clearly, the right to individual autonomy presupposes such control, or at least it does so unless some special case can be made that DNA is more akin to cars, houses, and other commodities than appears to be the case.

Accordingly, prima facie we should reject the claim that a person has property rights in the moral sense to his or her DNA, let alone that anyone else or any organization does. Therefore, we ought to reject the claim that intellectual property rights, in the legal sense, should be on offer (i.e., that human genes and genetic data should be allowed to be patented, or at least that the moral underpinning of any such legal right could be a prior moral right, as

---

[33] Koepsell, *Who Owns You*, Chapter 7.

in the case of legal property rights based on "mixing one's labour" with prior unowned material).[34]

Moreover, our DNA is central to our identity in ways in which, for example, our little finger is not. For DNA contains instructions that in large part make us who we are. As such, DNA is not simply a set of facts about us, such as the fact that we might have one finger that is longer than another one and shorter than a third. Rather, DNA is constitutive of us as human beings, and enables us to continue living and to generate our offspring. DNA consists of design instructions for the creation and functioning of a human being. So, the question that needs to be asked at this point is who possesses the moral rights to these constitutive elements of the identity of the morally valuable entity that is a human being.[35] Prima facie, the answer is the morally valuable entity itself—namely, the individual human being. However, there is a complication at this point.

On the one hand, there is individual human biological identity, the identity of a single human being, based on a DNA profile that is unique to that human being (or, in the case of twins, two human beings). Such a DNA profile enables each human being to be differentiated from all other human beings (other than one's identical twin if one has one). In short, there are individual human genomes. On the other hand, however, there is the biological identity of the human species—the identity of all human beings based on a DNA profile that is unique to human beings. Such a generic DNA profile enables human beings to be differentiated from members of other species. In short, there is the human genome.

So, there is the individual human genome and the human genome, and their respective DNA profiles overlap. Moreover, members of the same family have overlapping DNA profiles. What is the implication of this for the ascription of moral rights given that, as argued above, in the case of morally valuable beings, such as humans, they possess the moral rights to control their own DNA data? The following conception suggests itself.

In the case of DNA data constitutive of a single individual's biological and, therefore, personal identity, the individual in question possesses the moral right to control this data.[36] However, members of a family (or identical

---

[34] The philosopher, John Locke, famously argued for this moral basis for property rights in his *Second Treatise*, Chapter 5: Of Property (any edition).

[35] There are, of course, further questions regarding the nature and extent of these moral rights and the purposes for which they should be exercised (e.g., for the well-being of the individual human being). However, we cannot pursue these questions here, important though they are.

[36] We assume that individual biological identity is in part constitutive of personal identity.

twins) have a joint moral right to control overlapping DNA data (i.e., data concerning genetic features that members of the family in question share).

What of the human genomic data? Presumably, there is a joint moral right of all members of human race to control overlapping DNA data pertaining to the biological identity of each and every human being qua member of the human species.[37]

4.3.2 Autonomy/Privacy

Universal and quasi-universal forensic DNA databases have significant implications for individual autonomy, given the relationship between autonomy and personal identity discussed above (identity/autonomy) but also for individual privacy, given the relationship between autonomy and privacy (autonomy/privacy). In this section our concern is primarily with the use of DNA data by law enforcement for identification, and therefore with autonomy/privacy. Autonomy/privacy is understood primarily as informational privacy, and therefore as an aspect of individual autonomy: more specifically, the right to control one's personal information. The threat to autonomy/privacy from these databases is considerable since they can be used, or alternatively misused, by law enforcement for identification and investigation purposes. Moreover, as mentioned above, DNA and other genetic information is unlike, for instance, personal financial data, in that it is permanent, unalterable and indeed in part constitutive of one's personal identity. Thus, one can change one's bank account, phone number, driver's license, or tax file number, and so on, but one cannot change one's DNA. Therefore, DNA and other genetic information are reliable lifelong identifiers. This means they have greater utility for law enforcement than other forms of personal data. However, it also means there is much more at stake in terms of an individual's privacy and autonomy if this genomic data is acquired by criminals, or for that matter provided to law enforcement or other legitimate agencies (including private sector ones). That is, there are significant privacy/autonomy and related data security concerns, especially where large-scale DNA testing is proposed.[38]

---

[37] If this is correct, it has profound implications for genetic research and testing, and so-called human enhancement inter alia. However, we cannot pursue these difficult issues here.

[38] To try to ensure the data security of large DNA databases, one suggestion as a condition of collection might be that the personal data associated with each DNA sample be encrypted. The encryption could be done locally, say in a dedicated app on a person's phone, which uploads the encryption

Moreover, as stated above, the genome of a person is constitutive of that person's individual-specific (biological) identity.[39] Accordingly, the threshold for the infringement of an individual's right to control access to their genomic data is higher than for most other personal information. Further, universal, or quasi-universal DNA databases can be used in conjunction with other databases, including databases of other genomic data, biometric data (e.g., fingerprints and facial images), financial information, phone metadata, and so on. Again, while the integration of all these databases with DNA databases has greater utility for law enforcement, it also potentially undermines individual autonomy/privacy to an even greater extent than is the case if all these databases are not integrated with DNA databases.

As argued in Chapter 2, Section 2.1.1, privacy is a right that people have in relation to other persons and organizations with respect in part to the possession of information about themselves by other persons and organizations (e.g., DNA profiles and other genomic data stored in law enforcement, government, or commercial databases). DNA profiles can enable other information to be derived, such as their health status, paternity relationships, and so on—by other persons (e.g., by law enforcement analysis of biological material at a crime scene or at other sites that can link a person with that particular location).[40] Importantly, privacy rights are closely associated with the more fundamental moral value of autonomy. While privacy delimits an informational—and, for that matter, an observational—space (i.e., the private sphere), the right to autonomy consists of a right to decide what to think

---

key to an escrow service, independent of police. The lab technician collecting the data would receive only a code to write on the sample. If a match is found, the police would apply, with an appropriate warrant, to the escrow authority for the personal data. Once the investigation is complete, or, after some specified time interval, if no arrest is made, the data held by the escrow service would be destroyed.

[39] It might be thought, therefore, that the person owns one's genome. However, as already noted, the concept of ownership understood as private property is problematic when applied to one's own body, body parts, or other constitutive biological material, including one's DNA. One's relationship to one's body, for instance, seems importantly different to one's ownership of one's car, if for no reason other than that one's car is not literally constitutive of one's identity. Ownership of data about one's constitutive bodily material further complicates the matter. At any rate, we cannot address these complex philosophical issues here. Suffice it to say that ownership rights and privacy rights involve rights to control information (i.e., involve autonomy). Hence, we will assume that autonomy is certainly implicated whatever view one takes of ownership rights to biological material and data pertaining to such material—that is, we will assume that there are moral rights to control one's DNA data and this assumption is sufficient for our purposes here.

[40] There is a vast philosophical (and legal) literature on privacy, including in security contexts. Some useful influential works are: S. Warren and L. Brandeis, "The Right to Privacy," *Harvard Law Review* 4, no. 5 (1890): 193–220; Solove, *Understanding Privacy*; and Macnish, *Ethics of Surveillance*.

and do, and the right to control the private sphere. So, the right to privacy consists of the right to exclude organizations and other individuals (i.e., the right to autonomy) both from personal information, such as that included in DNA, and from monitoring of where they have been and whom they have been with. Hence it is a right to privacy/autonomy.

The right to privacy is not absolute. A person does not have a right not to be casually observed—as opposed to, for instance, followed around—in a public space, but arguably has a right for law enforcement agencies not to have access to their genomic data, even though this right can be overridden under certain circumstances—namely, if they have been convicted of a serious crime. Under such circumstances, their DNA profile will then be included in a forensic database. For instance, this right might be overridden if an individual is reasonably expected of being involved in a serious crime, and police have a warrant, approval from a judicial officer, legislative authority, and so on—and then only for the purpose of identifying persons who have committed a specific crime. If persons have committed a serious crime (e.g., murder) in the past, it would arguably (see below) be morally acceptable to utilize the retention of their genomic data—as it relates to identification, not health conditions—to include in a database and match against samples obtained from crime scenes. This is a specific and targeted measure to improve public safety, which even then can only be used in such a way that has been legislated for by a democratically accountable government. As discussed above, there already have been millions of individuals in countries, such as the United Kingdom and the United States, included in forensic DNA databases of this type since the early 2000s. However, this justification for retention of DNA profiles does not extend to innocent persons or even to suspects who are not subsequently convicted of crimes. Accordingly, it does not justify universal nor quasi-universal forensic DNA databases, insofar as the latter involves the accessing of the DNA of innocent persons (e.g., those who have submitted DNA to commercial ancestry and health testing services, and their relatives) who have not consented to the DNA profiles being accessed by law enforcement.

Privacy/autonomy is a moral right of an individual (as is identity/autonomy). However, the implications of an infringement of the privacy/autonomy and identity/autonomy rights of groups of people, and ultimately the citizens of an entire state or subgroup of a state must also be considered. Violations on a large scale can result in a power imbalance between the state and the citizenry, and thereby undermine liberal democracy itself. The universal collection of DNA from entire populations of millions or tens of millions of people in the Tibetan Autonomous Region and the Xinjiang Uyghur Autonomous Region, for identification and analysis by the Chinese government, are striking examples of large-scale violations of privacy and autonomy rights by an authoritarian state.

However, as will be discussed further, law enforcement being able to indirectly access the genomic data of the entire population in a liberal democracy, such as the United States or the United Kingdom, through familial searching or forensic genealogy, without any legal authority, would also be unacceptable. While quasi-universal forensic DNA databases may not be equivalent to the universal databases created in these regions of China, it is nonetheless unacceptable. For one thing, as we have just seen, it violates individual autonomy/privacy and identity/autonomy rights. For another thing, as the social credit system in China graphically illustrates, it threatens to generate an unacceptable power imbalance between the citizenry and the state, especially when the potential to integrate DNA databases with other databases is taken into account. A cornerstone of liberal democracy is that the citizens exercise control over the state rather than the reverse.

4.3.3 Right Not to Self-Incriminate

The privilege against self-incrimination entitles a person to refuse to answer any question, or produce any document, if the answer or the production would tend to incriminate that person—and is integrated into the legal systems of liberal democracies around the world, either in common law or statute, such as the Fifth Amendment in the US Constitution.[41] The rationale can be traced to the fact that the state has substantially more resources at its disposal in prosecuting crime than are available to those against whom that power is exercised, and there are potentially dire consequences at stake, such as imprisonment. A further consideration is respect for the dignity and privacy of individuals.[42]

---

[41] The privilege has been described in both the legal and philosophical literature. See e.g., E. Morgan, "The Privilege against Self-Incrimination," *Minnesota Law Review* 34, no. 1 (1949): 1–45; Robert Gerstein, "Privacy and Self-Incrimination," *Ethics* 80, no. 2 (1970): 87.

[42] See e.g., Andrew Ashworth, "Self-Incrimination in European Human Rights Law—A Pregnant Pragmatism?," *Cardozo Law Review* 30 (2008): 751; Mike Redmayne, "Rethinking the Privilege against Self Incrimination," *Oxford Journal of Legal Studies* 27, no. 2 (2007): 209–32; Ian Dennis, "Instrumental Protection, Human Right or Functional Necessity? Reassessing the Privilege against Self-Incrimination," *Cambridge Law Journal* 54, no. 2 (1995): 342–76.

It can be argued that legally requiring a person to provide DNA evidence that might inculpate themselves is a breach of the legal privilege not to self-incriminate, which is in turn based upon the moral right not to self-incriminate. Let us set aside the legal privilege and focus on the apparently underlying moral right not to self-incriminate.

The right not to incriminate oneself seems to be closely related to the right to self-defence. The notion is normally held to be that no matter how heinous the crime a person may have committed, they always retain the moral right to defend their life. So, a convicted murderer sentenced to death, or a terrorist sentenced to death, is morally entitled to try to prevent his or her executioner from performing the execution, even up to the last moment. Similarly, people always retain the right not to intentionally incriminate themselves, although of course they may choose to self-incriminate, just as they may choose not to defend themselves. On this view even people who have committed a heinous crime retain the right not to, in effect, speak against themselves or otherwise intentionally facilitate their own conviction. This view is consistent with the absence of any right not to incriminate oneself accidentally or inadvertently. It is also consistent with consenting, perhaps by way of implied consent, to self-incrimination in certain circumstances (e.g., in driving a car, drivers might be held to have consented to alcohol tests).

Notice that the moral right not to self-incriminate, as is the case with most, if not all moral rights, is not absolute, and so it can be overridden, at least in principle. Of course, things might be different with the legal right not to self-incriminate, although what the legal right ought to be is a matter to be settled at least in part on moral grounds. Moreover, the moral and legal right not to self-incriminate might not apply in certain circumstances, most obviously if the person in question has already been convicted of a crime and, consequently, the issue of incrimination for that crime, if not for other crimes, does not arise. Further, arguably, it is morally justifiable to collect the DNA profile (to the extent required for identification purposes) of a person convicted of at least some serious crimes (e.g., murder), and retain it after they have completed their sentence, on the grounds that they continue to pose such a serious risk that their right not to self-incriminate, if accused of a crime in the future, is overridden. Relatedly, it might be argued that it is morally justifiable to retain a DNA database of all those persons convicted of serious crimes on the grounds that their individual rights not to self-incriminate—now qualified because of their conviction—are outweighed by the contribution such a database makes to protecting the community at large. It is consistent with all this that a person, including one reasonably suspected of a serious crime, who voluntarily allows their DNA profile to be collected to enable their exculpation should have the right to have their DNA profile destroyed, perhaps after the investigation has been completed.

What of those who refuse to provide their DNA, and who have not been convicted of a serious crime? Some of these are straightforward cases of suspects who can refuse, based on the legal (morally based) right not to self-incriminate. Other cases are not so straightforward. Suppose there is a population or group, none of whose members is a uniquely identified suspect of the crime being investigated. Rather, there is merely the possibility that a member of the population or group is the offender, but it is not known which one. Now suppose that most of the members voluntarily provide DNA in order to assist the police and to remove any suspicion from themselves. Is this a violation of the right not to self-incriminate of the ones who refuse to provide their DNA? Of course, the exercise of a person's right to provide evidence to exculpate himself does not in such cases consist of a violation of another person's right not to self-incriminate herself. For the right of person A not to self-incriminate does not entail a duty on the part of another person B not to incriminate A. Moreover, if person A is to self-incriminate then A will have to perform an action—in this scenario, presumably the action of refusing to provide her DNA. Has the person who refuses to provide their DNA necessarily invoked their right not to self-incriminate? Not necessarily, given the other available moral justification for so refusing—namely, an innocent person's invocation of the exercise of their right to privacy/autonomy. However, the right to privacy/autonomy might be justifiably overridden in the circumstances (e.g., in the case of a serial killer). Nevertheless, could the person reasonably invoke their right not to self-incriminate? Presumably they could, at least on the view of this moral right outlined above. What of the matter of adverse inferences being drawn?

Firstly, the exercise of the right not to self-incriminate does not necessarily exclude the possibility of adverse inferences being drawn, so the protection it affords is incomplete. Secondly, if it is believed that the protection it affords should not be reduced in this way, then legislation could be enacted stating that failure to provide DNA in these circumstances cannot be taken to constitute reasonable suspicion in a formal sense (e.g., it cannot justify arrest, let alone constitute evidence at trial). If such legislation were enacted, then the person who refused to provide her DNA would not be incriminating herself, intentionally or otherwise. The right to silence operates in something like this manner (i.e., adverse inferences cannot be made, at least at trial). Either way, the moral right not to self-incriminate would include the right to refuse to provide one's DNA.

Note that whereas this right to refuse to provide one's DNA in these circumstances is an impediment to criminal investigations, its effect on an investigation is mitigated if the other members of the population or group voluntarily provide their DNA and, more generally, if innocent persons discharge what might be regarded as their collective moral responsibility to provide their DNA (albeit in the context of their DNA profiles being destroyed on completion of the investigation). We return to the issue of collective moral responsibility below. First, we must consider joint rights to genomic data.

4.3.4 Joint Moral Rights

The genome of a person is not only constitutive of that person's individual-specific (biological) identity, that same genome is *in part* constitutive of the individual-specific (biological) identity of the person's relatives—to a decreasing extent depending on the degree of relatedness (e.g., a sibling is more related than a second cousin). Evidently, therefore, genomic data involves joint rights—but what are joint rights?[43] Roughly speaking, two or more agents have a joint moral right to some good, including potentially some data or knowledge, if they each have an individual moral right to the good, if no one else has a moral right to it, and if the individual right of each is dependent on the individual rights of the others. Thus, the right of moral agent A to some good G (jointly held with moral agent B) brings with it an essential reference to the right of B to G (jointly held with A), and does so via the good, G. Moreover, being a joint right, neither A nor B can unilaterally waive it.

Joint rights need to be distinguished from universal individual rights. Take the right to life as an example of a universal individual right. Each human being has an individual right to life. However, since my possession of the right to life is wholly dependent on properties I possess as an individual, it is not the case that my possession of the right to life is dependent on your possession of that right. Notice that joint rights can be based at least in part on

---

[43] Miller, "Collective Rights"; Miller, *Social Action*, Chapter 7; Miller, "Joint Rights: Human Beings, Corporations and Animals."

properties individuals possess as individuals. The right to political participation is based in part on membership of a political community, and in part on possession of the property or right of autonomy.[44]

Joint rights can arise in a variety of ways. Joint rights can arise by way of promises. The owner of a house might confer joint ownership rights of the house on his two sons, for example. These joint rights might be joint moral rights and joint legal rights if the promise in question was legally binding. Another important moral basis for joint moral rights is joint action, specifically joint action that produces a good (i.e., a good to which there is a joint right). Consider, for instance, two business partners or the coauthors of a book. Again, these joint moral rights might also be joint legal rights, depending on the nature of the laws in the jurisdiction in question.

As stated above, the genome of a person is not only constitutive of that person's individual-specific (biological) identity, that same genome is *in part* constitutive of the individual-specific (biological) identity of the person's relatives. Accordingly, there is a species of joint right to control genomic data in play here, and not merely an exclusively individual right. The right to control one's genome data needs to be regarded, we suggest, as a (qualified) joint right (i.e., a right jointly held with the individual's relatives).[45] If these rights are, as we are suggesting, joint rights, then it follows that an individual may not have an exclusive individual right to provide his or her genomic data to consumer genetic testing providers or to law enforcement. Of course, when it comes to serious crimes, the consent of an individual to access his or her genome data is not necessarily required (e.g., if the individual is a past offender and hence his or her genomic data in the form of a DNA profile is held in a law enforcement database).[46] However, in cases where identifying the person who has committed a crime relies on the genomic data of relatives known to be innocent, and the relatives in question have a joint right to the data in

---

[44] Thus, the distinction between joint and collective rights or group rights is important. Miller, "Collective Rights and Minorities"; Peter Jones ed., *Group Rights* (Aldershot: Ashgate, 2009); Miller, *Social Action*, Chapter 7.

[45] It is a qualified joint right since the genomic data of any one of the persons is not identical to the genome data of the other persons (i.e., the sets of genomic data are overlapping). Moreover, there is a further question with respect to the degree of overlap that would underpin a joint right. Presumably, two persons (A and B), who are very distant relatives, and therefore have only have marginally overlapping genomic data (insofar as that data does not overlap with the genomic data of all humans) might not have a joint right to the data in question. The degree of overlap is very slight and their familial relationship too tenuous to underpin a *joint* right. Accordingly, the boundaries of joint rights are vague, and as a result fixing the limits of joint rights somewhat arbitrary.

[46] On the other hand, there is the potential collateral 'damage' (infringement of privacy/autonomy) to the relatives of criminals, given partially overlapping DNA profiles.

question, then it may be that *all* of these relatives need to have consented to the collection of the genomic data in question.[47] For in voluntarily providing one's DNA to law enforcement a person is, in effect, providing law enforcement with the partially overlapping DNA data of the person's relatives. But presumably a person does not have a moral right to decide to provide law enforcement with another person's DNA data. Accordingly, it seems that person A does not have a moral right to *unilaterally* provide law enforcement with his or her own data (i.e., A's DNA data, given that in doing so A is providing to law enforcement the partially overlapping DNA data of A's relatives, B, C, D, and so on). Rather, A, B, C, D, and so on have an (admittedly qualified) joint moral right to the DNA data in question, and therefore the right (being a joint right) has to be exercised jointly (i.e., perhaps all or most must agree). Naturally, as is the case with individual moral rights, joint moral rights can be overridden. For instance, A's individual right to know whether he is vulnerable to a hereditary disease might justify his providing his genomic data to health authorities and doing so without the consent of any of his relatives. In relation to our concerns here, the joint moral right of a group of persons to refuse to provide law enforcement with the DNA data in a murder investigation, for instance, may well be overridden by their collective moral responsibility (see next section) to assist the police. However, there is a residual question: what of the member of the group who committed the murder and who has, therefore, a moral right not to self-incriminate? We return to this important residual issue below. Let us first elaborate the notion of collective moral responsibility that is required.

4.3.5 Collective Moral Responsibility

As we have seen, the collection of and access to genomic information for law enforcement purposes has continued to expand over the past decade in both liberal democracies and authoritarian regimes. The public collection programs implemented in China can be contrasted with access to commercial databases in the United States. However, both result in what Smith has described as a quasi-universal forensic DNA database, enabling all citizens to potentially be identified in a criminal investigation, if necessary. This aspect

---

[47] This consent issue adds to other problems that exist with direct-to-consumer genetic testing, such as the accuracy of the tests and the fact that the results are not provided in a clinical setting by a healthcare professional.

of the discussion will examine whether there is a *collective moral responsibility* to investigate serious crime that overrides individual privacy and autonomy rights and makes these actions morally justified.[48] Here we are assuming that the DNA in question can be used only for identification purposes. We are not concerned with the far more morally problematic possibility of the provision of detailed genetic information associated with a person's externally visible physical traits, ancestry, ethnicity, inherited diseases, and so on.

Evidently, strategies for combating crime involve a complex set of often competing, and sometimes interconnected moral considerations (e.g., some privacy rights, such as control over personal data, are themselves aspects of autonomy). Hard choices must be made. However, the idea of a collective responsibility on the part of individuals to jointly suffer some costs (e.g., loss of privacy rights), in favour of a collective good (e.g., prosecuting serious crime) lies at the heart of all such effective strategies. Accordingly, we need an analysis of the appropriate notion of collective responsibility.

Central to collective responsibility is the responsibility arising from joint actions and joint omissions. We analyse this concept and defend this analysis in detail in Chapter 5, Section 5.3. Here we summarize the concept. A joint action can be understood as follows: two or more individuals perform a joint action if each of them intentionally performs his or her individual action, but does so with the (true) belief that in so doing each will do their part and they will jointly realize an end that each of them has and that each has interdependently with the others (i.e., a collective end).[49] On this view of collective responsibility as joint responsibility, collective responsibility is ascribed to individuals.[50] Moreover, if the joint action in question is morally significant (e.g., by virtue of the collective end being a collective good or a collective harm), then the individuals are collectively *morally* responsible for it. Each member of the group is individually responsible for his or her own contributory action, and (at least in the case of most small-scale joint action each is also individually (fully or partially) responsible for the aimed at outcome (i.e., the realized collective end) of the joint action. However, each is individually responsible for the realized collective end, *jointly with the others*; hence, the conception is relational in character. As already mentioned, if the

---

[48] Miller, "Collective Moral Responsibility"; Seumas Miller, "Collective Responsibility as Joint Responsibility" in *Routledge Handbook of Collective Responsibility*, ed. Saba Bazargan-Forward and Deborah Tollefsen (New York: Routledge, 2020), 38–50.
[49] Miller, "Collective Moral Responsibility."
[50] Ibid.

collective end of the joint action is a collective good or a collective harm, then these individual persons are collectively morally responsible for this good or harm.

Let us now apply this concept of collective moral responsibility to access to genomic information by law enforcement agencies to investigate and prosecute crime, and in particular to universal and quasi-universal DNA databases. Certainly, there is a collective good (see Chapter 5, Section 5.2) to which the use of this information will make a significant contribution—namely, the investigation and prosecution of serious crimes and the prevention of harm and preservation of the lives of those who may otherwise have been harmed if a serial killer or rapist is not brought to justice as swiftly as possible.[51] Naturally, those whose lives would not have otherwise been preserved receive a benefit—namely, their life—that those who would not have been impacted do not receive. Moreover, crime imposes economic and social costs for society that affect individuals more broadly than those who are directly victimized by crime.

Other things being equal, and if a universal or quasi-universal forensic DNA database operates effectively, there is a collective moral responsibility on the part of members of the state to submit their DNA. Of course, other things might not be equal. For instance, the data made available to authorities might be misused. Moreover, there are the moral rights to privacy/autonomy in play and, as we have seen, the moral right not to self-incriminate. As argued above, there is a collective moral responsibility of joint rights holders of DNA to provide this DNA to law enforcement, at least in the case of serious crimes. That is, their joint moral right is overridden by their collective moral responsibility. However, this collective moral responsibility applies in specific cases on a piecemeal basis. It is not a collective moral responsibility to provide their DNA data in a manner that contributes to a universal or quasi-universal DNA database. Moreover, it is not a collective moral responsibility to provide their DNA data on a permanent basis. Rather, they have a joint moral right that the data be destroyed upon the conclusion of the specific criminal investigation and associated trial.

What of the moral right not to self-incriminate? Arguably, the right not to self-incriminate overrides the individual responsibility of an offender or suspect to provide her DNA data to law enforcement. Note that this individual moral responsibility (overridden by the right not to self-incriminate)

---

[51] Ibid.

is the offender's or suspect's responsibility to contribute her DNA data to assist law enforcement, and as such is the offender's component responsibility (so to speak) of the group's collective (i.e., joint) moral responsibility to provide their DNA data to law enforcement. Accordingly, whereas most of the members of the group are, all things considered, morally required to provide their DNA data, the offender or suspect is not, all things considered, morally required to do so. Her right not to self-incriminate, should she choose to exercise it, affords her protection at this point. However, the protection is limited insofar as law enforcement will, nevertheless, have the benefit of the DNA data of the other members of the group, and that DNA data may overlap with the offender's if she is a relative and, if not, an adverse inference might still be able to be made with respect to the offender or suspect.

Notice that, this conception of collective responsibility as joint responsibility implies that each relevant person has an individual moral responsibility to provide a sample of their DNA, assuming the others do. So, it is not simply a matter of whether each wants to do so; rather, each has a moral obligation to comply, given the others, or most of the others, comply. However, it does not follow from this that each should be compelled to comply. It does not follow that compliance should be a matter of enforceable law. On the other hand, if the numbers who choose to comply under circumstances in which compliance is voluntary is not sufficient to meaningfully assist the criminal investigation in question, then it may well be that compliance ought to be enforced (i.e., the magnitude of the evil to be avoided outweighs any given individual's privacy/autonomy right, and indeed the aggregate privacy/autonomy rights or joint moral right).

And there is this further point. Given the increasing amount of data available to public and private sector agencies, such as smartphone metadata and location history, it is important that the use of this data is limited to what is under warrant for the investigation of serious crimes. It is important that this does not lead to normalization or more widespread use of sensitive data in cases where it is not appropriate.

4.3.6   DNA Databases: Ethical Guidelines

Let us now summarize the legal implications of our position. We do so in the light of the following general points. Universal DNA databases and, as we have seen, *existing* quasi-universal DNA databases, compromise individual rights—notably, privacy/autonomy rights, and identity/autonomy rights, and weaken the protection afforded by the right not to self-incriminate. In relation to DNA data, the privacy/autonomy rights in question turn out to be, or to be based on,[52] qualified joint moral rights. That said, citizens have a collective moral responsibility to assist law enforcement in relation to serious crime and in particular to provide their DNA data on a case-by-case basis, if required. Indeed, this moral responsibility may need to be enforced, since in relation to serious crimes it evidently overrides privacy/autonomy rights. However, arguably, the right not to self-incriminate overrides the individual moral responsibility—including when it occurs as a component of a wider collective moral responsibility—to assist law enforcement.

In summation: First, universal databases should not be permitted under law if they require compelling everyone to provide DNA.[53] This is not morally justified. Rather only the DNA of those convicted of serious crimes should be collected and retained permanently. The DNA of those arrested and charged with crimes, but not convicted, may be collected and retained for a reasonable period of time (e.g., three to five years).

Second, a person reasonably suspected by law enforcement of committing a serious crime, or who is among a group of familial relatives one or more of whom is suspected of committing a serious crime, has respectively an individual or joint (i.e., collective) moral responsibility, and ought to have a legal responsibility, to provide their DNA to law enforcement for exculpatory or inculpatory purposes. Those who voluntarily provide their DNA to assist law enforcement under these circumstances and are exculpated have a moral right, and ought to have a legal right, to have their DNA destroyed within a reasonable time (e.g., normally at the conclusion of the investigation).

Third, the individual moral and legal responsibility to provide one's DNA to law enforcement under the circumstances described above— whether it is jointly held with others—is overridden by the moral right not to self-incriminate, but not by any alleged duty not to incriminate others, in instances where it is likely to incriminate. The moral right not to self-incriminate should also be a legal right. Whether or not an adverse inference should be able to be drawn at trial, and the weight to be given to such an

---

[52] There might be a joint moral right to a DNA profile entailing joint rights to determine access but the latter might, nevertheless, be appropriately regarded as based on non-joint individual rights to privacy/autonomy rather than themselves joint individual rights to privacy/autonomy.

[53] This would be even more morally problematic if the DNA in question would not only enable identification but also the determination of individual characteristics.

inference, from a refusal on the part of someone charged with a serious offence to provide DNA on grounds of self-incrimination ultimately depends
on the overall security threat posed by the type of crime in question.

Fourth, law enforcement should not have the legal right to access DNA
databases collected for other purposes, except in two sorts of case. In the first
kind of case there is a particular already uniquely identified person, who is
reasonably suspected of having committed a serious crime, and access to
their DNA data is granted under warrant. In the second kind of case, there
is a particular already uniquely identified person, who is *not* suspected of
having committed a serious crime but who is a member of a group of familial
relatives one or more of whom are reasonably suspected by law enforcement
of having committed a serious crime, and access to the nonsuspect's DNA
data is granted under warrant.

Moreover, in this second kind of case the nonsuspect has an individual
moral responsibility, held jointly with their relatives (i.e., there is a collective moral responsibility), and ought to have a legal responsibility to provide their DNA to law enforcement to assist their investigations[54]. Those
who voluntarily provide their DNA to assist law enforcement under these
circumstances have a moral right—and a moral duty to their relatives—and
ought to have a legal right and duty, to have their DNA held in law enforcement databases destroyed within a reasonable period (e.g., normally at the
conclusion of the investigation).

Fifth, persons intending to provide their DNA for another purpose (e.g.,
to a health provider or to a commercial provider to determine their ancestry)
have a moral right, and should have a legal right, to be informed that their
DNA data might be accessed by law enforcement in the above-described
circumstances. Moreover, in providing their DNA for another purpose, they
have a moral right, that ought to be a legal right, to limit in advance the period during which their DNA is stored. Further, depending on the type of
database in question, it may be that they do not have the individual right to
provide their DNA to the provider in question. It is a joint right that perhaps
ought to be a legal joint right, and therefore the consent of family members
may be morally, and perhaps ought to be legally, required (e.g., in the case of
commercial databases established for the purpose of tracing ancestry).

---

[54] The DNA data with its identity label could be held in some sort of escrow. Police can trawl the
DNA records without access to the identity data. If they find a match, given appropriate warrants,
then they can apply to get the identity of the DNA. On this view, supposing this process is practicable,
an individual might not have an obligation to provide DNA unless this level of security were in place.

Finally, we note that if these conditions are met then the effect might be to eliminate quasi-universal databases, since people who have not been convicted of a serious crime might be reluctant to consent to have their DNA data stored in commercial, health, and other databases—or, at least, they might require it to be destroyed after relatively short periods. On the other hand, if these conditions are met then the effect might not be to eliminate quasi-universal databases. However, if the latter eventuality obtains then these quasi-universal databases would, according to this hypothesis, comply with the relevant moral constraints, and as such, they ought to be legally permissible.

4.4 Facial-Recognition Technology

As we saw in the case of DNA, one's face, as opposed to one's facial expression at a specific time (e.g., whether one is smiling or not right now), is an unalterable feature of a human being and constitutive in part of their personal identity. However, unlike DNA, a human being's face is expressive of their inner mental self (e.g., their emotions), and somewhat under their control.[55] As such, a human being's face is, constitutive in part of their personal identity in a different, and in some respects more profound sense than is their DNA. Moreover, a person's face, while it is more or less unalterable, does undergo gradual change over time—notably, as part of the ageing process. In this respect, it mirrors one's changing personal identity (e.g., the adult is the same person as the child he or she was but also importantly different). Given one's facial image is an image of a constitutive feature of one's personal identity (i.e., one's face), and given the tight connection between identity and autonomy, evidently control of one's facial image is importantly connected to individual autonomy. Specifically, a person has a moral right to control images of his or her face (e.g., digital photos).[56] This includes accurate facial images of persons embedded in larger false visual representations of them (e.g., a photo of a crime scene manipulated to include a photo of a suspect). It also includes fake facial images that are not detectable as fake by anyone

---

[55] We say *more or less* unalterable given the complicating factor of major cosmetic surgery.

[56] Moreover, there is a technique known as perceptual hashing, used by Apple in its contentious software proposal to scan for child porno on iPhones, which could be used by anybody to request that all images of their face be expunged. T. Holwerda, "Researchers Produce Collisions in Apple's Child Abuse Hashing System" *CSNews*, 18 August 2021. https://www.osnews.com/story/133835/rese archers-produce-collision-in-apples-child-abuse-hashing-system/ Accessed 5/11/2023.

without special equipment or training. However, unlike DNA profiles, facial images are easily and surreptitiously obtainable, and ubiquitous because of the widespread use of social media. Accordingly, control of one's facial image is increasingly difficult, and therefore, given the possibility of using facial-recognition technology to identify and track a person, the threat to individual privacy and autonomy posed by facial-recognition technology is considerable.

Biometric facial-recognition technology involves: (1) the automated extraction of facial images from passport photos, drivers' licences, social media sites (e.g., Facebook), and elsewhere; (2) the digitization of these facial images; (3) the conversion of these facial images into a contour map of the spatial and geometric distribution of the facial features on these images; (4) the storage of these facial features (thus extracted and converted); (5) the comparison, using algorithms of newly acquired facial images (thus extracted and converted), with those already stored in databases to identify individuals. Biometric facial-recognition systems can be integrated with the closed-circuit television systems that already exist in public spaces to search for, identify, and thereby track, people in real time.[57]

The expanding use of this technology in law enforcement, border protection, and national security contexts raises a number of pressing ethical concerns for liberal democracies in relation to individual privacy and autonomy, and democratic accountability.[58] Moreover, the fact that this technology is already in widespread use in authoritarian states, such as in China's social credit system, raises the spectre of its widespread use in liberal democracies like the UK. Thus, British police have used it at football matches.[59]

LONDON (AP)—When British police used facial recognition cameras to monitor crowds arriving for a soccer match in Wales, some fans protested by covering their faces. In a sign of the technology's divisiveness, even the head of a neighbouring police force said he opposed it. The South Wales police deployed vans equipped with the technology outside Cardiff stadium this week as part of a long-running trial in which officers scanned

[57] Marcus Smith, Monique Mann, and Gregor Urbas, *Biometrics, Crime and Security* (Abingdon: Routledge, 2018).

[58] Kleinig et al., *Security and Privacy*. See also Miller and Smith, "The Ethical Application of Biometric Facial Recognition Technology."

[59] https://apnews.com/article/7266356b2c244e3970afeabeaeb48e49. Accessed 30/10/2023.

people in real time and detained anyone blacklisted from attending for past misbehaviour. Rights activists and team supporters staged a protest before the game between Cardiff City and Swansea City, wearing masks, balaclavas, or scarves around their faces.

We also note that it is not simply a matter of the use of this technology by law enforcement agencies. Criminal organizations, firms, and private individuals now have access to this technology, as well as to a very large database of facial images and associated publicly available information. The latter database has been created by the private firm, Clearview AI. Clearview AI created a database of literally billions of facial images by scraping them off social media sites and related sources. Moreover, these facial images have links to websites from which the facial images were scraped, thus providing information about the identity of the persons whose facial images are in the database. Further, Clearview AI's software enables its buyer to compare a facial image the buyer might possess (e.g., as a result of taking a digital image of a stranger in a public space), with images in the Clearview AI's huge database and, thereby (via the publicly available information associated with these facial images) identify the stranger in question (e.g., who they are, what they do, and so on). Accordingly, criminal organizations could use facial-recognition technology (and Clearview AI's database and app, in particular) to thwart law enforcement (e.g., to identify an undercover operative by taking his photo and comparing it with his facial image that is on a police college graduation photo),[60] and to engage in crimes, such as stalking (e.g., by using facial image to gain more information about a person being stalked) and identity theft (e.g., using a facial image in conjunction with other personal data with a view to defrauding potential victims). There is also the danger of hacking into Clearview AI's databases, and thereby not only accessing the facial images and associated publicly available information, but also information about the users of the Clearview AI product. Photographic identification, such as a driver's licence, can be altered by criminals to create a new identity or assume an existing one. This may then be used to open new accounts and build up debt.

Moreover, criminals can apply ML techniques to existing facial images to create new images, and thereby thwart law enforcement. Already hackers

---

[60] Kelly W. Sundberg and Christina M. Witt, "Undercover Operations: Evolution and Modern Challenges," *Journal of the AIPIO* 27, no. 3 (2019): 3–17.

have attacked security systems to make them think a person is someone else, thus circumventing no-fly lists.

> By using machine learning, they created an image that looked like one person to the human eye, but was identified as somebody else by the face recognition algorithm—the equivalent of tricking the machine into allowing someone to board a flight despite being on a no-fly list.[61]

Facial-recognition technology is increasingly being used in the private sector for authentication purposes. For instance, Amazon is pushing ahead with facial recognition to improve efficiency, enhance customer convenience, and reduce costs in its stores. In stores so equipped, a customer walks in, grabs whatever, and walks out. Facial recognition is then used to debit their registered cards. Such contactless processes have been given a boost by the COVID-19 pandemic.

> Earlier this year, Amazon said it would start licensing its cashier-less shopping software to other stores as retailers seek to limit face-to-face contact and cut wait times in line. The "Just Walk Out" system uses computer vision to track customers throughout stores. When they're done shopping, customers walk out of the store without ever scanning and paying for items. That technology has sparked more interest from retailers as grocery and other store workers face danger and exhaustion during the pandemic.[62]

Facial recognition is being used within some schools in the US but is being challenged.

> The New York legislature today passed a moratorium on the use of facial recognition and other forms of biometric identification in schools until 2022. The bill, which has yet to be signed by Governor Andrew Cuomo, comes in response to the launch of facial recognition by the Lockport City School District and appears to be the first in the nation to explicitly regulate or ban use of the technology in schools.[63]

---

[61] https://www.technologyreview.com/2020/08/05/1006008/ai-face-recognition-hack-misidentif ies-person/.

[62] https://www.washingtonpost.com/technology/2020/09/08/robot-cleaners-surge-pandemic/. Accessed 30/10/2023.

[63] https://venturebeat.com/2020/07/22/new-york-bans-use-of-facial-recognition-in-schools-statewide/. Accessed 30/10/2023.

Portland, Oregon, has gone further:

> Portland, Ore., City Council on Wednesday unanimously adopted two
> landmark ordinances banning city and private use of facial recognition
> technology. . . . The first bars all city bureaus from acquiring or using the
> controversial technology with minimal exceptions for personal verifica-
> tion. . . . The second blocks private entities from using the software that
> scans faces to identify them in all public accommodations. . . . That second
> ordinance goes beyond the steps other cities, like Boston, San Francisco
> and Oakland, Calif., have taken to limit government applications of facial
> recognition. . . . "What makes Portland's legislation stand out from other
> cities is that we're prohibiting facial recognition technology use by private
> entities in public accommodations," Mayor Ted Wheeler (D) said during
> Wednesday's deliberations. "This is the first of its kind of legislation in the
> nation," he added.[64]

There is legislation proposed in 2021 in the EU which may constrain facial
recognition.

> Facial recognition and other high-risk artificial intelligence applications
> will face strict constraints under new rules unveiled by the European Union
> that threaten hefty fines for companies that don't comply. The European
> Commission, the bloc's executive body, proposed measures on Wednesday
> that would ban certain AI applications in the EU, including those that ex-
> ploit vulnerable groups, deploy subliminal techniques or score people's
> social behaviour. The use of facial recognition and other real-time remote
> biometric identification systems by law enforcement would also be prohib-
> ited, unless used to prevent a terror attack, find missing children or tackle
> other public security emergencies.[65]

Notwithstanding new legislation in the US, EU, and elsewhere, regulation
of facial-recognition technology is playing catch-up. For facial recognition
is entering a new and potentially more morally problematic phase. It took a
long time to get facial recognition reasonably accurate and sufficiently robust,

---

[64] https://thehill.com/policy/technology/515772-portland-adopts-landmark-facial-recognition-
ordinance/. Accessed 30/10/2023.
[65] https://www.bloomberg.com/news/articles/2021-04-21/facial-recognition-other-risky-ai-set-
for-constraints-in-eu. Accessed 30/10/2023.

unaffected by beards, spectacles, scars, or other appearance variables. At the time of writing, there are still errors made because of decoration, like brightly coloured dots or patches. The challenge now is getting recognition of emotion from faces, which will add further profiling information for surveillance and concomitant advertising. Prominent in this regard is a company, Affectiva. Rosalind Picard in the MIT Media lab published a study in 1997 describing how there are emotions we recognize in ourselves, such as fear, and others which are below conscious awareness, yet manifest themselves as micro-, transient, facial expressions. Her view then was that use of such data should be carefully monitored and restricted. Yet two decades later a company, Affectiva, was spun-off to do exactly this. Thus, facial-recognition technology, during an internet video conference, could be monitoring the emotions of participants and gathering information of which even they themselves are unaware.[66]

4.4.1 Personal Identity, Autonomy, and Privacy

We have invoked a distinction between individual autonomy and personal identity.[67] In respect of personal identity, we need to distinguish between numerical identity and qualitative identity. Two peas in a pod might be qualitatively identical; they share all their properties in common and are, therefore, indistinguishable from one another. Nevertheless, they are numerically distinct; there are, after all, *two* peas. Moreover, with respect to the qualitative identity of human beings, we need to distinguish between what we might refer to as generic identity and personal identity. The qualitative identity of human being consists, let us assume, in possession of whatever properties are definitive of human beings (e.g., rational animals). Let us refer to this as generic human identity. The qualitative identity of a single human person consists in part in the properties that are definitive of human beings.

---

[66] Recent EU laws on AI will likely add restrictions on emotion recognition systems in law enforcement, border management, the workplace, and educational institutions. https://www.europarl.eur opa.eu/news/en/press-room/20230609IPR96212/meps- ready- to- negotiate- first- ever- rules-for-safe-and-transparent-ai. Accessed: 30/10/2023.

[67] What we are referring to as personal identity might be referred to as individual identity for persons. Moreover, persons are sometimes distinguished from human beings (e.g., perhaps there are rational agents who are persons but not humans). Moreover, problems of personal identity include determining what counts as the constitutive features of a person such that they are no longer a person if they lose those features, or they are no longer the same person if they lose those features. We do not need to pursue these philosophical issues here but see the entry on personal identity in the *Stanford Encyclopedia of Philosophy* https://plato.stanford.edu for an overview.

However, it also includes whatever additional properties, or combination, extent, and degree of those properties that are constitutive of generic human identity, which are definitive of that person. Let us refer to this as personal identity. Note that some properties, or combination thereof, taken to be definitive of a person's personal identity *by that person* might not, in fact, be definitive (e.g., a person might falsely believe himself to be Napoleon). Note also that properties that might enable a person to be differentiated from another person, such as for instance their history, might be definitive of that person's personal identity, whereas other properties that enable the person to be differentiated from other persons, such as their unique birthmark, might not necessarily be definitive of that person's personal identity.

Individual autonomy (as opposed to, for instance, collective autonomy) is a definitive property of fully functional, adult human beings and (in different degrees) of the personal identity of any one of these. Moreover, in acting autonomously a human person acts to a greater or lesser extent in accordance with other features of their personal identity. In acting autonomously, a person cannot simply ignore their human and personal identity. Further, at least for social beings such as humans, personal identity consists in part in relationships with other human beings (e.g., relationships with family, friends, enemies and/or members of one's community). We note that one's face and its expressions is a fundamental communicative and expressive element in these relationships, and indeed an aspect of one's personal identity, at least for most people.

We need to distinguish between moral rights to things that are constitutive of personal identity and moral rights to things that are not so constitutive (e.g., one's car or other private property). That said, the boundaries of the distinction are far from clear cut.

What are some of the constitutive features of the personal identity of a fully functional, adult human being, above and beyond individual autonomy? Such features include the following ones: an individual body, including a unique face; a highly specific and integrated set of mental, social, and physical skills; a highly specific set of relationships with other individuals, including family members, friends, and peers; and an individual history, including memories thereof. Doubtless, an individual person would not necessarily lose their identity if they lost an arm or a leg, if they gained a new set of friends, if they switched professions, if they immigrated to a new country, if they learnt a new language, if they converted to a new religion, or if they suffered a loss of hearing or a minor loss of memory.

We cannot, and need not, precisely delineate the defining features of the personal identity of human persons.[68] Rather, we need merely to acknowledge its existence, gesture at some of its elements, and assert that at least some elements of it have an intrinsic moral importance worthy of protection by rights. Here we offer again the example of one's face. It would surely be an egregious rights violation to alter another person's face (e.g., to deface it with acid, even aside from the infliction of suffering involved). Naturally, we should not confuse one's face with an image of one's face, nor should we confuse accessing or circulating one's facial image with defacing that image. Nevertheless, the centrality of one's face to one's identity implies a right to control one's facial images, which in turn implies a privacy right, due to the close relationship between individual autonomy and privacy (as argued in Chapter 2, Section 2.1.1). On the other hand, given the necessarily public nature of one's face and its expressions, the right to control one's facial images is by no means absolute—as, for instance, one's right not to express one's thought might be. Indeed, one has a duty, for instance, to show one's face for purposes of identification, and therefore a derived duty to allow access to one's facial image for like purposes (e.g., to present one's passport with a photo of oneself at border posts).

We saw in Chapter 2, Section 2.1.1, firstly, that privacy is a right that people have in relation to other persons, the state, and organizations with respect to: (a) the possession of information (including facial images) about themselves by other persons and by organizations (e.g., personal information and images stored in biometric databases); or (b) the observation/perceiving of themselves—including of their movements, relationships and so on—by other persons (e.g., via surveillance systems, including tracking systems, that rely on biometric facial images).[69] Biometric facial recognition is obviously implicated in both informational and observational concerns.

Secondly, we saw that the right to privacy is closely related to the more fundamental moral value of autonomy. Roughly speaking, the notion of privacy delimits an informational and observational space (i.e., the private sphere). However, the right to autonomy consists of a right to decide what to think and do and the right to control the private sphere and, therefore, to decide *whom to exclude and whom not to exclude* from it.[70] So the right to

---

[68] Moreover, they may be a need to distinguish between stronger and weaker senses of personal identity such that, for instance, one ceases to be the same person in a weak sense if one loses most of one's memories.

[69] Kleinig et al., *Security and Privacy*.

[70] Ibid.

privacy consists of the right to exclude organizations and other individuals (i.e., the right to autonomy) both from personal information and facial images, and from observation and monitoring (i.e., the private sphere). Thus, a degree of privacy is necessary for people to pursue their personal projects, whatever those projects might be. For one thing, reflection is necessary for planning, and reflection requires a degree of freedom from the distracting intrusions, including intrusive surveillance, of others.[71] For another, knowledge of someone else's plans can lead to those plans being thwarted (e.g., if one's political rivals can track one's movements and interactions then they can come to know one's plans in advance of their implementation), or otherwise compromised, (e.g., if whom citizens vote for is not protected by a secret ballot, including a prohibition on cameras in private voting booths, then democracy can be compromised).

Naturally, as we also argued in Chapter 2, Section 2.1.1, the right to privacy, including one's control of facial images, is not absolute; it can be overridden. Moreover, its precise boundaries are unclear. A person does not have a right not to be observed in a public space, but arguably does have a right not to be photographed in a public space, let alone have an image of their face widely circulated on the internet, even though this right can be overridden under certain circumstances. For instance, this right might be overridden if the public space in question is under surveillance by CCTV to detect and deter crime, and if the resulting images are only made available to police—and then only for the purpose of identifying persons who have committed a crime in that area. What of persons who are present in the public space in question and recorded on CCTV, but who have committed a serious crime, such as terrorism, elsewhere—or at least are suspected of having committed a serious crime elsewhere and are, therefore, on a watch-list?[72] Presumably, it is morally acceptable to utilize CCTV footage to identify these persons as well. If so, then it seems morally acceptable to utilize biometric facial-recognition technology to match images of persons recorded on CCTV with those of persons on a watchlist of those who have committed, for instance, terrorist actions, or are suspected of having done so, as the SWP were arguably seeking to do in the *Bridges* case.

---

[71] Ibid.

[72] Drawing on the legislative example (s 5(1) *Crime Commission Act 2012* [NSW]) stated above, we will define a serious crime as an offence punishable by imprisonment for life, or for a term of three or more years.

In addition to the rights of a *single* individual, it is important to consider the implications of the infringement, indeed violation, of the privacy and autonomy rights of the whole citizenry by the state and/or by other powerful institutional actors, such as corporations. Such violations on a large scale can lead to a power imbalance between the state and the citizenry, and thereby undermine liberal democracy itself.[73] The surveillance system imposed on the Uighurs in China, incorporating biometric facial-recognition technology, graphically illustrates the risks attached to large scale violations of privacy and related autonomy rights.

Accordingly, while it is morally acceptable to collect biometric facial images for necessarily circumscribed purposes, such as passports for border control purposes and drivers' licences for safety purposes, it is not acceptable to collect them to establish vast surveillance states, as China has done, and to exploit them to discriminate based on ethnicity. However, images in passports and driving licences are, and arguably ought to be, available for *wider* law enforcement purposes (e.g., to assist in tracking the movements of persons suspected of serious crimes that might be unrelated to border control or safety on the roads). The issue that now arises is the determination of the point on the spectrum at which privacy and security considerations are appropriately balanced.

Privacy can reasonably be overridden by security considerations under some circumstances, such as when lives are at risk. After all, the right to life is, in general, a weightier moral right than the right to privacy.[74] Thus, utilizing facial-recognition technology to investigate a serious crime such as a murder, or to track down a suspected terrorist, if conducted under warrant, is surely ethically justified. On the other hand, intrusive surveillance of a suspected petty thief might not be justified. Moreover, given the importance of, so to speak, the aggregate privacy/autonomy of the citizenry, threats to life on a small scale might not be of sufficiently weighty to justify substantial infringements of privacy/autonomy (e.g., a low-level terrorist threat might not justify a citizen-wide biometric facial-recognition database). Further, regulation and associated accountability mechanisms need to be in place to ensure that, for instance, a database of biometric facial images created for a legitimate purpose (e.g., a repository of passport photos) can be accessed by

---

[73] Walsh and Miller, "Rethinking 'Five Eyes' Security Intelligence Collection Policies and Practice Post Snowden."

[74] Ibid.

border security and law enforcement officers to enable them to prevent and detect serious crimes, such as murder, but not used to identify protesters at a political rally.

We have argued that privacy rights, including in respect of biometric facial images, are important in part because of their close relation to autonomy, and although they can be overridden under some circumstances—notably, by law enforcement investigations of serious crimes—there is obviously a point where infringements of privacy rights is excessive and unwarranted. Access on a case-by-case basis under judicial warrant to a national biometric facial recognition database (e.g., of passport holders), for use in relation to serious crimes and subject to appropriate accountability mechanisms, may well be acceptable, but general law enforcement access to billions of facial images scraped by Clearview AI from social media accounts without the consent of those to whom they belong in order to detect and deter minor offences is unacceptable. And, of course, establishing a surveillance state (e.g., to the extent that has been achieved in China) would be beyond the pale. Clearly the devil is in the detail. Let us now turn directly to security.

4.4.2 Security and Public Safety

Security can refer to national security (e.g., harm to the public from a terrorist attack), community security (e.g., disruptions to law and order), and organizational security (e.g., breaches of confidentiality and other forms of misconduct and criminality). At other times it is used to refer to personal security, including personal physical security. Physical security, in this sense, is security in the face of threats to one's life, freedom of movement, or personal property—the latter being goods to which one has a human right. Violations or breaches of physical security obviously include murder, rape, assault, and torture.[75] Biometric facial recognition systems could assist in multiple ways to enhance security in each of these senses. Thus, a biometric facial recognition system could help to prevent fraud by better establishing identity (e.g., falsified driver's licences) and facial recognition data would be likely to help to investigate serious crimes against persons, such as murder and assault (e.g., via CCTV footage).

---

[75] Kleinig et al., *Security and Privacy*; Miller, *Dual Use Science and Technology, Ethics and Weapons of Mass Destruction*.

As stated in Chapter 1, Section 1.1.2, security should be distinguished from safety, and therefore cybersecurity from cybersafety, although the concepts of security and safety are related and the distinction somewhat blurred. We tend to speak of safety in the context of wildfires, floods, pandemics, and the like, in which the harm to be avoided is not intended harm. By contrast, the term *security* typically implies that the threatened harm is intended. At any rate, it is useful to at least maintain a distinction between intended and unintended harms and, in relation to unintended harms, between foreseen, unforeseen, and unforeseeable harms. For instance, someone who is carrying the COVID-19 virus unknowingly because they are asymptomatic, is a danger to others, but nevertheless might not be culpable if, for instance, they had taken reasonable measures to avoid being infected, had an intention to test for infection if symptoms were to arise, and if when aware of being infected would take all possible measures not to infect others. While biometric facial-recognition systems can make an important contribution to security, their utility in relation to safety is less obvious, even though they could assist in relation to finding missing persons or ensuring unauthorized persons do not unintentionally access dangerous sites.[76]

A number of potential ethical problems arise from the expanding use of biometric facial recognition for security purposes, especially in the context of interlinkage with nonbiometric databases, data analytics, and artificial intelligence. First, the security contexts in which their use is to be permitted might become both very wide and continuing (e.g., as when the counterterrorism security context becomes the war without end against terrorism, which then becomes the war without end against serious crime, which then becomes the war without end against crime in general).[77]

Second, data, including surveillance data, originally and justifiably gathered for one purpose (e.g., taxation or combating a pandemic) can be interlinked with data gathered for another purpose (e.g., crime prevention) without appropriate justification. The way metadata use has expanded from initially being used by only a few agencies to now being used quite widely by governments in many western countries is an example of function creep and illustrates the potential problems that might arise with the introduction of biometric facial recognition systems.[78]

---

[76] Miller and Smith, "The Ethical Application of Biometric Facial Recognition Technology."
[77] Miller and Gordon, *Investigative Ethics.*
[78] Monique Mann and Marcus Smith, "Automated Facial Recognition Technology: Recent Developments and Approaches to Oversight," *UNSW Law Journal* 40, no. 1 (2017): 121–45.

Third, various general principles taken to be constitutive of liberal democracy are gradually undermined, such as the principle that an individual has a right to freedom from criminal investigation or unreasonable monitoring, absent prior evidence of violation of laws. As noted above in 4.1, in a liberal democratic state, it is generally accepted that the state has no right to seek evidence of wrongdoing on the part of a particular citizen, or to engage in selective monitoring of that citizen, if the actions of the citizen in question have not otherwise reasonably raised suspicion of unlawful behaviour and if the citizen has not had a pattern of unlawful past behaviour that justify monitoring. Moreover, in a liberal democratic state, it is also generally accepted that there is a presumption against the state monitoring the citizenry. This presumption can be overridden for specific purposes, but only if the monitoring in question is not disproportionate, is necessary or otherwise adequately justified, kept to a minimum, and is subject to appropriate accountability mechanisms. Arguably, the use of CCTV cameras in crime hotspots could meet these criteria if certain conditions were met (e.g., police access to footage was granted only if a crime was committed or if the movements of a person reasonably suspected of a crime needed to be tracked). However, these various principles are potentially undermined by certain kinds of offender profiling, and specifically ones in which there is no specific (actual or reasonably suspected) past, imminent, or planned crime being investigated. Biometric facial recognition could be used to facilitate, for instance, a process of offender profiling, risk assessment, and subsequent monitoring of people who by virtue of fitting these profiles are considered at risk of committing crimes, even though the only putative 'offences' committed were fitting these profiles.

We conclude with several general points that ought to guide policy in this area. Firstly, privacy in relation to personal data, such as facial images, consists in large part in the right to control the access to, and use of, that data. Moreover, security consists in large part in individual rights—notably, the right to life, as well as to institutional goods, such as law and order. Biometric facial-recognition technology gives rise to security concerns, such as the possibility of identity theft by a sophisticated malevolent actor, even as they resolve old privacy and confidentiality concerns, such as reducing unauthorized access to private information and thereby strengthening privacy protection. This security problem is particularly acute given that one's face is a more or less unalterable feature of oneself. Therefore, one cannot change one's facial image as a unique identifier in the manner that one can change one's password. In short, the problems in this area cannot be framed in terms of a simple weighing of, let alone trade-off between, individual privacy rights versus the community's interest in security.

Secondly, the establishment of comprehensive, integrated biometric facial recognition databases and systems by governments (and now the private sector), and the utilization of this data to identify and track citizens (e.g., via live CCTV feeds) has the potential to create a power imbalance between governments and citizens that risks undermining important principles that are constitutive of the liberal democratic state (e.g., privacy).

Thirdly, the expanding use of biometric facial-recognition databases and systems must be clearly and demonstrably justified in terms of efficiency and effectiveness in the service of *specific* security and/or safety purposes, rather than by general appeals to community security or safety.

Finally, insofar as the use of facial-recognition and other biometric identification systems can be justified for specific security and safety purposes, and therefore privacy and other concerns mitigated, it is nevertheless imperative that their use be subject to accountability mechanisms to guard against misuse. Citizens should be well informed about biometric facial-recognition systems and should have consented to the use of these systems for the specific, justified purposes in question. Their use should be publicly debated, backed by legislation, and their operation subject to judicial review.

5

Public Health, Pandemics, and Cybertechnology

Individual Rights and Collective Goods

5.1 Introduction

Public health and cybersecurity intersect at a number of points. Most obviously, health databases and medical equipment that rely on cybertechnology are subject to cyberattacks. Personal health information stored in electronic databases can be stolen, modified, or destroyed by hackers, and the cyber-based medical equipment in hospitals can be disabled by viruses. Perhaps most importantly, as the WannaCry attack on the UK National Health System (NHS) dramatically demonstrated, such databases and computer-based equipment can be subject to ransomware attacks in which the attacker encrypts the data and will only decrypt it if paid a ransom.

On 12 May 2017 the NHS was hit by the devastating WannaCry ransomware attack. Five hundred and ninety-five out of 7545 general practitioners' (GP) offices (8 percent) and eight other NHS and related organizations were infected. Eighty out of 236 NHS trusts, which include both GPs and hospitals, across England were affected by the WannaCry attack. Services were impacted even if the organization was not infected by the virus (e.g., if they took their email offline to reduce the risk of infection). The extent of the attack was enormous, with around seven thousand appointments cancelled, and a possible loss of another nineteen thousand follow-on appointments.[1] The NHS responded well to what was an unprecedented incident, with no reports of harm to patients or of patient data being compromised or stolen.

However, as COVID-19 has also dramatically demonstrated, data security is not the only cybersecurity issue in public health. Importantly, inadequate

[1] https://www.england.nhs.uk/wp-content/uploads/2018/02/lessons-learned-review-wannacry-ransomware-cyber-attack-cio-review.pdf. Accessed 15 June 2021.

Cybersecurity, Ethics, and Collective Responsibility. Seumas Miller and Terry Bossomaier, Oxford University Press.

regulation of cyberspace, and of social media platforms in particular, has enabled antivaxxers to spread disinformation, propaganda, and conspiracy theories (see Chapter 3) in order to discredit vaccination programs and, typically, play down the severity of the virus. This has had the effect of facilitating the spread of the pandemic by virtue of the large numbers of unvaccinated people who, nevertheless, continue to interact with others as though being infected was not a serious health risk. In some cases, antivaxxers have engaged in violent protests, like the so-called freedom convoy to Ottawa in early 2022, which led to the shutdown of the downtown area of a major city, the closure of a major US/Canada border crossing for a couple of weeks, and the declaration of a state of emergency by the premier of Ontario.[2]

Moreover, if left unchecked, pandemics by their very nature lead to security problems (e.g., security breaches at quarantine facilities and overcrowded hospitals, violence at retail outlets due to food shortages, and violent protests against lockdowns). Accordingly, the lack of regulation in cyberspace has not only had the effect of enabling the views of antivaxxers and the like to flourish, thereby undermining vaccination programs, but it has also caused security problems to arise as a consequence of an uncontrolled pandemic.

A third cybersecurity issue—or set of issues—arises from the fact that cybertechnology is in the forefront of efforts to combat COVID-19 and future pandemics (e.g., by collecting, analysing, and disseminating public health data, such as infection rates and the tracking and tracing (via, for instance, metadata) of those who are infected. That is, the use of various types of cybertechnology is an important countermeasure to many security threats posed by pandemics, whether the spread of pandemics is exacerbated by inadequate regulation of cyberspace or not. However, such public health databases have moral costs—notably, a reduction in privacy and consequential data security risks, as the WannaCry ransomware attack on the NHS demonstrates. Accordingly, the moral benefits of such databases need to be weighed against these moral costs. In the case of public health databases relevant to combating pandemics, the moral benefits are, no doubt, likely to outweigh the moral costs. However, this needs to be demonstrated, given that in other cases the balance might go in the opposite direction. For instance, many commercial databases of personal information (e.g., some of those used for advertising purposes, may well not be justified given the

---

[2] M. Woods and J. Pringle, "Ontario Premier Says, 'Ottawa under Siege' and Declares State of Emergency." https://ottawa.ctvnews.ca/freedom-convoy-2022. Accessed 30/10/2023.

privacy and data security risks, notwithstanding the commercial benefits, especially to shareholders). Again, governments may be driven, particularly during pandemic emergencies, to collect more data than they strictly need, or to use the data collected for purposes other than combating the pandemic, thereby endangering public confidence in government. This potential moral cost should at least be considered. For instance, Singapore authorities admitted that data from a centralized digital contact-tracking system, called TraceTogether, could also be accessed by the police, contrary to previous assurances.[3]

A fourth ethical issue that is relevant to cybersecurity, although indirectly, is that of vaccination against pandemics. The creation of vaccines and the implementation of vaccination programs involves big data, AI, and various other cybertechnologies, and therefore gives rise to privacy/autonomy and related data-security issues. However, the issue of the moral justifiability of vaccination programs to counter pandemics is germane to cybersecurity because it is presupposed in the discussions of data security issues and also the cybersecurity issues pertaining to antivaxxer propaganda, conspiracy theories, and the like disseminated in cyberspace. Accordingly, it is important to get clear on this prior question of the moral justifiability of vaccination programs to combat pandemics. Here, as elsewhere, settling issues in the ethics of cybersecurity presupposes settling prior ethical issues. The ethical analysis of privacy undertaken in Chapter 2 and of disinformation, propaganda, and hate speech undertaken in Chapter 3, prior to the resolution of the issues of data security and of computational propaganda (respectively) are other instances of this.

5.1.1 Pandemics and Cybersecurity

Pandemics are a potential security problem (as opposed to a safety problem). In the case of the COVID-19 pandemic, an *actual* serious security problem arose by very large numbers of people *knowingly* or, at least, *negligently*— that is, by refusing to take reasonable precautions against its spread, such as by getting vaccinated or, in the case of political leaders, refusing to implement policies enabling and requiring people to take reasonable

---

[3] https://www.nature.com/articles/d41586-023-02130-6?utm_source=Nature+Briefing&utm_c ampaign=303d9b9929-briefing-dy-20230704&utm_medium=email&utm_term=0_c9dfd39373-303d9b9929-42343011. Accessed 30/10/2023.

precautions—infecting other people, leading to the severe illnesses and deaths of very large numbers of people. It is, of course, impossible to precisely determine the extent to which the COVID-19 virus has been spread knowingly or negligently. However, the number of people, including national leaders, such as former US President Donald Trump and former President of Brazil Jair Bolsonaro, who have through their public stance on vaccination inadvertently facilitated the spread of the virus is presumably in the hundreds of thousands, if not millions. Moreover, pandemics, and certainly COVID-19, create tremendous burdens on critical health infrastructure (e.g., hospitals), cause large-scale economic downturns, and potentially lead to political instability. In short, while pandemics are in the first instance a personal security problem, as well as a safety problem, they are also a potential national security problem. In both cases cybertechnology is implicated in multiple ways, including as part of the solution.

Governments around the world have used a variety of measures to respond to the COVID-19 pandemic. These measures include identifying the occurrence and spread of an infectious disease in a community—notably, by means of disease reporting requirements, which have traditionally taken place by phone, mail, or fax, but more recently by means of digital disease reporting, including syndromic surveillance (e.g., observance of disease categories identified by means of clusters of symptoms). Other measures seek to stop the spread of the disease by means of economic lockdowns, curfews, border shutdowns, quarantines, contact tracing, social distancing, hand washing, mask wearing, and so on. They also include, crucially, the creation of vaccines to generate a degree of immunity.

Cybertechnology, such as metadata and phone apps, has been used for contact tracing. Moreover, there is an important role for big data and associated analytics, including machine-learning (ML) techniques, in the prevention and mitigation of pandemics, especially when used in conjunction with whole genome sequencing (WGS). WGS is a microbiological strain typing method which can be used to "produce universally understood data expressed in the sequence of nucleotides which can be used to reconstruct transmission pathways, highlight missing cases, and, potentially, locate an individual at the time of exposure."[4] This pathogen WGS data can be linked to administrative and other big data, like mobile phone tracking data (GPS)

---

[4] Chris Degeling et al., "Community Perspectives on the Benefits and Risks of Technologically Enhanced Communicable Disease Surveillance Systems," *BMC Medical Ethics* 21, no. 31 (2020): 1–14.

and social media use, to provide very early warning and accurate monitoring during the early stages of disease outbreaks. Moreover, the digitized health and genetic data of populations can be used more generally to identify groups that are particularly vulnerable to specific pathogens, especially when such data is linked with social media data, as well as data about age, location, occupation, and so on. It can also be used for research purposes (e.g., creation of vaccines).

We note that the potential utility of sharing health and genetic data stored in databases, sometimes referred to as biobanks, has given rise to the idea of a health and medical information commons (HMIC), an idea which has been given further impetus by developments in big data analytics and ML techniques. The idea of a health and medical information commons is associated with notions of common-pool or shared resources elaborated by Elinor Ostrom,[5] and refers to a species of *epistemic* (i.e., knowledge-based) commons. Below we provide an analysis in terms of Miller's concept of a collective (epistemic) good.

The use of cybertechnology to combat pandemics raises several important ethical questions about privacy/autonomy (e.g., control of one's health data) and data security (e.g., accountability for data breaches). However, it also raises other important ethical questions such as justice issues. For instance, questions of justice arise when data associated with certain groups is more susceptible to hacking because of insufficient resources to implement data encryption, or (less directly) if there are serious risks to the vulnerable from the unvaccinated who are themselves not at serious risk. The latter typically refuse to bear any costs (such as being vaccinated) to protect the vulnerable and, in some cases, have been influenced by antivaxxer disinformation spread on social media.

A fundamental moral concept that needs to be in play in these discussions is that of collective moral responsibility (see Chapter 4, Section 4.3.5). This notion comports with the concept of an epistemic commons in at least two respects. Firstly, there is a collective moral responsibility to combat a pandemic. Secondly, there might be a derived collective moral responsibility to contribute to the creation of such a commons by providing relevant personal data, including not only health and genetic data linked to social media data for research purposes, but also location data from one's smartphone for

[5] Ostrom, *Governing the Commons*. See also Seumas Miller "Collective Responsibility and Information and Communication Technology," in *Information Technology and Moral Philosophy*, ed., J. van den Hoven and J. Weckert (Cambridge University Press, 2008), 226–50.

contact tracing purposes. However, it is important to note that the contribution to the commons should have a defined time horizon and defined connectivity with other data. Thus far, there has not been a great deal of progress in relation to a global health and medical information commons to assist in the control of pandemics.[6] On the other hand, there have been significant developments in relation to technologically enhanced communicable disease surveillance systems and contact tracing.[7]

Governments have been quick to use metadata, apps, social media, and messaging services to respond to the COVID-19 pandemic. Most notably, South Korea has taken phone metadata tracking to another level and used it to directly inform community messaging about the virus. The government publishes anonymized data of the locations of individuals who have contracted COVID-19, making it available to the public via websites and apps. Text messages are sent to citizens in a specific locality by the health authority. These are very specific and can include anonymized maps of individuals' location history. Depending on the population size of the locality, the specificity of these messages may allow those individuals to be identified, and therefore may infringe upon privacy rights. However, under certain conditions, the collective moral responsibility to combat the pandemic may override individual privacy rights—and, therefore, the right to autonomy, which is constitutive of privacy (see Chapter 2, Section 2.1).

For, as argued in earlier chapters, the right to privacy is not absolute; it can be overridden. Moreover, its precise boundaries are unclear. A person does not have a right not to be observed in a public space, but arguably does have a right not to have their movements tracked by their smartphone, even if this right can be overridden under certain circumstances. For instance, this right might be overridden if a person has been directed to self-isolate in a hotel because they have recently returned from overseas—and then only for the purpose of identifying other members of the public who may have been exposed to the virus.[8] What of persons who are carrying the COVID-19 virus and are at a risk of passing it on to other members of the community? Presumably, it is morally acceptable to utilize available data to identify these persons. If so, then it seems morally acceptable to utilize metadata to identify whom these

---

[6] Mary Majumber, Juli Bollinger, Angela Villanueva, Patricia Deverka, and Barbara Koenig, "The Role of Participants in a Medical Information Commons," *Journal of Law, Medicine and Ethics* 47, no. 1 (2019): 51–61.

[7] Degeling et al., "Community Perspectives."

[8] Moreover, their right to freedom of movement might also be overridden, given the need for quarantine and, if necessary, enforced quarantine.

individuals may have contacted, to isolate them, and to provide treatment as early as possible to reduce the chance that they will become ill and possibly die. This will reduce the number of people to whom they will pass the disease.

Providing more specific information to the community in response to the pandemic, as South Korea has done, can be contrasted with the initial approach of extensive lockdowns that were implemented in Europe, the United States, and Australia. Of course, quarantine, enforced lockdowns, and the like compromise individual autonomy. However, arguably, the collective moral responsibility to combat the pandemic overrides individual autonomy rights, albeit restrictions on autonomy, such as lockdowns, also have deleterious economic effects.

Evidently, strategies for combating COVID-19 involve a complex set of often competing, and sometimes interconnected (e.g., some privacy rights, such as control over personal data, are themselves aspects of autonomy) moral considerations. Hard choices have to be made. However, the idea of a collective responsibility on the part of individuals to jointly suffer some costs (e.g., loss of privacy rights) in favour of a collective good (e.g., eliminating or containing the spread of COVID-19) lies at the heart of all such effective strategies. This idea provides the theoretical framework for this chapter on cybersecurity and public health, including, but not restricted to, pandemics. Accordingly, this chapter provides an analysis of the appropriate notions of collective goods and collective responsibility. This theoretical framework is applied to the notion of a health and medical information commons and a variety of surveillance tools used to combat COVID-19, including phone metadata tracking.

Moreover, even if the collection, storage and analysis of health and medical data in an information commons, and the use of metadata or similar tracking and tracing methods, can be morally justified in principle as necessary to avert the threat to public health posed by COVID-19, nevertheless, ethical problems arise from the expanding use of such data for public health surveillance and other security purposes. This is especially the case because of its interlinkage with other data available to governments, such as data from social media, biometrics, and the rapidly developing capabilities of data analytics and artificial intelligence. First, the security contexts in which their use is to be permitted might become both very wide and ongoing. For example, the COVID-19 ('biosecurity emergency') context becomes the need to prevent future pandemics and maintain public health more generally; just as, arguably, the 'war' (without end) against terrorism became the war (without end) against serious crime; which, in turn, could at the limit become the 'war' (without end) against crime in general and, in doing so, result in unnecessary and disproportionate curtailment of civil liberties. Second, data, including surveillance data, which was originally and justifiably gathered for one purpose (e.g., taxation or combating a pandemic) is often interlinked with data gathered for another purpose (e.g., crime prevention), for which there is no appropriate justification. Metadata use, in particular, has expanded in some countries, from initially being used by only a few police and security agencies to wide use by governments in many western countries. This is an example of function creep and illustrates the potential problems that might arise as the threat of COVID-19 eases.[9]

5.2 Security, Public Health, and Collective Goods

The notion of security is somewhat vague (as we saw in Chapter 2, Section 2.1). Sometimes it is used to refer to a variety of forms of collective security, such as national security (which may be undermined by terrorist attacks), community security (which may be undermined by high levels of street crime), and biosecurity (which may be undermined by pandemics). At other times, it is used to refer to personal physical security. Physical security in this sense is security in the face of threats to one's life, freedom of movement, or personal property—the latter being goods to which one has a human right. As has been discussed, metadata could help to identify individuals that need to be tested for COVID-19 because they may have been in close contact with a person with the disease. Naturally, metadata might have a limited role to play here if, for instance, there was an alternative means with fewer privacy risks, such as a Bluetooth application using the DP3T protocol or its Apple-Google collaborative extension.[10] More generally, any cyber application designed for pandemic tracking and tracing purposes might fail because it is not be taken up by a sufficient proportion of the population to be effective, or because there might be performance issues reducing its efficacy. The emergency nature of pandemics, such as COVID-19, means that multiple

[9] Seumas Miller and Marcus Smith, "Ethics, Public Health and Technology Responses to COVID-19," *Bioethics* 35, no. 4 (2021): 364–71.;

[10] "COVID-19 Digital Tracing Worked," *Nature* Briefing, 3 July 2023. https://www.nature.com/articles/d41586-023-02130-6?utm_source=Nature+Briefing&utm_campaign=303d9b9929-briefing-dy-20230704&utm_medium=email&utm_term=0_c9dfd39373-303d9b9929-42343011. Accessed 30/10/2023.

mitigation strategies are often invoked in parallel since there is inadequate time to test each one thoroughly before release. There is a scientific and moral imperative to evaluate such strategies down the line. The UK has done just that with the contact tracing app, estimated to have saved 9,600 lives and prevented around one million infections.[11]

Personal (physical) security is a more fundamental notion than collective security. Indeed, collective security in its various forms is in large part derived from personal security. Thus COVID-19, for example, is a threat to public health and national security precisely because it threatens the lives of individual citizens. However, collective security is not simply aggregate personal (physical) security. For example, COVID-19 might be a threat to the stability of a government and, as such, a national security threat.

As mentioned above, security should be distinguished from safety, although the two concepts are related, and the distinction somewhat blurred. We tend to speak of *safety* in the context of natural disasters, pandemics, and other crises in which the harm to be avoided is not intended (and does not involve culpable negligence). By contrast, the term *security* typically implies that the threatened harm is intended (or is a matter of culpable negligence). At any rate, it is useful to at least maintain a distinction between intended and unintended harms—and, in relation to unintended harms, between foreseen, unforeseen, and unforeseeable harms. For instance, someone who is unknowingly carrying the COVID-19 virus because they are asymptomatic, is a danger to others, but nevertheless might not be culpable if they had taken reasonable measures to avoid being infected, intended to test for infection if symptoms arose, and then take all possible measures not to infect others, if infected. On the other hand, in a general sense, a pandemic is a personal security, as opposed to a safety, issue, insofar as members of a population are aware that, if infected, they pose a threat to the lives of themselves and others, which they can choose to minimize (e.g., by complying with infection reporting, contact tracing, quarantine requirements, hand washing, social distancing, mask wearing, and, ultimately, vaccination) but choose not to minimize it.[12] While the danger emanates from the transmissible, virulent pathogen, nevertheless, it is the irresponsible behaviour of human beings that transforms this safety issue into a security issue.

[11] Kendall et al., *Nature Communication* 14, 858 (2023). http://scholar.google.com.au/scholar?q=Kendall,+M.+et+al.,+Nature+Communication.+14,+858+(2023).
[12] Miller, *Dual Use Science and Technology, Ethics and Weapons of Mass Destruction*, 9.

Notions of collective security, collective safety, public health and so on are collective goods. There is evidently a family resemblance between notions such as *common good, collective good, public good, common interest, collective interest, public interest* and so on. Such goods or interests are attached to, or are enjoyed by, groups and other collectives, such as the Australian people, members of the Wagga Wagga local community, or the pharmaceutical industry. The contrast here is between common goods, common interest (or, in our parlance, collective goods), on the one hand, and a single person's interest or a benefit that is or could be produced and/or enjoyed by a single person. Historically, notions of the common good, common interest and the like in the political sphere are associated with philosophers such as Aristotle, Aquinas, Hobbes, and Rousseau.

There is a distinction to be made between the common good and specific common goods. Security, clean air, and an efficient transport system are all examples of common goods. We can presumably, at least in principle, offer a definition of the notion of a common good and draw up a list of such goods. By contrast, the common good—which is often, but not always, what is in the common interest—is something to be determined anew in a multiplicity of ever-changing circumstances. The common good in our sense of a collective good is an unspecified, or rather underspecified, state to be realized by collective action.[13]

Economists typically speak of a species of common goods that are *public* goods. They define public goods as being nonrival and nonexcludable.[14] If a good is nonrival, then my enjoyment of it does not prevent or diminish the possibility of your enjoyment of it (e.g., a street sign is nonrival since my using it to find my way has no effect on you likewise using it). Again, a nonexcludable good means that, if anyone is enjoying it, then no one can be prevented from enjoying it (e.g., national defence). The public goods in question are typically relativized to the nation-state but increasingly to the global economy.[15]

---

[13] Jean-Jacques Rousseau, *The Social Contract*, trans. H. J. Tozer (Ware: Wordsworth Editions, [1762] 1998); T. H. Green, *Lectures on the Principles of Political Obligation* (London: Longmans, Green and Co., 1895). See also Miller, "Collective Rights"; Seumas Miller, "Institutions, Collective Goods and Individual Rights," *ProtoSociology: An International Journal of Interdisciplinary Research* 18–19 (2003): 184–207; Miller, *Social Action*, Chapter 7; *Moral Foundations of Social Institutions*, Chapter 2.

[14] John G. Head, *Public Goods and Public Welfare* (Durham: Duke University Press, 1974).

[15] Georges Enderle, "Whose Ethos for Public Goods in the Global Economy?" *Business Ethics Quarterly* 10, no. 1 (January 2000): 131–44.

Nonrivalness and nonexcludability are relevant to the characterization of common goods, although the notion of a common good is not necessarily defined in terms of them. Other properties relevant to the notion of a common good include equality and jointness of production. Many common goods are jointly produced, maintained, or renewed. And perhaps, if a common good is enjoyed, then it is enjoyed equally by all; or if not, it ought to be. There are further distinctions to be made in relation to common goods and, specifically, collective goods in our sense.

As Joseph Raz points out, there are necessarily common goods and ones that are merely contingently common (we retain Raz's term, "common good" in this section).[16] A right of access to a water supply might only be contingently common. This would be so if, when the water supply is cut off, everybody's supply is cut off. But under a different system, selective cutting off is possible. By contrast, a tolerant society is necessarily a common good. The tolerance of the society is not something that could be channelled to certain individuals only. The public-health good of achieving herd immunity in the face of a pandemic such as COVID-19 is a necessarily common good, since it is not something that could be channelled to certain individuals only. Here we assume that there are a range of measures that are required to achieve herd immunity, including compliance with the requirement to inform health authorities if one is infected, subjecting oneself to contact tracing requirements, quarantine for those infected, general compliance with hand-washing and mask-wearing protocols in certain settings, and a high level of vaccination rates, none of which is sufficient on its own. That is, no one could guarantee not being infected or avoiding illness, even if fully vaccinated, absent these conditions. On the other hand, if there was a vaccine that was 100 percent effective against all variants of a virus and, as might be the case, some had access while others did not, then the right of access to the vaccine would not be a common good, not even a contingently common good.

Following Denise Reaume with respect to necessarily common goods, we can further distinguish between those that an agent can choose not to enjoy himself, and those that he cannot choose not to enjoy.[17] Perhaps clean streets are of the former kind and a law-abiding society of the latter. A recluse could

---

[16] Joseph Raz, "Rights-based Moralities," in *Theories of Rights*, ed. J. Waldron (Oxford: Oxford University Press, 1984), 187.

[17] Denise G. Reaume, "Individuals, Groups and Rights to Public Goods," *University of Toronto Law Journal* 38, no. 1 (1988): 438–68. See also Miller, "Collective Rights" and "Institutions, Collective Goods and Individual Rights."

not be prevented from enjoying clean streets or a law-abiding society, but he could choose not to enjoy clean streets by never going out. On the other hand, even by staying at home he cannot choose not to enjoy a law-abiding society. Again, if herd immunity is achieved and, consequently, a pandemic is eliminated, then no one could choose not to enjoy this public health good. Therefore, such herd immunity would be a necessarily common good.

Let us now explore common goods that are jointly produced, maintained, or renewed.[18] Perhaps the territory occupied by the people of the Netherlands is a common good in this sense, since much of it would be under water were it not for the elaborate system of dykes put in place, maintained, and extended over hundreds of years by its citizens. Again, the achievement of the elimination of a pandemic by means of the joint compliance of members of the community with public health protocols would be a jointly produced and jointly maintained common good.

What is the relationship, if any, between moral rights and jointly produced common goods? Presumably, the participants in the joint enterprise in question have a joint moral right to the common good. Accordingly, even if the good considered in itself is not a common good, each of the individual (jointly held) rights to that good is a right to it *qua* common good. For example, using two boats and a single large net we could jointly catch a hundred fish. By prior agreement we could possess individual rights to fifty fish each. But this agreement is something additional to the joint right, and indeed presupposes it. Imagine that unexpectedly the good produced (i.e., the fish caught) could not be parcelled out in the manner envisaged in the agreement. Perhaps we caught only one very large fish, or no fish, but instead a rare and valuable old ship. If so, each individual could still claim an individual (jointly held) right to the good, and therefore legitimately insist on making some different agreement, or perhaps no agreement. By analogy, if each member of a community contributed to the elimination of a pandemic by means of his or her compliance with the public health protocols then each would have a right to enjoy this common good and have this right jointly with the others.

Jointly produced common goods give rise to the so-called free-rider problem. The problem arises for goods that can be produced even if some members of the group do not contribute (i.e., they ride freely). The problem is that each knows they can ride freely, and it is in their interest to do so. However, if all act on their self-interest, then the good will not be produced,

---

[18] Miller, *Moral Foundations of Social Institutions*, Chapter 2.

something that is ultimately in no one's interest.[19] Compliance with COVID-19 public health protocols, including those involving cybertechnology (e.g., checking in to restaurants using a COVID-19 app and contributing one's COVID-19 status to a database), does give rise to a partial free-rider problem—namely, for those who are not at risk of illness, let alone death, should they contract COVID-19. If everyone else complies with COVID-19 protocols but such a person does not, then he or she will receive the benefits conferred by herd immunity (e.g., a well-functioning economy and public health system) without bearing any of the costs, assuming there are no sanctions for noncompliance. However, there are a significant number of people, notably the elderly and those with underlying health issues, who are at great risk of serious illness, and perhaps death, if they are infected and their self-interest, therefore, is to comply with the public health protocols—and, especially, to be vaccinated—even if they could avoid compliance in a context of general compliance.

The free-rider problem is particularly acute for rational egoist theories,[20] since such theories assume that rational human action is always self-interested, or at least that where self-interest and the common good conflict, self-interest always wins. The two generally proposed solutions to the free-rider problem are: (1) top-down state structures that enforce compliance with rules that exclude free riding; (2) free-market arrangements in which the pursuit by each of his self-interest is in the common interest. A problem that arises for (1) is that of 'Who guards the (self-interested) guards?' A problem for (2) is that in the case of many collective goods (e.g., scarce fishing resources), the assumption that the pursuit of individual self-interest will serve the common interest is simply empirically false (e.g., as in the well-known example of the tragedy of the commons).[21] However, fortunately, as Elinor Ostrom has persuasively argued,[22] these are not the only two options, and the rational self-interest model, which tends to underpin them, is a serious over-simplification of human motivation (e.g., human beings can be motivated by a concern for fairness or for the well-being of others). Accordingly, while

[19] Michael Taylor, *The Possibility of Cooperation* (New York: Cambridge University Press, 1987); Mancur Olson, *The Logic of Collective Action* (Cambridge, MA: Harvard University Press, 1965).

[20] Thomas Hobbes, *Leviathan* (Cambridge: Cambridge University Press, [1651] 1996).

[21] Seumas Miller, "The Global Financial Crisis and Collective Moral Responsibility," in *Distribution of Responsibilities in International Law*, ed. Andre Nollkaemper and Dov Jacobs (Cambridge: Cambridge University Press, 2015), 404–33.

[22] For an introduction to her ideas, see Elinor Ostrom, *The Future of the Commons: Beyond Market Failure and Government Regulation* (Indianapolis: IEA, 2012).

top-down state structures and free markets solve many collective action problems—although in doing so they should not be understood, we suggest, as being wholly reliant on the motive of individual self-interest—they do not solve all of them, and other options are available. 'Bottom-up' governance systems are also possible in which, in effect, the individual members of a community jointly own the resources at their disposal and their jointly produced outputs (i.e., there are shared property rights) and also jointly enforce compliance with rules excluding free riders.[23] Importantly, Ostrom argues in favour of the possibility of 'mixed regimes' that have elements of top-down, market-based, and bottom-up institutional forms, and solutions to collective action problems that are a matter of 'horses for courses.'

We note that on Miller's account,[24] the collective ends pursued by organized joint activity are common goods ("collective goods," in his parlance) by virtue of their possession of the following three properties: (1) they are goods that are produced, maintained, or renewed by means of the *joint activity* of members of organizations or communities (e.g., schools, hospitals, police services, the Australian population, or New Yorkers); (2) they are *available to the whole community* (e.g., clean drinking water, clean environment, basic foodstuffs, public health, and public security); and (3) they *ought* to be produced, maintained, or renewed and made available to the whole community since they are desirable—as opposed to merely desired—and as such members of the community have a *joint moral right* to them.

One question that arises here is whether those that do not chose to participate in such joint action to produce or maintain collective goods, such as public health, nevertheless have a moral right to them. For instance, does the free rider who refuses to comply with public health protocols, including being vaccinated, have a moral right to the herd immunity generated by others. Perhaps not, but she might necessarily enjoy this benefit (i.e. she cannot be deprived of it). Moreover, such a person presumably should not be deprived of the benefit of a vaccine, supposing she wants it, since at the very least her being vaccinated affords protection to others—and, in any case, she does have a right to protection if it is available and at little cost to others. On the other hand, such an individual's compliance with some protocols might justifiably be enforced (e.g., she might reasonably be subjected to forcible

[23] Margaret McKean and Elinor Ostrom, "Common Property Regimes in the Forest," *Unasylva* 180, no. 46 (1995): 3–15.
[24] Miller, *Moral Foundations of Social Institutions*, Chapter 2.

quarantine if he refuses to be tested to determine whether she is infectious and refuses to be vaccinated). That is, she has a right to be offered a vaccine and a right to refuse it. However, if she refuses to be vaccinated and to be tested, then others may have the right to exclude her from public areas—and, for that matter, from private areas inhabited by others, such as members of her family—or even to quarantine her (forcibly, if necessary), since she constitutes a potential danger to them and is culpable in this regard.

5.3 Collective Moral Responsibility

As we have seen above, the notion of collective responsibility lies at the heart of the ethics of pandemic control, including the use of cybertechnology for contact tracing and to collect, store and analyse health and genetic data for surveillance purposes but also in the development of vaccines. However, this notion (see Chapter 4, Section 4.3.5) needs further analysis.

One of the central notions of collective responsibility is responsibility arising from joint actions (and joint omissions). Roughly speaking, a joint action can be understood thus: two or more individual persons perform a joint action if each of them intentionally performs his or her action but does so with the (true) belief that, in so doing each will do their part, and they will jointly realize an end that each of them has, and which each has interdependently with the others (i.e., a collective end).[25] On this view of collective responsibility as joint responsibility, collective responsibility it is ascribed to individuals.[26] Moreover, if the joint action in question is morally significant (e.g., by virtue of the collective end being a collective good or a collective harm), then the individuals are collectively *morally* responsible for it. Each member of the group is individually responsible for his or her own contributory action and, at least in the case of most small-scale joint action, each is also individually (fully or partially) responsible for the aimed at outcome (i.e., the realized collective end) of the joint action. However, each is individually responsible for the realized collective end, *jointly with the others*. Hence, the conception is relational in character. As already mentioned, if the collective end of the joint action is a collective good or a collective harm, then

[25] Seumas Miller, "Joint Action," *Philosophical Papers* 21, no. 3 (1992): 275–97 and Miller, *Social Action*, Chapter 2.
[26] Miller, *Social Action*, Chapter 8; Miller, *Moral Foundations of Social Institutions*, Chapter 3.

these individual persons are collectively morally responsible for this good or harm.

Here we need to make a number of important points. Firstly, this account of collective responsibility as joint responsibility pertains not only to joint actions but also to joint *omissions* (e.g., cases in which members of a group decide not to jointly act to avoid a harm to themselves or others such as a group of antivaxxers). Secondly, it is possible that while each participant in a morally significant joint action makes a causal contribution to the aimed at outcome of the joint action, none of these contributing actions considered on its own is either necessary or sufficient for this outcome. This is especially so in the case of large-scale joint actions involving a large number of participants, as in the case of a health or medicine information commons. Thirdly, large-scale morally significant joint actions and omissions, such as fighting the COVID-19 pandemic, introduce a range of issues that are often not present in small-scale, morally significant joint actions and omissions. (See Chapter 7, Section 7.1) for further analysis of large-scale joint actions.) For one thing, large-scale cases often involve hierarchical organizations, and hence there is the potential for those in subordinate positions to have diminished moral responsibility. For another thing, the extent of the contribution to the outcome of a joint action or omission can vary greatly from one participant to another (e.g., one person might contribute by staying at home while another is a front-line health worker). Indeed, some of those who make a causal contribution to a joint action—and especially to large-scale joint actions—might, nevertheless, not be genuine participants in that joint action, because in performing their contributory action, they were not aiming at the outcome constitutive of the joint action; they did not have its collective end as their end.

Evidently, at least in those nation-states affected, there is a collective moral responsibility to comply with reasonable measures to combat the COVID-19 pandemic. Naturally, compliance with COVID-19 protocols (e.g., providing health and location data, wearing a mask, being vaccinated, going into quarantine) involves a cost to each individual in terms of his or her loss of autonomy, for instance. However, there is a collective good to which compliance with these measures can make a significant contribution—namely, the preservation of not only the health of many, but also the lives of those who would otherwise have died as a result of the pandemic. Moreover, compliance indirectly contributes to the preservation of the livelihoods of many who might lose their jobs and businesses in the economic downturn resulting from an out of control pandemic. For without this compliance, the pandemic would become widespread and ongoing, leading to overrun hospitals and government-imposed lockdowns that result in a severe economic downturn. Naturally, those whose lives would not have otherwise been preserved receive a benefit—namely, their life—that those who would have survived had they been infected do not receive. However, it is by no means certain who would survive, if infected, and who would not. Moreover, the death of large numbers of the community as a result of a pandemic imposes personal and economic costs on those who survive the pandemic. Further, and most importantly, the good health of many, and the survival of large numbers of the members of a nation-state (or other community) is surely a good that outweighs the privacy costs imposed on the members. Consider, for instance, that at the time of writing over one million people have died in the United States from COVID-19 compared to just under three thousand in the 9/11 terrorist attacks. Evidently, the good health of millions and the survival of hundreds of thousands of the members of a community outweighs the autonomy and privacy cost to each individual member, including the autonomy and privacy costs to those who would survive, and remain in good health, even if infected. Moreover, it also outweighs the aggregate autonomy and privacy costs of the members of the community. But what of the costs in terms of livelihoods? The argument might be slightly different—namely, that if the pandemic is not brought under control, then livelihoods, as well as lives, will be lost on a vast scale. In other words, the proposition that lives need to be traded against livelihoods misunderstands the nature of the relationship. This confusion arises because at some low levels of infection rates, there might need to be a trade-off between lives and livelihoods. But to control the pandemic the notion of a cooperative scheme involving a collective good and collective moral responsibility, understood as joint moral responsibility, needs to be invoked.[27]

However, in respect of cooperative schemes, we need to distinguish between those in which there is a benefit to those who participate and those in which this is not the case. The obligation might seem less clear in the latter case; indeed, it might seem unfair. But consider this argument:

> Sometimes an agent or agents have an obligation to conform to a scheme
> which burdens that agent or agents, but which significantly benefits another

---

[27] Miller, *Moral Foundations of Social Institutions*, 70–76.

agent or agents. But such an obligation has little to do with the fairness of a co-operative scheme. Rather it concerns the importance or moral value of the collective end realised by the co-operative scheme. Such obligations arise, especially, in cases of need—as opposed to desire for a benefit—and the greater the need, the greater the disadvantage one ought to be prepared to suffer in order to help fulfil that need. The need in question may belong to a majority or a minority of the participants in the scheme.[28]

And there is this further point regarding to the greater costs that might be imposed on some members of the community than on others in relation to COVID-19. Here we need to invoke the concept of a web of interdependence.[29] In any nation-state there is a complex structure of direct and indirect interdependence including across time. For instance, there is direct interdependence between employers and employees, police officers and citizens, food producers and food consumers, and so on. There is indirect interdependence between health workers and their patients, because, firstly, patients rely on health workers, but health workers rely also on their patients who may be their employers, food producers, and/or police officers. Moreover, this interdependence exists across time, and even across generations insofar as the older generation is now dependent on the younger, while the younger was once dependent on the older, and so on.

Of course, this web of interdependence is not such that meeting the needs of a single person is a necessary or sufficient condition for the meeting of the needs of any other single person, let alone of all or even most other persons in a given nation-state (or larger community). Rather there is a complex web of partial interdependence between individuals, between subsets, and between individuals and subsets.

This de facto web of interdependence undermines the proposition that those who are not vulnerable to COVID-19 *only* have moral obligations to those who are vulnerable by virtue of the needs of the latter, although these needs do in fact also generate obligations, as we saw above. For those who are not themselves vulnerable to COVID-19 also have needs, even if not for health protection from the virus. Their past, present, or future needs (e.g. for an education or for present and future employment in a tourist sector decimated by COVID-19) are, will be, or have been met in the past,

---

[28] Miller, *Social Action*, 148.
[29] Miller, *Moral Foundations of Social Institutions*, Chapter 2.

directly or indirectly, by members of the group who are vulnerable to the virus. Accordingly, the web of interdependence generates reciprocal moral obligations among members of a nation-state and these obligations obviously entail preserving the lives of their fellow members. Naturally, there is a limit to these obligations in cases where those called upon to meet the need are required to incur significant, potentially disproportionate, costs. This raises complex moral questions that we cannot pursue further here beyond making the point that, in the case of health workers confronting COVID-19, there are stringent moral obligations on the part of members of governments, in particular, as well as citizens, to ensure that the risks to these health workers are minimized. In the case of members of governments, there is the obligation—indeed collective responsibility—to provide adequate personal protective equipment (PPE) to health workers, and to design and implement public health policies and guidelines for citizens that ensure hospitals and other facilities are not overwhelmed by COVID-19 cases. In the case of the citizens, there is the obligation—indeed collective responsibility—to comply with these public health guidelines.

5.4 Health and Medical Information Commons

As mentioned above, an important part of controlling infectious disease pandemics, such as COVID-19, is personal data (e.g., health and medical data) collection, storage, and analysis (e.g., by means of ML techniques) for health and medical research purposes (e.g., vaccine development), for clinical interventions (e.g., treatment of respiratory distress), and to inform public health policy making, (e.g., mandatory wearing of facemasks, use of smartphone metadata for contact tracing, construction of purpose-built quarantine facilities). Moreover, the data in question may need to be interlinked with other data (e.g., social media data) and shared across different institutions and jurisdictions. Let us refer to such large-scale data collection, storage, and analysis facilities as an HMIC. Roughly speaking, it is a commons by virtue of two main properties: (1) the personal data in question is provided by a large number of the members of whole populations (i.e., the aggregated data is jointly provided); (2) the aggregated data constitutes a structured knowledge base of big data (a collective epistemic entity) that, since it is ultimately in the service of public health (e.g., via the use of data analytics), is itself an instrumental good (i.e., an epistemic collective good).

Some examples of the above include utilizing electronic health records (including clinical notes on these records made by doctors and nurses), electronic disease reporting data, immunization data, and demographic data for public health surveillance of infectious diseases (e.g., to determine the occurrence and spread of a disease, to identify immunization rates, and to identify via big data analytics adverse vaccine events and determine groups vulnerable to pathogens or vaccines). Other examples include ones that do not utilize health care records but utilize social media data or metadata (e.g., for contact tracing) or health self-reports, which aggregate online reports of symptoms, known as participatory surveillance (e.g., the Flu Near You system).[30]

We note that the goal of a unitary global HMIC may well prove illusory. Rather what might be achievable is a "collection of many different health-related commons (or common pool resources) that would benefit from the widespread adoption of a group of high-level but flexible principles."[31]

Since the health and medical data in question is personal, it is apparently individually 'owned' by the members of the population in question, at least in the sense that each has a moral right to control access to this information and to the uses to which it is put.[32] Accordingly, each must consent to the collection of this data or, if not, there must be an overriding moral consideration in favour of its collection (e.g., a public health emergency, such as a pandemic). In the case of a pandemic, the overriding moral consideration is ultimately global security and, as constitutive elements, the national security of individual nation-states, given the following considerations. Firstly, the threat posed by the pandemic is to the lives of the members of the global population (i.e., to aggregate personal security). Secondly, there is a threat to the national security of individual nation-states, if the outbreak is not contained and, as a result, there is an economic collapse and social disorder. However, nation-states are currently the fundamental institutional custodians of security in its main forms.

Framed in this manner, there are individual rights to the data and to the collective goods of public health and the security dependent on the

[30] Effy Vayena and Lawrence Madoff, "Navigating the Ethics of Big Data in Public Health," in *The Oxford Handbook of Public Health*, ed. Anna Mastroianni, Jeffrey Kahn, and Nancy Kass (Oxford: Oxford University Press, 2019), 354–66.

[31] Robert Cook-Deegan, Mary Majumder, and Amy McGuire, "Introduction: Sharing Data in a Medical Information Commons," *Journal of Medicine, Law and Ethics* 47, no. 1 (April 2019): 7–11.

[32] Amy McGuire, Jessica Roberts, Sean Aas, and Barbara Evans, "Who Owns the Data in a Medical Information Commons?," *Journal of Law, Medicine and Ethics* 47, no. 1 (April 2019): 62–69.

maintenance of public health. However, matters are more complex than this, since there is an intervening collective *epistemic* good—namely the aggregated data. This collective good is epistemic since it consists in knowledge, and it is an instrumental good by virtue of being a means to realize the goods of public health and security. The question that now arises pertains to the nature of the ownership, or moral rights, to this collective epistemic good. After all, any data element pertaining to a single person is unimportant considered on its own. It is only the aggregate of data that has utility. Big data is the means to the end. We suggest that this collective epistemic good is jointly owned by those who contributed their personal data.

Of what does this joint ownership consist,[33] given that the data in question needs to be analysed if it is to be useful for public health and security purposes? Presumably, the joint owners of the data contribute their data on condition that it is analysed in certain ways, to realize certain purposes (e.g., to prevent and mitigate pandemics). However, it would not follow from this that their joint rights to the data and, for that matter, the underlying individual rights to it, had been extinguished. Here there are a number of possibilities.

Perhaps the personal data of, say, Peter Jones, is individually owned by him insofar as it is in a form that enables him to be uniquely identified but not if it does not enable him to be uniquely identified. (i.e., once anonymized it is simply a numerical unit in a statistical generalization, albeit there is the problem of re-identification.[34] Perhaps also the joint rights to the data,[35] held by the members of the population, are extinguished once they consent to have it collected (albeit there are problems with the practicality of informed consent because of indirect links to databases and uses far removed from the original database and uses consented to).[36] On the other hand, since Peter Jones might have consented to provide his personal data only on condition it was used in an anonymized form, and since the members of the population possessed of a joint right to the epistemic good (i.e., the aggregated data) might have agreed to contribute to the data bank only on

---

[33] There are complications here in respect of the precise relationships of privacy, autonomy and ownership rights that we do not need to discuss here. See Chapter 2, Section 2.1.1, and Chapter 4, Section 4.3.1. Suffice it to say that ownership rights and privacy rights involve rights to control information (i.e., involve autonomy rights).

[34] Vicki Xafis et al., "An Ethics Framework for Big Data in Health and Research," *Asian Bioethics Review* 11, no. 1 (2019): 231.

[35] On collective rights, see Miller, *Social Action*, Chapter 7.

[36] Vicki Xafis et al., "An Ethics Framework," 231.

condition it was used to combat pandemics, then neither Peter Jones' individual right nor the members of the population's joint right would have been extinguished. Rather, these rights would have been waived under certain limited conditions, or perhaps they would have given up certain rights among the cluster of rights that constitutes ownership but not the entire cluster (e.g., in the manner that an author might retain her moral rights while giving up her economic rights). If the latter is the case, which rights are retained, and which alienated?

5.4.1 Privacy Preservation and Machine Learning

As we have seen, while HMICs produce collective goods they do have costs in terms of control of personal data, privacy (or, at least, privacy/autonomy—see Chapter 2, Section 2.1),[37] and, therefore, data security. Ideally, one's personal health data would only be available to one's doctor. Accordingly, if the database is hacked, then this is a violation of one's privacy rights. Moreover, if a researcher accesses this data for research purposes, this is an intrusion into one's sphere of privacy, even if it is morally justified by virtue of the collective epistemic good it enables and by virtue of being consented to. Further, granting such additional access to researchers may well increase the risk of breaches of data security. However, these privacy/autonomy and related data security costs can be mitigated by, for instance, privacy-preserving health mining software, such as OpenSafely, which enables health records to be accessed in situ and in their anonymous form. The anonymization process is such that it does not enable the researchers to access the prior personal data in unanonymized form. Instead of extracting sensitive patient records from the databases of the company which manages them on behalf of general practitioners, the OpenSafely research team has developed a suite of software that lets them run their massive analysis on the data in situ.

According to Williamson et al.,

Opensafely provides a secure software interface that allows detailed pseudonymized primary care patient records to be analysed in near-real time where they already reside—hosted within the highly secure data centre

---

[37] And perhaps also in terms of ownership understood as a species of property right, supposing there is ownership in this sense.

of the electronic health records vendor—to minimize the reidentification risks when data are transported off-site; other smaller datasets are linked to these data within the same environment using a matching pseudonym derived from the NHS number. Naturally, there is still the potential for the data centre of the electronic health records vendor to be hacked, although this data centre is highly secure.[38]

A closely related idea is the use of *swarm learning*. Rather than access a single large dataset, as in the Goldacre study, it uses a distributed learning model, comprising a number of nodes. In this respect, it is a further advance on the OpenSafely model in that the initial analysis and findings are done at each node prior to integration at the central level. A node could be a hospital, health trust, GP practice, and so on, which ideally has a meaningfully large dataset. The system then comprises:

- An ML model, where each node downloads and uses the same model and generates the same set of parameters therefrom, such as the weights in a neural network.
- The parameters computed are shared by each node and an averaging process is carried out.
- Addition of new nodes is done automatically through a blockchain-based (Chapter 7, Section 7.1.5) credential system (this is not specified in any great detail, and what really matters in terms of the system operation is that each node has to meet some sort of fidelity criteria).
- Large datasets on leukaemia, tuberculosis and COVID-19 all showed comparable performance of the distributed model to that of an aggregated model. (Similar to the OpenSafely scheme, the medical data never leaves its home node.)[39]

Again, the point can be made that there is still the potential for the data centre of the electronic health records vendor to be hacked, although it is highly secure. That said, the OpenSafely model, conducting research on anonymized data, and especially the distributed model, significantly

[38] E. Williamson et al., "Factors Associated with COVID-Related Death Using OPENSafely," *Nature* 584 (2020): 430–36.

[39] S. Warnat-Herresthal et al., "Swarm Learning for Decentralised and Confidential Clinical Machine Learning," *Nature* 594 (2021): 265–70.

mitigates data security problems. And there is further good news regarding data security and privacy/autonomy protection.

Ideally, the owner of sensitive health or other data, say a hospital, would prefer that the data only ever left the owner database in an encrypted form. At first sight, it would seem counterintuitive to carry out ML or data mining on encrypted data. It turns out that there is a way. The idea of homomorphic encryption is not new, and a lot of progress has been made in the last decade. The idea is that you have some private data (e.g., health data). You encrypt it with one of a class of homomorphic encryption functions and ship it off somewhere for ML. The learning is done on the encrypted data and the results are shipped back encrypted. The ML machine never sees unencrypted data. The bottleneck has been speed—the process is time-consuming—but a lot of recent work has looked and getting more practical algorithms. Thus: ". . . Our solutions are efficient in terms of both execution time and communication, e.g., completing a GWAS over 20K patients and four million variants in <5 h. . . (GWAS is Genome Wide Association)."[40]

As this book goes to press, the Viterbi engineering school at the University of Southern California has announced a hardware acceleration device, TREBUCHET.[41] It reduces the homomorphic speed gap from around one hundred thousand times slower to closer to a factor of ten. Thus, privacy preserving ML on health or other data *is* possible. The issue has become one of tractable computational overhead.

5.5 Data Security

While our focus thus far has been on the collective security benefits (collective goods) of an HMIC as an important element in a global system of pandemic control versus the individual costs in terms of control of personal data (and, to this extent, privacy), there are potential security downsides to HMIC. Specifically, there is the problem of data security. If data security is breached it may result in an individual harm (e.g., infringement of individual privacy), including individual harms on a large scale (e.g., infringement of the privacy of very large numbers of individuals). Moreover, these individual

---

[40] David Froelicher et al., "Truly Privacy-Preserving Federated Analytics for Precision Medicine with Multiparty Homomorphic Encryption," *Nature Communications* 12, no. 5910 (2021): 1–10.

[41] https://viterbischool.usc.edu/news/2023/02/trebuchet-a-high-powered-processor-for-cutting-edge-encryption/. Accessed 28 February 2023.

harms could be very serious (e.g., if electronic health records are destroyed and, as a result, there are delays in treatment and potential death). Some of these harms are so-called group harms; that is, they are harms to members of a group *qua* members of a particular group. For instance, a group harm could consist of an increased death rate among indigenous people because of deliberate falsification of the appropriateness of a vaccine for members of this group.

Ransomware can also result in group harms as well as individual harms. Ransomware uses encryption as a weapon to extract ransom money and, unfortunately, it has proven to be particularly effective in the health sector. For instance, UK's National Health Scheme was attacked in this way, as was outlined in the introduction to this chapter. As we saw in Chapter 2, encryption offers enormous benefits. For instance, the whole of e-commerce depends upon being able to feed credit card numbers safely into a web site, relying on the https encryption protocol.

However, it is possible to use encryption as a weapon. Consider the victim of ransomware, whose data on a hard disc is maliciously encrypted for financial or other gain. In a ransomware attack, the target computer becomes infected by a piece of malware. It may arrive from an email attachment, a dubious website, a Trojan horse app, or by other means. Once installed, it then autonomously encrypts the hard disc, rendering it unusable to its owner. Decryption requires a key, for which a ransom must be paid.

Consider the case of a large hospital. Patient records, treatment procedures and schedules are now kept online. A terrorist hacker could gain access to this system and encrypt the contents of patient records, or perhaps encrypt records of only select, unspecified patients. Many patients could suffer harm as a result.

Deleting large amounts of data would not be so effective, since it would usually be possible to restore the data from backups. But a similar effect could be achieved by maliciously altering selected records, such as changing drug dosage. Such mechanisms may be released on the Internet to attack autonomously and indiscriminately.

Lest it be thought that ransomware is a minor issue consider the following. Ransomware is thought to cost the Australian economy $1 billion per year, but it usually operates on a single machine at a time. A user opens an email attachment, which contains the ransomware, encrypting the computer. However, in the recent case of WannaCry, it was attached to a worm, which spread very rapidly, dropping the ransomware on each computer it affected.

Many organizations were compromised, including Deutsche Bahn (German railway), Telefonica (Spanish telecom) and importantly, given our concerns here, the National Health Service in the UK, raises the policy issue in a significant way, as described above in the introduction to this chapter.[42]

Let us deal first with the morality of hacking. In many countries, hacking is illegal, and in some countries it carries very severe penalties. Yet hacking is not necessarily unethical or immoral. Suppose an imprisoned journalist from Western Europe who was reporting on an autocratic regime with a very poor human rights record is sentenced to death. If a hacker can free her, by forging documents, altering computer records, even gaining control of security systems, then the actions of the hacker would rightly be regarded as morally permissible, even heroic—notwithstanding that hacking is illegal both in the authoritarian state and in the journalist's and hacker's liberal democratic home country. In contrast to our heroic hacker, in the NHS case, the actions of the hackers were highly immoral. Patients could have died or had their condition worsen because of missed medicine, wrong doses, delayed surgery, and so on. Thus, the actions of the hackers endangered not a few but possibly many human lives.

The report into the incident notes that:

In July 2016, the National Data Guardian published 10 data security standards been designed to address basic cyber vulnerabilities. Adherence to these standards by the health and care system could have significantly mitigated the impact of the WannaCry attack on our services.

Thus, since a cyberattack was a possibility some people in the NHS may have acted negligently, and therefore, unethically, in failing to implement such recommendations; alternatively, their failure may have been due to lack of resources. At any rate, the first and primary failing was not installing a security patch to Microsoft Windows 7, a supported operating system. None of the eighty NHS trusts affected had installed this patch. The older, unsupported, Windows XP, was used to control devices such as MRI scanners, which were compromised too. In this case, presumably, the patch should have been installed by the machine vendors or whoever was contracted to provide maintenance. This is unlikely to have fallen into the legacy software

---

[42] Tobias A Mattei, "Privacy, Confidentiality, and Security of Health Care Information: Lessons from the Recent Wannacry Cyberattack," *World Neurosurgery* 104, no. 1 (August 2017): 972–74.

category since the operating system remained the same. It is also not likely to have taken very long to do, thus the downtime would have been quite short.

The attack exploited a vulnerability in the Windows operating system, rather than spreading it by email phishing attacks, so very few NHS staff would have had any culpability. The attack was not just on the NHS. The vulnerability was part of the mechanism for sharing files between machines, and thus it spread rapidly without user error. It infected 230,000 machines in 150 countries, but was fortunately terminated the same day by a cybersecurity researcher who identified and triggered a kill switch. Exactly what this switch was for is not obvious.

As it happens there had been warning of such an attack: NHS Digital's CareCERT bulletin on 25 April 2017 following the receipt of intelligence of a specific threat from British Telecom (BT) on 24 April 2017. In other words, BT had identified the threat three weeks earlier.

One of the major difficulties with cybersecurity is assessing the level of risk. Since cyberattacks are evolving and multiplying so rapidly, risk assessment is difficult and the level of resources that need to be devoted to each and every risk is not clear. Machine learning may play a role in prioritizing risks in the future. The National Audit Office report into the incident noted that two NHS trusts had already suffered cyberattacks, thus they knew the risk was tangible. In fact, prior to the attack NHS Digital had assessed eighty-eight trusts for cybersecurity and *all* had failed.[43]

Installing a security patch is not a relatively routine task. Thus, somewhere in the chain of control, somebody decided that it had a lesser priority than other tasks, taking the risk that patients could be compromised. The recommendations of the NHS report emphasized lines of responsibility, information flow, training in looking out for cyber vulnerabilities and handling an attack when it does occur.

By way of mitigation, the NHS has been under financial pressure for some time. The cost of upgrading software systems was one that could be delayed. There was a further intricacy, in that when an operating system upgrade is a major one, specialized software may not work properly, if at all. Hence, it is not just the cost of the operating system upgrade, but the testing and updating of other software. In some cases, such legacy software may be very old and extremely difficult to update without introducing new bugs. However, the

---

[43] https://www.nao.org.uk/report/investigation-wannacry-cyber-attack-and-the-nhs/. Accessed 16 June 2021.

point to be stressed here is that health and medical data is particularly vulnerable to ransomware attacks, given the threat to lives it poses.

5.6 Contact Tracing and Phone Applications

As we saw above, cybertechnology, such as metadata and apps, have been used for contact tracing. Moreover, pathogen data based on whole genome sequencing (WGS) can be used to reconstruct transmission pathways, highlight missing cases, and potentially locate an individual at the time of exposure.[44] This pathogen WGS data can be linked to mobile phone tracking data (GPS) and social media use, to provide, very early warning and accurate monitoring during the early stages of disease outbreaks. As we also saw above, the provision of more specific information to the community in response to the pandemic, as South Korea has done, can be contrasted with the initial approach of extensive lockdowns implemented in Europe, the United States and Australia. Of course, quarantine, enforced lockdowns, and the like compromise individual autonomy. However, evidently, the *collective moral responsibility* to combat the pandemic overrides the individual privacy rights that might be compromised using these cyber technologies, and contact tracing applications in particular.[45] That is, the collective responsibility to combat the pandemic overrides the data security risks and, specifically, the potential consequences of access to personal data by malevolent actors.

As we have seen, strategies for combating COVID-19 involve a complex set of often competing, and sometimes interconnected moral considerations (e.g., some privacy rights, such as control over personal data, are themselves aspects of autonomy). Hard choices must be made. However, the idea of a collective responsibility on the part of individuals to jointly suffer some costs in favour of a collective good (i.e., eliminating or containing the spread of COVID-19) lies at the heart of all such effective strategies. The use of contact tracing technology is no different in this respect. It involves a loss of privacy in particular for the sake of realizing the collective good.

Accordingly, in those nation-states affected by COVID-19 where phone applications are widely available, there is a collective moral responsibility to make use of them for contact-tracing purposes. However, there is also a moral

---

[44] Degeling et al., "Community Perspectives."
[45] Miller and Smith, "Ethics, Public Health and Technology Responses to COVID-19."

responsibility to ensure the apps have minimal loss of privacy consistent with their efficient functioning. The use of one or more of these applications involves, at least potentially, a cost to each individual in terms of his or her loss of privacy since his or her movements are or might be tracked. Moreover, there is the additional potential cost of this location data being misused by the government. However, there is a collective good to which the use of one or more of these applications can make a significant contribution—namely, the preservation of the lives of those who would otherwise have died as a result of the pandemic, and, indirectly, the preservation of the livelihoods of those whose jobs and businesses would otherwise have been lost as a result of the severe economic shutdown.

So, there is a collective responsibility of ordinary citizens in relation to phone applications, in particular. Other things being equal, and assuming that the phone applications in question are effective, there is a collective moral responsibility to download an application and utilize it. Naturally, other things might not be equal. For instance, the data made available to authorities might be misused (as in the Singapore example mentioned above). Moreover, the set of persons who are collectively morally responsible might not include all citizens, or all the members of the community (e.g., those who are unable to use a smartphone or who cannot afford one should perhaps be excluded, depending on whether a smartphone is available to them at little or no cost.

Notice that, as mentioned above, this conception of collective responsibility as joint responsibility implies that each relevant person has an individual moral responsibility to download a tracing application (assuming the others do). So, it is not simply a matter of whether each wants to do so. Rather, each has a moral obligation to comply, given that others have complied. However, it does not follow from this that each should be compelled to comply. It does not follow that compliance should be a matter of enforceable law. On the other hand, if the numbers of those who choose to comply under voluntary circumstances is not sufficient to enable a Bluetooth application to be effective, then it may well be that compliance ought to be enforced by the state. For the magnitude of the harm to be avoided outweighs any given individual's autonomy in respect of using the application (as well as his or her privacy), and indeed the aggregate autonomy (and privacy) in respect of using the application. Moreover, the moral weight attached to the reciprocal obligations generated by the web of interdependence can also be placed on the scale in favour of enforced compliance by the state.

Further, given the questions about the functionality of tracing applications and the seriousness of the threat posed by COVID-19, governments may need to resort to an option more invasive of privacy, such as analysis of metadata. Compliance with this option might need to be enforced. Depending on the extent of the COVID-19 infection and the number of lives at risk, this should only be done in specific cases, where individuals known to have the disease have placed others in the community at risk. If so, then greater, yet morally justified, moral costs to privacy and associated autonomy would be imposed on members of the community. However, the government's policy in this regard would ultimately be underpinned by the collective moral responsibility of members of the community to save the lives and livelihoods of those threatened by the pandemic—and the latter is ultimately dependent on the former. However, in the light of the example of counterterrorist legislation enacted post-9/11, and the fact that the government already has access to a wide range of data sources about individuals, it is important that this does not lead to a normalization of the use of these phone applications, which were introduced as an emergency measure to combat the pandemic—that is, it is important that the infringements of privacy and autonomy in question do not continue in circumstances in which they are no longer necessary.

5.7 COVID-19 Vaccination

As we saw above, cybertechnology, such as metadata and phone apps, are used for contact tracing, and there is an important role for big data and associated analytics, including ML techniques, in the prevention and mitigation of pandemics and, therefore, the maintenance of global and national security, especially when used in conjunction with whole genome sequencing (WGS). Further, the digitized health and genetic data of populations can be used to identify groups that are vulnerable to specific pathogens, especially when such data is linked with social media data, regarding age, location, occupation and so on, and can also be used for research purposes (e.g., in the creation of vaccines). In short, cybertechnology is deeply implicated in combating pandemics, especially in the creation and use of vaccines, which are the most important tool in combating pandemics such as COVID-19. Moreover, as was argued earlier, cybertechnology is unfortunately implicated in the spread of pandemics, or at least in obstructing attempts to combat COVID-19. Thus, inadequate regulation of cyberspace and of social media platforms, in particular, has enabled antivaxxers to spread disinformation, propaganda, and conspiracy theories in order to undermine vaccination programs and facilitate the spread of the pandemic. This has led to the security problems that arise as a consequence of an uncontrolled pandemic.

To this point we have discussed a variety of ethical issues that arise from the use of cybertechnology in combating pandemics, as well as its role in facilitating their spread—or, at least, obstructing pandemic countermeasures. However, we have not discussed, but have tended to assume, the answer to the question of the moral obligation, or lack thereof, to be vaccinated in the first place. Hitherto we have tended to assume, that compliance with safe and effective vaccination programs against pandemics is in fact a moral obligation (other things being equal). However, this assumed answer requires elaboration in the light of a strong and continuing antivaxxer movement. Moreover, the question of the moral obligation to be vaccinated against pandemics has a number of elements—notably, the moral basis, conceptual structure, and strength of this supposed obligation. Therefore, we turn now to address this issue.

The notion of collective responsibility has clear relevance to the ethics of vaccination against the COVID-19 virus in the context of a global pandemic. In the following discussion, we need to keep in mind the threefold distinction between the decision maker, the vaccine recipient, and the beneficiary.[46] Thus, the decision maker might be the government rather than the vaccinated person, and the primary beneficiary of vaccinating a young healthy person might be an elderly person. We also need to keep in mind the distinction between the microlevel (e.g., the individual Jones and his contact), and the macrolevel (e.g., what happens at the aggregated national level if everyone or most, with the exception of Jones, do or do not get vaccinated).

Evidently, vaccination is a necessary condition to prevent infection to oneself and avoid transmission to others. Accordingly, there is an individual moral obligation to be vaccinated, at the very least in order to avoid becoming infected and, thereby, infecting others (e.g., the moral obligation not to risk harm to others). In addition, there is the derived moral obligation not to impede the vaccination process by, for instance, spreading disinformation as antivaxxers have done, and continue to do.

---

[46] Steven Kraaijeveld, "Vaccinating for Whom? Distinguishing Between Self-Protective, Paternalistic, Altruistic and Indirect Vaccination," *Public Health Ethics* 13, no. 2 (July 2020): 190–200.

Here we need to note that most young, healthy individuals infected by COVID-19 are unlikely to be seriously adversely affected by at least the early strains of the virus whereas the elderly and those with certain underlying health conditions are, by contrast, likely to suffer serious adverse effects, including death, if they are infected. Accordingly, the aged or unhealthy members of a population have a strong self-interest in being vaccinated to protect themselves. By contrast, those who are young and healthy might not have a strong self-interest in being vaccinated. On the other hand, members of both groups have a moral obligation to be vaccinated in order not to infect others, especially the aged and unhealthy who might die as a consequence. Incidentally, given the risk to others of serious adverse effects, including potentially death, if a young healthy person infects an old and/or unhealthy person, the decision on the part of the young healthy person to be vaccinated should not be regarded as altruistic. Rather, it is ultimately based on the moral obligation not to harm others.

In addition, there is a need to achieve herd immunity to control the pandemic and, thereby secure the lives, livelihoods, and political stability of the world's populations. Pandemics pose a threat to aggregated personal security, national security, and indeed global security. Therefore, there is a collective moral responsibility on the part of members of any relevant population to achieve herd immunity. However, since herd immunity does not require 100 percent immunity, there is the threat of free riding on the part of some of those who do not want to be vaccinated, especially given some vaccines have a small risk of serious side-effects (e.g., blood-clotting in the case of the AstraZeneca vaccine), and young, healthy persons might not suffer serious adverse effects, if infected by the virus. There is a contrast here with those whose thinking operates only at the microlevel and who get vaccinated only to protect themselves—believing, perhaps falsely, that they have full immunity—and, as a result, do not support the vaccination of others (presumably because they don't much care whether others become infected and suffer the disease). For this latter group are behaving somewhat irrationally, even in terms of their own self-interest, since if COVID-19 gets out of control, even as they are protected by their immunity, then they will inevitably suffer indirect consequences (e.g., initially an overwhelmed health care system and, eventually, economic collapse).

We can conclude from this discussion that there is both an individual moral responsibility at the microlevel and a collective (i.e., joint) moral responsibility at the macrolevel to be vaccinated against COVID-19. However, there is the residual problem of free riders, motivated only by their individual rational self-interest (i.e., the young and healthy who refuse the vaccine, given herd immunity has been reached). The problem here is the well-known one of the pursuit of individual rational self-interest running counter to collective rational self-interest. If, for example, the number of young and healthy persons refusing to be vaccinated reaches a threshold of 33 percent of the total population, then herd immunity might not be reached. Assuming the cohort of 33 percent cannot be convinced by moral suasion, social pressure, or 'nudges' (e.g., public information campaigns and ubiquitous vaccination centres), then coercive means might have to be deployed, given what is at stake (i.e., aggregated personal security and national security), and given the relatively small moral cost to be paid by the young and healthy persons in question, both individually and collectively (i.e., in aggregate). Individually, that cost consists in almost all cases in taking a small risk of an adverse response to the vaccine, resulting in no adverse outcome. Collectively, that cost consists in the aggregate of these adverse responses (e.g., a handful of deaths). These coercive means might exist on a spectrum, at one end of which might be exclusionary policies, such as refusal of entry to crowded venues, and at the other end of which might be criminal charges for failing to be vaccinated—unless, for instance, one was known to have an anaphylactic response to the vaccine.

It is sometimes argued that notions of collective moral responsibility are not applicable to very large, unstructured sets of individuals. Accordingly, so the argument runs, the members of the population of a nation-state, let alone the global community of states, confronting a pandemic cannot have a collective moral responsibility to participate in the mass vaccination program needed to achieve herd immunity. Why not? The claim is that since the causal impact on herd immunity of any single person being vaccinated is infinitesimal, then a single person does not have a responsibility jointly with others to be vaccinated. This is, of course, consistent with an individual person having a moral obligation to be vaccinated himself to avoid infecting another. In this respect, the vaccination scenario is different from the climate change scenario. In the latter scenario, a single person's action of restricting his or her carbon emissions cannot possibly make any difference to global warming or another's well-being, including his or her own.

In the vaccination scenario, as in the climate change scenario, it would be absurd to claim that each of us is fully morally responsible for the large-scale harm caused by (respectively) the failure to achieve herd immunity and the totality of luxury carbon emissions (e.g., an individual is not fully morally responsible for the failure to achieve herd immunity; or, likewise, the loss of habitats and lives consequent upon human-induced global carbon emissions). Rather, each of the millions of individuals has at most a radically diminished moral responsibility for the large-scale harm resulting from (respectively) the failure to achieve herd immunity and reduce carbon emissions.

Doubtless, the reason for the absurdity of the claim of full, individual, moral responsibility for the massive harm lies in the large numbers involved in both scenarios, and the fact that each can only make a tiny causal contribution to the outcomes in question. However, the question that needs to be asked is whether each member of a very large group can come to have as their end an outcome to which each individual can only make, and only needs to make, a very small contribution. Self-evidently, the answer is in the affirmative. Consider, for example, an appeal to members of a population for funds for disaster relief or for a political candidate (e.g., Bernie Sanders' appeal of a few dollars from each contributor). The end in question is a collective end and the aggregate of actions performed to realize that (collective) end is a joint action.

Moreover, the collective end in question—namely herd immunity—is a collective moral good, and therefore the joint action required to achieve it is a collective (i.e., joint) moral responsibility. That is, each has a responsibility jointly with the others to adopt the end of mass vaccination to achieve herd immunity, and to do so by being vaccinated. Notice that, as with many joint actions, this is consistent with the fact that no one acting alone could achieve this collective end of mass vaccination and, therefore, no one has *full* moral responsibility to achieve that end. Rather, each has a (small) *partial* responsibility, jointly with others, to achieve the end of mass vaccination, and thereby, herd immunity.

It might be argued against this that, as is the case with the individual responsibility to perform an individual action, there is only a collective moral responsibility to perform a joint action if it is possible to do so. Moreover, no one has a responsibility to contribute to a joint action if the others do not do likewise. This is correct. However, in the initial stages of vaccination, there is every chance that herd immunity, as the collective end, can be achieved. It is entirely possible, and indeed it is a reasonable bet that, if one or a few are vaccinated, then others will follow. Many will be willing to go first. This has been borne out in places such as Australia where, after an initial bout of so-called vaccine hesitancy, up to 95 percent of the eligible populations of various states have been vaccinated. Here are the reasons for this. Firstly, in the case of the mass vaccination program, unlike the climate scenario, it is entirely clear what the action in question must be—namely, to be vaccinated. The collective good of herd immunity is likewise both clear and highly morally and rationally desirable. Secondly, the costs of being vaccinated are very small. Thirdly, unlike in the climate change scenario, if each person acts on his individual moral obligation to be vaccinated to avoid infecting his or her small number of contacts, then the macrolevel collective end of herd immunity will also be achieved. That is, there is the likelihood that many will act on this individual obligation, and thereby achieve their collective responsibility. Fourthly, the government is likely to introduce measures to urge people to be vaccinated and, if these measures prove ineffective, even to coerce them to do so, in order to achieve the moral imperative of herd immunity. Accordingly, in these circumstances, there is a collective moral responsibility to achieve herd immunity by means of mass vaccination, notwithstanding the impediments to doing so—notably, disinformation about alleged dangers of vaccines, and collective irrationality, more generally.

6

Cyberconflict

Covert Political Action, Cognitive Warfare, and Cyberweapons

Contemporary nation-states, and for that matter, nonstate actors, like corporations, now suffer and inflict ongoing cyberattacks on a large scale. Whether all or any of these attacks constitute war rather than conflicts short of war[1] or mere breaches of security (criminal or otherwise) is not always entirely clear, and is in any case contested.[2] Here we distinguish between the sometimes overlapping categories of cyberwar, cyberconflict short of war, cognitive warfare (i.e., a species of nonkinetic warfare waged in cyberspace, utilizing computational propaganda—see Chapter 3), cyberterrorism, cybercrime, cyberespionage, and what we refer to as covert political cyberaction (i.e., a species of covert political action). While the category of cybercrime is now well established in law, the other categories are not, or at least it is controversial whether they have been satisfactorily worked out in detail.[3] Moreover, as mentioned in Chapter 4, the line between cyberwar and cybercrime has been somewhat blurred by the state use of criminal cybergangs, such as the Russian state's use of ransomware gangs (e.g., Conti).[4] At the very least, this

---

[1] See S. B. Ford, "*Jus Ad Vim* and the Just Use of Lethal Force-Short-of-War," in *Routledge Handbook of Ethics and War: Just War in the 21st Century*, ed. F. Allhoff et al. (Abingdon: Routledge, 2013), 63–75.

[2] See, for instance, Larry May, "The Nature of War and the Idea of "Cyber War,"" and James L. Cook, "Is There Anything Morally Special about Cyber War?," in *Cyber War: Law and Ethics for Virtual Conflicts*, ed. J. D. Ohlin, K. Govern, and C. Finkelstein (Oxford: Oxford University Press, 2015).

[3] The Tallinn Manual is a recent attempt to define cyberwar adequately (Schmitt, ed., *Tallinn Manual on The International Law Applicable to Cyberwar*). However, whether it has succeeded or not is controversial. See, for instance, Lucas, *Ethics and Cyber War*. See also Gross and Meisels, *Soft War*; J. Galliot, *Force Short of War in Modern Conflict*; Herb Lin, "Overview of Relevant IHL Rules and Principles That May Be Challenged by Cyberwar" (paper presented at the *Cyberwarfare, Ethics and International Humanitarian Law workshop*, Geneva, Switzerland, 21–22 May 2014).

[4] According to the UK's National Cyber Security Centre, it is highly likely that key group members of Conti, a Russian ransomware gang, "maintain links to the Russian Intelligence Services from whom they have likely received tasking. The targeting of certain organizations, such as the International Olympic Committee, by the group almost certainly aligns with Russian state objectives." See "UK Cracks Down on Ransomware Actors," 23 February 2023. https://www.gov.uk/government/news/uk-cracks-down-on-ransomware-actors.

Cybersecurity, Ethics, and Collective Responsibility. Seumas Miller and Terry Bossomaier, Oxford University Press.
© Oxford University Press 2024. DOI: 10.1093/oso/9780190058135.003.0007

involves state crimes, but might, under some circumstances constitute war.[5] Our own view is that many instances of offensive cyberactions, if conducted under the authority or effective control of governments, should be regarded as instance of covert political action. Our general stance regarding issues in this area, and specifically whether they require a new normative framework, is as follows.[6]

The claim that there is a distinct new category of war—namely, cyberwar— that sits alongside conventional kinetic war and nuclear war is questionable. Roughly speaking, conventional war is held to necessarily involve 'killing people and breaking things' in the service of taking and holding territory (ultimately one's own territory in the case of a war of self-defence). However, cyberconflict does not necessarily involve any of these things. But perhaps cyberwar is a species of cyberconflict involving organized groups engaged in an ongoing series of cyberattacks in which there is massive destruction of critical infrastructure, leading to large-scale loss of life (e.g., one of many cyberattacks destroys the physical components of an electricity power grid in the middle of winter, indirectly causing numerous deaths). In addition, of course, conventional war in contemporary settings uses cyberweapons, and more generally, has an important cyber dimension. Consider, for instance, the February 2022 Russian invasion of the Ukraine. It has involved a wide range of cyberattacks, including on Ukraine's banks and government departments.[7] However, arguably, the cyber dimension in an otherwise conventional kinetic war would have to become the dominant dimension for the war to be reasonably described as a cyberwar.

A mode of warfare or, at least, a mode of conflict, which cuts across the distinction between kinetic war (including conventional and nuclear war) and cyberwar, or at least cyberconflict, is so-called information warfare. The definitions of information warfare and related notions, such as psyops, are multiple and somewhat unclear.[8] Suffice it to say that most kinetic wars involve the use of disinformation, propaganda, and psychological manipulation strategies, or at least as tactical components of strategies. More recently,

---

[5] James Martin and Chad Whelan, "Ransomware Through the Lens of State Crime," *State Crime*, 12, no. 1 (2023): 1–25. https://www.scienceopen.com/document_file/c292fcee-3ae3-4f60-98cf-11a933700fba/ScienceOpen/SCJ_12_1_Martin%20and%20Whelan.pdf. Accessed 30/10/2023.

[6] This view was initially expressed in Miller, "Cyber-attacks and 'Dirty Hands.'"

[7] M. Alazab, "Russia Is Using an Onslaught of Cyber Attacks to Undermine Ukraine's Defence Capabilities," *The Conversation*, 24 February 2022. https://theconversation.com/russia-is-using-an-onslaught-of-cyber-attacks-to-undermine-ukraines-defence-capabilities-177638.

[8] But see, for instance, A. Bernal, C Carter, I Singh, K Cao and O. Madreperla "Cognitive Warfare," NATO Report, 2020.

the use of these kinds of strategies has been waged in cyberspace. However, in doing so, it has used computational propaganda (see Chapter 3) and its associated techniques of big data analytics, machine learning (ML), psychological manipulation (indeed, potentially, neurophysiological manipulation), and the like. This emerging form of warfare is now referred to as cognitive warfare (whether or not the term *war* is really appropriate, given its connotations of the use of kinetic force).[9] As such, cognitive warfare is an important strategy in the context of both kinetic wars, such as the war currently being waged by Russia against the Ukraine, and cyberwars—or, at least, cyberconflicts short of war, supposing there are no such things as cyberwars, strictly speaking). We discuss cognitive warfare below (see 6.2 below).

Conventional war typically has an ultimate political (and, in some cases, moral) purpose and that purpose might not necessarily *consist* of taking and holding territory on a permanent basis. For instance, an armed humanitarian intervention might reach the threshold of war, but without involving the permanent occupation of territory taken and held during the successful intervention. Perhaps the 1979 Uganda-Tanzania war is a case in point. This involved the armed intervention by a Tanzanian army against Uganda leading to the replacement of the Ugandan dictator, Idi Amin. In this case, as in other cases of humanitarian armed intervention, taking and holding territory is typically a temporary means to the ultimate political (and moral) purpose. Nuclear war also involves 'killing people and breaking things', even though potentially on a scale at which the population formerly occupying the territory struck by the nuclear weapons may well have been decimated and the territory rendered uninhabitable.

The terms *conventional* and *nuclear*, used in relation to wars, qualify the nature of the wars to which they are applied, and in doing so typically refer to the dominant form of weaponry used to conduct those wars. Here we note that cyber*weapons* are somewhat different from conventional weapons, such as guns and bombs. For cyberweapons used in malware and related cyberattacks do not directly kill people or destroy buildings; rather they destroy or steal data or render it unavailable (perhaps temporarily) and/or deny or disrupt data processing. Accordingly, let us refer to them, somewhat paradoxically, as *soft* cyberweapons to distinguish them from kinetic weapons,

---

[9] A. Backes and A. Swab, "Cognitive Warfare: The Russian Threat to Election Integrity in the Baltic States," (paper, Belfer Center for Science and International Affairs, Harvard Kennedy School, Harvard, November 2019).

such as guns and bombs, that directly kill people and destroy buildings.[10] That said, so-called *autonomous* armed drones, for instance, do directly 'kill persons and break things', so they are a species of *hard* weaponry—and their autonomy, which is a defining feature, is dependent on cybertechnology. So let us refer to these as *hard* cyberweapons (or, cyber-based hard weapons). However, let us exclude from our consideration at this point a kinetic war in which hard cyberweapons, such as autonomous drones, are the dominant form of weaponry used and focus on a notional war in which soft cyberweapons, like malware, are the dominant forms of weaponry used. Insofar as soft cyberweapons are the dominant form of weaponry used in a conflict that is otherwise justifiably defined as a war, then the war might reasonably be regarded as a cyberwar. However, it might be argued that soft cyberweaponry is unlikely to be the dominant form of weaponry used in what is indisputably a war. Or, at least, soft cyberweaponry has not thus far been the dominant form of weaponry used in such a war.[11] Of course, cyberattacks have been (by definition) the dominant form of attack in cyberconflicts. However, these cyberconflicts in which soft cyberweaponry has been the dominant form of weaponry have evidently not risen to the threshold of conflict reasonably characterized as war; rather they have been instances of conflicts short of war, depending on how the latter notion is understood. They have not done so because they have not in fact involved large-scale killing people and breaking things, even if they *potentially* could have done so, indirectly, and nor have they involved taking and holding territory. Rather these forms of cyberconflict seem more appropriately regarded as instances of conflict short of war—or as an ancillary means of fighting a conventional war.[12] We note that psyops operations, information warfare, and cognitive warfare are species of the latter. Aside from their nonkinetic character, these kinds of conflict often occur in what are acknowledged on all hands to be peacetime conditions (e.g., Russian interference in the 2020 US Presidential election). Moreover, in many cases they might be appropriately regarded as species of covert political action (see 6.1 below).

Here we note that conflict short of war does not necessarily exclude the use of lethal weapons. For instance, an assassination by State A of an enemy

---

[10] We wish to resist the view that words or other modes of communication can literally constitute weapons. Accordingly, we are excluding so-called cognitive warfare from the discussion at this point.

[11] May, "The Nature of War and the Idea of 'Cyber War.'"

[12] May, "The Nature of War and the Idea of 'Cyber War'"; Gross and Meisels, *Soft War*; Galliott, *Force Short of War in Modern Conflict*.

spy of State B in the territory of a third party, State C, during peacetime, and credibly denied by State A, could constitute an act short of war. Accordingly, even if a one-off, credibly denied, cyberattack using a soft cyberweapon did indirectly lead to a person's death, or a small number of deaths, and intentionally so, it would not necessarily constitute an act of war.

What of the applicability of just war theory to cyberconflicts involving predominantly soft cyberweapons? The latter, we have claimed, is de facto, even if not necessarily, a species of conflict short of war. Accordingly, whereas the historically most influential normative theory of war—namely, just war theory—is, let us assume, an acceptable normative theory for waging conventional kinetic war (if not all-out nuclear war) and might, therefore, reasonably be thought to be serviceable, appropriately modified, for use in relation to conventional wars using hard cyberweaponry (other things being equal),[13] it would not follow that it is serviceable for use in relation to cyberconflicts in which soft cyberweapons are the dominant form of weaponry. Thus, the use of soft cyberweapons (and, for that matter, propaganda used in cognitive warfare by means in part of cybertechnology) makes it extremely difficult to respect the principle of discrimination (e.g., to avoid also targeting civilians when targeting combatants—see 6.2 below). On the other hands, some of its constitutive principles, if appropriately modified, such as the principles of necessity and proportionality, are likely to have some degree of applicability.

And there is this further point. Cyberconflict involving predominantly soft cyberweapons—or, for that matter, as we saw above, merely the use of the epistemic or other psychological techniques of cognitive warfare by means in part of cybertechnology—is not only de facto a species of conflict short of war, it is also, we suggest, at least characteristically, a species of covert political action. But just war theory is evidently not serviceable as a normative theory of covert political action, if only for the reason that covert political action is typically not conducted in accordance with a number of the principles of just war theory enshrined in international law (e.g., the law prohibiting perfidy, which requires combatants to fight openly and without treachery).[14]

---

[13] Things might not be equal if many forms of cyberattack necessarily target innocent persons, as well as combatants and morally culpable noncombatants. See Seumas Miller, "Civilian Immunity, Forcing the Choice and Collective Responsibility," in *Civilian Immunity*, ed. I. Primoratz (Oxford: Oxford University Press, 2007), 137–66, for an account of justifiably targetable culpable noncombatants.

[14] Heather Roff, "Cyber Perfidy, Ruse and Deception," in *Binary Bullets*, ed. F. Allhoff, A. Henschke, and B. J. Strawser (Oxford: Oxford University Press, 2016), 201–27.

Therefore, arguably, just war theory is not serviceable for most, if not all, forms of cyberconflict. Nevertheless, to reiterate, some of its constitutive principles, appropriately modified, are likely to be applicable.[15] Accordingly, there is a pressing need to provide a normative theory of covert political action that is applicable to cyberconflict (i.e., to cyberconflict short of war).

So much for soft cyberweapons and the epistemic techniques of cognitive warfare. What of hard cyberweapons and so-called autonomous weapons in particular? These seem to give rise to genuinely new moral questions such as, for instance, concerning the so-called responsibility gap (see 6.2.3 below). Hence there is a need for a serviceable normative theory regarding the use of autonomous weapons both as a means of waging war and as a weapon in use in cyberconflict and other forms of conflict short of war.

Accordingly, our focus in this chapter is, firstly, on providing a normative theory of cyberconflict short of war in which soft cyberweapons and/or the epistemic techniques of cognitive warfare are the dominant means of attack—that is, a normative theory of what is characteristically, we suggest, a form of covert political action; or, if not, merely a dimension of kinetic wars, and therefore conflicts in which, we assume, just war theory is applicable, at least in principle.

Our focus, secondly, is on so-called autonomous weapons, a species of hard weaponry. In focusing on these two issues, we are not suggesting that there cannot be cyberwars or that there cannot be forms of cyberconflict short of war that are not instances of covert political action.

We argue in section 6.1 that while the notion of a cyberwar (appropriately analysed) is coherent and potentially applicable, nevertheless, most cyberattacks that are not conducted against the enemy in a kinetic war, but which are conducted by nation-states—perhaps using criminal organizations that are effectively under their control, at least for these purposes—against other nation-states for political reasons, are best understood neither as acts of war (not even acts of economic war), nor necessarily as crimes (at least in the jurisdiction mounting the cyberattack, although they may constitute state crimes under international law or under the laws of the state attacked),[16] but rather as a new species of covert political action, that of covert political *cyberaction*.[17] Moreover, many

---

[15] And still others might be problematic, even if they ought to be applied, at least in principle (e.g., the principle of discrimination). See section 6.2 below.

[16] And if we assume that terrorist actions are crimes.

[17] This claim was argued for in Miller, "Cyber-attacks and 'Dirty Hands.'"

cyberattacks conducted against the enemy in a kinetic war might also be instances of covert political cyberaction (e.g., if they target civilian infrastructure and are credibly denied). We also discuss some of the principles, including some of those constitutive of just war theory, that might have application to cyberconflict in general and to covert political cyberaction in particular.

In section 6.2, we discuss so-called cognitive warfare. Since cognitive warfare involves computational propaganda, we help ourselves to some of the conclusions we drew regarding computational propaganda in Chapter 3. However, cognitive warfare, and the use of computational propaganda in the context of cognitive warfare, raises additional ethical issues such as, for instance, the moral rights to freedom of communication (or lack thereof) of the members of nation-states in these settings.

In section 6.3, we consider the issue of autonomous cyber-based weapons. Cybertechnology is playing an increasing role in kinetic wars, but also potentially in state terrorism, counterterrorism, and covert political cyberaction, not only by way of soft cyberweapons but also by way of hard cyberweapons—notably, autonomous weapons. Autonomous weapons have given rise to a plethora of ethical problems. Indeed, the issue of autonomous weapons is currently one of the most controversial and hotly contested in the field of military ethics. If the term *autonomous weapons* is being used to refer to weapons that are not under meaningful human control, then, we argue, the use of autonomous weapons should be prohibited, at least under most circumstances.

6.1 Cyberconflict and Covert Action

Covert political cyberaction does not include purely defensive measures, such as building firewalls, layers of password protection, and the like.[18] Rather, it is offensive action, including offensive action undertaken in response to a present or future attack. Covert political cyberaction is multifarious and includes covert political cyberattacks (lethal and nonlethal), as well as certain forms of cyberespionage.

---

[18] An earlier version of the material in this section appeared in Miller, "Cyber-attacks and 'Dirty Hands.'"

Recent high profile cyberattacks, including acts of cyberespionage, include the following[19] (in addition to at least some of the above-mentioned cyberattacks by Russia on the Ukraine in the context of its invasion of the Ukraine):[20] the denial of service cyberattack on Estonian banks, media, and government websites in 2007, presumably perpetrated by Russia; the Stuxnet attack, in which malware was used to disrupt Iran's nuclear enrichment ICT (information and communication technology) infrastructure through a US-Israeli operation (Olympic Games) to disrupt Iran's nuclear program; the Israeli bombing of a Syrian nuclear facility after they had penetrated Syrian computer networks and disabled their air defence systems (Operation Orchard); [21] the US computer-security firm (Mandiant), which has documented ongoing Chinese cybertheft and disruption of the websites and other ICT infrastructure of US corporations and government agencies;[22] the release by the US National Security Agency (NSA) contractor Edward Snowden of a large amount of confidential intelligence data to the international press, including in relation to projects Verizon and PRISM;[23] the collection by the NSA of the metadata from the calls made within the US, and between the US and any foreign country, of millions of customers of Verizon and other telecommunication providers (Verizon); [24] and the agreements between NSA and various US-based Internet companies (Google, Facebook, Skype, and so on) to enable NSA to monitor the online communications of non-US citizens based overseas (PRISM).[25]

In the case of the cyberattack on Estonia, it is important to note that Estonia is a member of NATO, and that Russia was the chief suspect.

[19] See Peter W. Singer and Allan Friedman, *Cybersecurity and Cyberwar: What Everyone Needs to Know* (Oxford: Oxford University Press, 2014), for an outline of these various cyberattacks. On the Stuxnet and Estonia cases, see also Thomas Rid, *Cyber War Will Not Take Place* (New York: Oxford University Press, 2013), 32–34.

[20] There is a question here as to whether to classify a cyberattacks in the context of a war as an act of covert political action, as opposed to a clandestine military operation aimed that the attacker wants to keep secret for tactical or strategic military reasons. If the attack is on a military installation in the context of war, then it should likely not be regarded as an act of covert political action since, for instance, it would (presumably) have been conducted within the laws of war.

[21] David E. Sanger, *Confront and Conceal: Obama's Secret Wars and Surprising Use of American Power* (New York: Broadway Books, 2013).

[22] See Mandiant Intelligence Centre, *APT1: Exposing One of China's Cyber Espionage Units* (Washington, DC: Mandiant Intelligence Centre, 2013). http://intelreport.mandiant.com/Mandiant_APT1_Report.pdf. Accessed 23/6/2014.

[23] For a sympathetic account, see Luke Harding, *The Snowden Files: The Inside Story of the World's Most Wanted Man* (London: Guardian Books, 2014).

[24] Metadata is the unique phone/email numbers of the caller and recipient, the time and duration of the call, and the location of the caller and the recipient, but not the content of the communication.

[25] Moreover, US-based citizens (and US-based foreigners) were being surveilled as part of the PRISM program, which is against US law and their constitutional rights.

However, there were no deaths or destruction of property and computer technicians unblocked the networks relatively quickly, thereby ensuring the disruption was minimal. Moreover, Russia denied responsibility for the attack and NATO did not declare war. It seems, therefore, that this cyberattack did not constitute an act of war, but rather something short of war.[26]

By contrast, Operation Orchard was presumably an act of war since it involved the Israeli bombing of a Syrian nuclear facility immediately after an Israeli cyberattack on Syrian air defence systems. More specifically, the cyberattack itself was undertaken in the context of an act of war (the bombing), and only for the specific purpose of enabling the bombing.[27]

Other kinds of cases are more difficult to classify. Presumably a cyberattack in which both the aggressor and the victim were nation-states and which not only disabled ICT infrastructure, but which destroyed it and caused many deaths, could well count as an act of war, even if credibly denied), especially if conducted in the context of an existing kinetic war, as in the case of many of the cyberattacks conducted by Russia against Ukrainian civilian ICT infrastructure (e.g., disrupting electricity services during extreme winter temperatures).[28] But in such cases it is the destruction of *physical* infrastructure, and especially, the loss of *human life* that elevates these cyberattacks to cyberwar, or at least acts of cyberwar—and, for that matter, acts of cyberterrorism, if civilian infrastructure is targeted—even if the overall characterization of the war in question was that of a kinetic war.[29] Here we note that such acts of cyberwar might also be instances of covert political cyberaction, if credibly denied.

What of a denial of service cyberattack that did not destroy physical infrastructure, and did not immediately cause any loss of life, but did cause, and

[26] Rid, *Cyberwar Will Not Take Place*, 30–32; BBC News, "Estonia Hit by 'Moscow Cyber War'," *BBC News*, 17 May 2007. http://news.bbc.co.uk/2/hi/europe/6665145.stm, accessed 24/6/2014; Peter Finn, "Cyber Assaults on Estonia Typify a New Battle Tactic," *Washington Post*, 19 May 2007, http://www.washingtonpost.com/wp-dyn/content/article/2007/05/18/AR2007051802122.html, accessed 24/6/2014.

[27] Rid, *Cyberwar Will Not Take* Place, 42–43; Erich Follath and Holger Stark, "The Story of 'Operation Orchard': How Israel Destroyed Syria's Al Kibar Nuclear Reactor," *Spiegel Online International*, 2 November 2009, http://www.spiegel.de/international/world/the-story-of-operation-orchard-how-israel-destroyed-syria-s-al-kibar-nuclear-reactor-a-658663.html. Accessed 24/6/2014.

[28] M. Miller, "Russian Cyberattacks Aim to Terrorise Civilians," *Politico*, 1 November 2023. https://www.politico.com/news/2023/01/11/russias-cyberattacks-aim-to-terrorize-ukrainians-00077561. Accessed 30/10/2023.

[29] This definition of cyberwar as necessarily involving destruction of physical property and/or loss of human life appears to be the one favoured by the authors of the *Tallinn Manual*. See Schmitt, ed., *Tallinn Manual on The International Law Applicable to Cyberwar*.

was intended to cause, a prolonged period in which government-funded and administered welfare and other services were unable to be provided, leading to severe hardship among large sections of the population, even though there was no loss of life? And what are we to make of so-called collateral damage by contagion? Should those nation-states severely affected, even unintentionally, regard themselves as at war? Stuxnet, for example, while targeted at Iranian ICT infrastructure, also caused collateral damage by contagion; it infected and shut down computers and computer networks in places such as Indonesia and India.[30]

Whether or not collateral damage by contagion constitutes war partly depends on the nature and extent of the collateral damage in question, and on whether it was foreseen or reasonably foreseeable. Presumably, neither India nor Indonesia ought to have regarded themselves as being at war with the US or Israel because of Stuxnet. On the other hand, if a nation-state hellbent on prosecuting war with an enemy nation-state released a highly virulent form of malware, knowing that it would disable and destroy key components of the ICT infrastructure—including the components of life-supporting medical facilities—of various neutral nation-states, leading to substantial loss of life, then those erstwhile neutral states might well, and justifiably, regard themselves as being in a state of war with the aggressive state in question.

An important question to be addressed at this point is the nature and extent of the harm culpably caused. Presumably, thresholds of harm can be delineated, or at least described, to serve as benchmarks in determining what is, and what is not, an act of war. The authors of the *Tallinn Manual* and some others have argued that such thresholds must be specified, at least legally, and presumably also morally, in terms of the nature and/or extent of the injury, loss of human life, and/or physical destruction caused.[31] Apparently, the idea informing such proposals is that cyberattacks cannot in and of themselves constitute war (i.e., war properly understood).[32] Rather, a cyberattack could only count as an act of war if it had consequential effects in terms of

---

[30] Jared Anwer, "India Caught in Crossfire of Global Cyber War," *Times of India*, 20 August 2012, http://timesofindia.indiatimes.com/tech/it-services/India-caught-in-crossfire-of-global-cyber-war/articleshow/15567180.cms, accessed 24/6/2014; Judy Bachrach, "The Stuxnet Worm Turns," *World Affairs Journal*, 30 January 2013, http://www.worldaffairsjournal.org/blog/judy-bachrach/stuxnet-worm-turns, accessed 24/6/2014.

[31] See, for example, Christopher J. Eberle, "Just War and Cyberwar," *Journal of Military Ethics* 12 (2013): 54–67.

[32] May, "The Nature of War and the Idea of 'Cyber War' "; and Cook, "Is There Anything Morally Special about Cyber War?"

human injury, loss of life, and/or substantial damage to physical objects (e.g., buildings).

An important point to be stressed here is that whatever international and domestic law might have to say on this matter, the moral justification for any specific threshold setting is very much context dependent. In the context of the possibility of nuclear war between superpowers, such as the US and Soviet Russia during the Cold War, between Russia and NATO if Russia were to invade one of NATO's Eastern European allies (e.g., the Baltic States), or between the US and China if China were to launch a military attack on Taiwan—the threshold setting at which a cyberattack on the part of one of these powers ought to count as an act of war must be set very high indeed, perhaps so high it can never be met in practice.[33] This has important implications for the practice of covert political action, as we argue in the next section.

At any rate, here we need to distinguish four kinds of harm or damage. First, there is harm (physical or psychological) done to human beings per se. Second, there is damage done to buildings, ICT hardware and other human artefacts (as well as to the natural environment insofar as it supports individual and collective human life). Third, there is, as Randall Dipert notes, cyberharm or rather soft damage, in our terminology (e.g., damage to software and data as opposed to the physical ICT hardware itself).[34] Fourth, there is institutional damage or harm—that is, the undermining of institutional processes and purposes (e.g., major breaches of confidentiality in a security agency or loss of institutional control of territory).[35]

The point to be made here is that the third and fourth kinds of harm (cyberdamage and institutional damage/harm) might have thresholds at which war might be justified, independently of the level of the first two kinds of harm caused (i.e., the level of physical or psychological harm caused to humans per se and the level of destruction of physical property and the like).[36] More importantly, for our purposes here, the third and fourth kinds

---

[33] We are assuming that nuclear war should be off the table and that conventional war should also be avoided, given the possibility of escalation to nuclear war. However, there remains the possibility of proxy conventional wars. Perhaps the current war in the Ukraine is or will become a de facto proxy war between NATO and Russia.

[34] Randall R. Dipert, "Ethics of Cyberwarfare," *Journal of Military Ethics* 9 (2010): 400.

[35] Since institutions are constituted by roles that are occupied by human beings, they can be damaged or harmed (or both) depending on whether the human beings in question are harmed—and, if so, they would be harmed *qua* members of the institutions in question.

[36] Or at least that cyberharm and/or institutional harm could conceivably reach such a threshold independently, *to some extent*, of the first two kinds of harm.

of damage/harm might have thresholds at which a seriously destructive or *harmful* response short of war is morally, and perhaps legally, justified. Such responses might include economic sanctions and the like, but they might also include various forms of covert political action—notably, covert political cyberattacks. We return to this question in detail below.

In this context, it is important to distinguish cybertheft from other forms of cyberattack. As with theft in general—and fraud, defined as theft by means of deception—cybertheft does not necessarily involve damaging any human person or artefact. Nor does it necessarily include cyberharm as such. Conversely, other forms of cyberattack cause cyberharm, but do not necessarily involve theft. Thus, a so-called logic bomb might destroy data and algorithms, do so without damaging the actual physical hardware, and in a manner that does not enable the attacker himself to possess the data or algorithms in question.

Of course, in the case of cybertheft, what is stolen is typically intellectual property (e.g., data or algorithms). Being theft of intellectual property, cybertheft does not necessarily deprive the owner of the use of the property, although the owner may well be deprived of many of the rights and benefits of ownership, such as *exclusive* use, as well as the economic benefits that flow from exclusive access.

Cybertheft needs to be distinguished from cyberespionage. The latter refers to the theft by some computer-based means—as opposed to, for example, by physical removal of paper-based documents: (i) of data or other intellectual property stored in an ICT system; (ii) which is reasonably regarded as *confidential from a national security perspective*; (iii) in order to realize some *political or military purpose*. Here the Snowden case is salient.[37] Edward Snowden was a low-level private contractor to the NSA who breached legal and moral confidentiality obligations by engaging in unauthorized accessing, retrieving and/or releasing of a large volume of confidential data from the NSA to the international press. Snowden's activities are a major, indeed stunning, breach of institutional confidentiality and were enabled by ICT, and specifically, the existence of vast amounts of communicable, searchable, analysable, stored data on a networked computer. Given the importance of compliance with confidentiality requirements to the integrity of security agencies, and given the large volume of confidential data released, Snowden's actions surely did considerable institutional damage to the NSA, in particular. Nevertheless,

---

[37] As mentioned earlier, for more on this case, see Harding, *The Snowden Files*.

perhaps the release of some of this data to the press was morally justified by the public's right to know, for example, that the NSA was engaged in an extremely large-scale collection process of the metadata of US and other citizens (see the brief accounts of Verizon and PRISM above in this section). In response to this argument, it might be claimed that this large-scale collection process was morally justified, even if unlawful, by the need to protect the US against the national security cyberthreats emanating from China (see below) and Russia (see above and Chapter 2). However, even if this argument is accepted, there remains the problem of the unlawfulness, and relatedly, the absence of democratic consent to this large-scale collection process—and the lack of democratic accountability.[38]

Admittedly, according to this somewhat stipulative definition, there would also be a distinction between cyberespionage and what might be referred to as cyberindustrial espionage. The latter refers to the theft by some computer-based means of: (i) data or other intellectual property stored in an ICT system; (ii) that is reasonably regarded as confidential from a commercial perspective; (iii) to realize some commercial purpose.

The above-mentioned investigations by the US computer-security firm, Mandiant, indicate that China is a major cyberthief.[39] For there are multiple acts of cybertheft originating from the headquarters of the China's People's Liberation Army Unit 61398. Indeed, according to Mandiant, most cyberattacks on US corporations, US infrastructure (e.g., power grids), and US government agencies originate from China, and China's large scale cybertheft comprises vast amounts of data from 140 countries (e.g., personal data in the files of US Federal Government employees held by the Office of Personnel Management).[40]

Much of this stolen information is apparently commercial in character. However, much of it is politically and military sensitive, at least potentially (e.g., knowledge of covert operations or of weapons systems). Moreover, some of this stolen information might simultaneously be both politically and militarily sensitive information, as well as being confidential commercial information. As such, the acts of cybertheft in question might be both

---

[38] A point stressed by George Lucas in his *Ethics and Cyber War*, Chapter 7. Lucas also forcefully makes the preventive self-defence argument on behalf of the NSA.

[39] Mandiant Intelligence Centre, *APT1*.

[40] E. Nakashima, "Chinese Breach Data of 4 Million Federal Workers," *Washington Post*, 4 June 2015. https://www.washingtonpost.com/world/national-security/chinese-hackers-breach-federal-governments-personnel-office/2015/06/04/889c0e52-0af7-11e5-95fd-d580f1c5d44e_story.html. Accessed 30/10/2023.

acts of cyberespionage and acts of cyberindustrial espionage. Consider, for example, the design and performance details of a fighter aircraft being developed by a commercial company for the exclusive use of the air forces of a particular nation-state and its allies, as in the case of the US F-35 Joint Strike Fighter plane,[41] which was evidently the victim of a Chinese cybertheft operation.[42] Further, such cybertheft activities might indicate that the Chinese state has the ability to penetrate, and manipulate in harmful ways, supposedly very secure critical infrastructure and related systems, such as power grids, air traffic control operations, financial systems, and so on.

6.1.1 Covert Political Action, Covert Political Cyberattacks, and the Problem of Attribution

One problem in relation to cyberattacks by nation-states on other nation-states is the so-called problem of attribution, although developments in cyberforensics are evidently mitigating this problem.[43] Unlike most attacks in conventional wars—or, for that matter, conventional crimes of assault or theft—there is a major *epistemic* problem in cybersecurity: the problem of *reliably* attributing responsibility, and conversely the *credibility of denial* of responsibility on the part of culpable aggressors (at least, if these attacks are not undertaken as part of a conventional war, since in the latter case they might not be denied). Because harmful cyberactivity is difficult to distinguish from benign cyberactivity, and because actors in the cyberworld are densely interconnected by indirect pathways, it is often extremely difficult to pinpoint the source of a cyberattack, or even to know that an attack rather than a malfunction has taken place. The attribution problem is compounded when state actors use criminal organizations, so to speak, to do their dirty work (i.e., the criminal organizations are under the effective control of the state but only in respect of certain cyberoperations and not others, as in the case of some of

[41] See Singer and Friedman, *Cybersecurity and Cyberwar*, 93.

[42] Michael Mullins, "China F-35: Secrets Stolen from US Show Up in Its Stealth Fighter," *The Wire*, 14 March 2014. http://www.newsmax.com/TheWire/china-f-35-secrets-stolen/2014/03/14/id/559556/. Accessed 24/6/2014.

[43] George R. Lucas Jr, "Just in Silico: Moral Restrictions on the Use of Cyberwarfare," in *Routledge Handbook of Ethics and War: Just War in the 21st Century*, ed. F. Allhoff et al. (Abingdon: Routledge, 2013), 371; Neil C. Rowe, "Perfidy in Cyberwarfare," in *Routledge Handbook of Ethics and War: Just War in the 21st Century*, ed. F. Allhoff et al. (Abingdon: Routledge, 2013), 401; Office of the Director of National Intelligence, *A Guide to Cyber Attribution* (September 2018). https://www.dni.gov/files/CTIIC/documents/ODNI_A_Guide_to_Cyber_Attribution.pdf. Accessed 31/10/2023.

the above-mentioned Russian ransomware gangs). In such cases, the criminal organization's culpability might be determined, but the state is not necessarily orchestrating the activity in question.

Moreover, the attribution problem is not simply a technical issue, not a matter of technical computer forensics. As with the determining of culpability for crimes in general, or ascribing responsibility for covert acts of aggression in wartime, there is a complex mix of rational and evidential considerations in play.[44] These include: (i) elements of the framework of rationality, such as motive, ability and opportunity; (ii) physical evidence, for example as a basis for computer forensics; (iii) testimony. There is also the question of weighing the different kinds of evidence in play and the internal coherence of the overall narrative attributing responsibility to this or that actor.[45]

However, a key problem in the case of cyberattacks emanating from foreign nation-states, as opposed to from within a domestic jurisdiction, is the problem of access. It is not possible, for example, for the US to send a team of investigators, replete with computer forensics specialists to China, to the People's Liberation Army building from which Mandiant claims cyberattacks have emanated, for the purpose of interviewing relevant personnel, removing the computers for forensic scrutiny, and so on. China can both deny responsibility for the crimes in question and (on grounds of national sovereignty) deny access to investigators; yet, without access to such evidence, criminal responsibility may be extremely difficult to prove, and consequently, denial of criminal responsibility may well be credible. The same point holds in relation to cyberattacks emanating from Russia.

At any rate, the existence of the 'problem' of attribution, and consequently the credibility of denial, makes cyberattacks an extremely useful tactic for nation-states seeking to avoid outright war or to mask their attacks during a war on, for instance, civilian infrastructure because these attacks violate international law.[46] Nation-states responsible for such cyberattacks are

---

[44] See Miller and Gordon, *Investigative Ethics*.

[45] Some of these points are also covered by Danks and Danks, "Beyond Machines."

[46] Distinctions might be made between espionage and covert political operations—the latter being kinetic—and between covert political actions undertaken during war and clandestine actions (e.g., special forces military operations), undertaken during war. The former are dirty hands actions (see below 6.1.2). See Mitt Regan and Michele Poole, "Accountability of Covert Action in the US and UK," in *National Security Intelligence and Ethics*, ed. Seumas Miller, Mitt Regan, and Patrick Walsh (London: Routledge, 2021), 232–48. We will assume that espionage during peacetime can constitute covert political actions and lawful special forces military operations undertaken during war are not instances of covert political action. However, in doing so, we accept that these assumptions are stipulative; the boundaries in this area are vague.

typically engaged in the age-old strategy of covert political operations.[47] Historically, the tactics deployed in covert political operations have included assassination of the political leaders of such enemy states, targeted killing of terrorist leaders outside theatres of war, the financing of coups and other insurrectionary movements, and destabilizing enemy states by spreading disinformation and propaganda, deploying agent provocateurs, and so on.[48]

Some covert political operations, if they were done overtly, may well constitute acts of war and be taken as such. Assassinations of foreign leaders and orchestrations of coups are cases in point. On the other hand, some covert political operations, such as political espionage during peacetime, would probably not be regarded as acts of war, even if acknowledged by the offending nation-state.[49] Consider, for example, the revelations of US spying on the Chancellor of Germany, Angela Merkel. While it has certainly soured US-German relations it is not even close to triggering war between the US and Germany.[50] Moreover, espionage has not typically been regarded as a *casus belli* from a legal perspective.

Covert political operations are typically, but perhaps not necessarily, unlawful—at least in the nation-state against which they are directed, if not in international law. This is one reason why they are not conducted openly, though arguably not the main reason for covert political operations conducted in peacetime. Covert political operations outside war, while they may involve killings and the destruction of property, are typically designed to stop short of war. The whole point of such *covert* political operations is to weaken an enemy state, or defend oneself from being weakened, while plausibly denying that one is doing so, thereby averting outright war. It is, therefore, no accident that during the Cold War, in the shadow of nuclear war, the

---

[47] Loch Johnson, "The "Third Option," in *American Foreign Policy*," in *National Security Intelligence and Ethics*, ed. S. Miller, M. Regan, and P. Walsh (London: Routledge, 2021), 169–185.

[48] David L. Perry, *Partly Cloudy: Ethics in War, Espionage, Covert Action and Interrogation* (Lanham, MD: Scarecrow Press, 2009).

[49] The *Tallinn Manual* restricts espionage and cyber-espionage to activity conducted during war behind enemy lines (see Schmitt, *Tallinn Manual*, 158). Therefore, it holds that covert *remote* cyber information gathering in war is not cyberespionage but rather computer network exploitation (CNE). This definitional move seems somewhat artificial. Moreover, the term *exploitation*, which is frequently used in these contexts seems to us to be unhelpful. Why should not, for example, A exploit B's vulnerability by attacking B rather than merely stealing from B. For example, state A can exploit state B's internal dissension and invade B or exploit B's lack of cyberdefence by destroying B's physical ICT infrastructure. Perry excludes espionage from his definition of covert political actions (Perry, *Partly Cloudy*, 163). However, it is not entirely clear why he does so when other covert, politically motivated, *nonviolent* actions (e.g., telling lies, spreading disinformation or propaganda, and so on) are included.

[50] BBC News, "Angela Merkel: Spy Claims Test US Ties with Germany," *BBC News*, 18 November 2013. http://www.bbc.com/news/world-europe-24992485. Accessed 24/6/2014.

covert political operation was a favoured tactic of both the Soviet Union and the US.

Thus, with the assistance and under the influence of Moscow, if not under its direction, communist-controlled labour unions orchestrated a series of violent strikes in the late 1940s in key industries in France and Italy to destabilize the democratically elected governments. For their part, the CIA responded with various covert operations, including financing, and otherwise supporting anticommunist groups in these countries.[51]

In South America in the 1960s and 1970s—notably, in Chile, the CIA went much further and actively supported the overthrow of democratically elected President Allende, presumably on the grounds of his links with Cuba and with the Soviet Union. A further interesting case, mentioned by Perry, involved the British duping the US into believing Nazi Germany had a secret plan to attack the US.[52] The British did so by allegedly discovering a secret Nazi map. However, the map was a forgery by British Intelligence, made with a view to influencing the US to go to war against Germany.

While covert political operations are, and always have been, used by nation-states against terrorist groups, they are not so often used *by* terrorist groups. Terrorist groups are primarily interested in drawing attention to their crimes. Publicity is, as they say, the oxygen of terrorism. Accordingly, we suggest that many, if not most, cyberattacks by state actors against other state actors, especially cyberattacks by nuclear powers against other nuclear powers or their allies, can typically be appropriately regarded, not as acts of war, but rather as covert political operations—specifically, covert political cyberattacks—that stop short of war. We suggest that many of the cyberattacks emanating from China against the US (Mandiant), emanating from the US against, for example, Iran (Stuxnet), or from Russia against Estonia and other European states, can be so regarded. On the other hand, cyberattacks conducted by states against the civilian infrastructure of foreign states can be instances of state-terrorism, as in the case of some of the Russian cyberattacks against Ukrainian civilian electricity services mentioned above. One might question whether they were instances of *covert* political action. However, a state actor might wish to both explicitly deny that it was responsible for an unlawful attack, while simultaneously implicitly communicating that it was in fact responsible, to terrorize. Consider, in this connection,

---

[51] Perry, *Partly Cloudy*, 167f.
[52] Ibid, 166.

Russia's poisoning of ex-Soviet citizens now living in foreign states who spied for these foreign states (e.g., Sergei and Yulia Skripal).[53] In the latter case, Russia denied that it was responsible. Arguably, however, it was sending an implicit message to its own citizens not to spy on behalf of foreign states or risk being murdered. Accordingly, a terrorist action might also be a covert political action, although admittedly the two kinds of action stand in some tension.

Even though many cyberattacks are, as we have seen, covert political actions that are short of war, this is not to deny that there are cyberattacks that are, in fact, acts of war such as. Operation Orchard is an example. We are not denying the possibility of what would be quite literally cyberwar. Nor are we settling the difficult questions concerning the threshold settings for war. Moreover, there are many cyberattacks that are neither acts of war nor can be plausibly characterized as covert political operations. For example, many cyberattacks are simply crimes directed at corporations, carried out by criminals or criminal organizations for financial gain.

6.1.2 The Morality of Covert Political Cyberaction: 'Dirty Hands'

The actions that constitute the core of covert political actions are multifarious. As already mentioned, they include assassination of political leaders, targeted killings of terrorist leaders outside theatres of war, support for coups, sabotage, theft, spreading of disinformation, use of agents provocateurs, espionage, and so on. Aside from their political motivation, they have another thing in common: they are harmful actions, normally regarded as immoral. Moreover, covert political action, and therefore covert political cyberaction are typically illegal, either in terms of international or domestic law (or both).

In short, covert political actions, and therefore covert political cyberactions, are morally justified, if at all, by the greater good that they serve—specifically, the greater good that consists of the realization of their motivating political purposes. Naturally, the political purposes served by covert political actions do not necessarily morally justify these actions, and indeed in many cases the political purposes themselves are not morally

---

[53] Gordon Corera, "The Salisbury Poisoning," *BBC*, 4 March 2020. https://www.bbc.com/news/uk-51722301. Accessed 31/10/2023.

acceptable (e.g., covert operations conducted to further the expansionist political interests of the Soviet Union under Stalin).

However, the most appropriate moral category, or general description in the philosophical tradition, under which to categorize most covert political actions,[54] and therefore many, if not most, covert political cyberaction is, we suggest, that of so-called dirty hands.[55] Covert political action is typically a paradigm of dirty hands (although obviously many instances of dirty hands actions are not instances of covert political action), which means doing what is pro tanto morally wrong (and, typically, unlawful) in order to achieve some putative greater moral good, and in the case of covert political action—including covert political cyberactions—the greater moral good (it is assumed) of the relevant nation-state. This greater moral good of the nation-state is presumably its national security—as opposed to, for instance, its national interest, which might in some instance not be a good, objectively speaking (e.g., subjugation of a foreign country). The pro tanto moral wrongness of a dirty hands action typically consists in the fact that the action either: (1) deliberately inflicts serious harm on an innocent person or persons; or (2) deliberately inflicts serious harm on a culpable person or persons, but the harm is grossly disproportionate to their culpability; and/or (3) violates a morally justified law.[56] An instance of (1) might be police officers torturing an innocent girl in order to get her terrorist brother to provide information concerning a planned terrorist attack. A cyber example might consist of the following lethal cyberattack by State A as a deterrent in response to State B's prior lethal cyberattack: State A's intelligence agency launches a cyberattack on the computer-based control system of one of State B's dams, causing the floodgates to open, resulting in serious downstream flooding of villages and loss of *innocent* life. An instance of (2) might be police officers torturing a thief to discover the whereabouts of the proceeds of his crime without which proceeds a conviction would be unlikely. A cyber example might be a life-threatening denial of service cyberattack by State A's

---

[54] Though not all—not, for example, the 1981 US covert operation to rescue the US diplomats and other US citizens held hostage by Iran—its breach of Iranian sovereignty notwithstanding.

[55] For an influential treatment see Michael Walzer, "Political Action: The Problem of Dirty Hands," *Philosophy and Public Affairs* 2 (1973): 160–80. See, also, Seumas Miller, "Noble Cause Corruption in Politics," in *Politics and Morality*, ed. I. Primoratz (Basingstoke: Palgrave Macmillan, 2007), 92–112; and Seumas Miller, "Noble Cause Corruption in Policing," in *Corruption and Anti-Corruption in Policing: Philosophical and Ethical Issues* (Dordrecht: Springer, 2016), 39–52.

[56] Roughly speaking, a morally justified law is one that is promulgated by a legitimate legislature in a procedurally correct manner and is not morally unacceptable (e.g., by virtue of violating a fundamental moral right).

intelligence agency on the electrical power grid of State B, in the middle of winter to deter State B's public and private sector agencies from continuing their widely supported (among citizens of B) cybertheft of State A's national security and industrial secrets. An instance of (3) might be the targeted killing by the agents of State A of a terrorist leader, outside a theatre of war, in the jurisdiction of a legitimate state, State B. A cyber example might be the Stuxnet malware attack mentioned above. Notice that, in dirty hands scenarios, the dirty action might or might not be morally justified, all things considered. Either way, the dirty action is pro tanto a legal or moral wrong and the person seriously harmed has been wronged,[57] at least by virtue of having his or her legal rights violated.[58] Indeed, this being so, dirty hands actions are typically unlawful. Accordingly, an important question arises as to how those who engage in covert political action in a liberal democracy are to be held accountable.[59]

Here we need to distinguish dirty hands actions from lawful and morally justifiable, but nevertheless harmful actions. Presumably, the lethal and other harmful actions of soldiers in wartime, insofar as they comply with just war theory (both the jus ad bellum and the jus in bello), are not instances of dirty hands actions.[60] Nor are the harmful actions of police officers (e.g., the use of coercive force to effect an arrest), instances of dirty hands insofar as they comply with legally enshrined, community accepted, and objectively correct moral principles.[61]

If this is correct, then covert political action, and therefore covert political cyberaction, poses particular challenges, both for the standard law enforcement model and just war theory. On the one hand, covert political cyberaction is (more or less) by definition harmful action short of war; its

---

[57] And the law in question is a law that ought to exist (e.g., the 'dirty' action is a violation of sovereignty and sovereignty is morally desirable).

[58] So this person did not consent to being harmed; nor is it a harm of a kind and degree that the person could reasonably be expected to suffer in order to realize the greater good to which it is an effective and necessary means (e.g., as in the case of the use of coercive force by police to arrest a suspect who later turns out to be innocent). Moreover, the action was not in the person's interest, all things considered.

[59] Regan and Poole, "Accountability of Covert Action in the United States and the United Kingdom."

[60] Arguably, combatants on both sides are governed by a particularist principle of reciprocity according to which each combatant of State A is entitled to use lethal force against each combatant of State B, on condition each combatant of State B is entitled to use lethal force against each combatant of State A. See Miller, *Shooting to Kill*, Chapter 6; and Seumas Miller, "War, Reciprocity and the Moral Equality of Combatants," *Philosophia* (August 2023). https://link.springer.com/article/10.1007/s11406-023-00678-1.

[61] Miller and Blackler, *Ethical Issues in Policing*, 5–30.

raison d'être is typically to harm an enemy state without triggering war, and especially in the case of nuclear powers to avoid triggering nuclear war. Moreover, its remit in terms of national security might be somewhat wider than that of national defence understood as the territorial integrity and political independence of the nation-state. So, the application of just war theory is somewhat inappropriate. It largely misses its mark.

On the other hand, covert political cyberaction is (more or less) by definition unlawful—at least in the nation-state against which it is directed, and potentially under international law. Accordingly, there is a strong moral presumption against its use. Yet, for reasons elaborated below, it does seem morally justified on some occasions and in some areas—for example, cyberespionage (although espionage is not necessarily unlawful under international law). Moreover, its raison d'etre is not the enforcement of the law, as in the case of police work conducted by law enforcement agencies. It might be replied that some instances of offensive cyberaction against criminal organizations and nation-states engaged in state crimes might be reasonably regarded as law enforcement. Perhaps so. However, consider a scenario in which state A mounts a cyberattack against state B, as a deterrence to B conducting unlawful cyberattacks against A; or, alternatively, A mounts cyberattacks against criminal organization C, based in B as a deterrence to C conducting unlawful cyberactivities against A, and doing so in the context of B refusing to assist A to bring C to justice. This does not much look like law enforcement, since there might not be any relevant international law recognized by B, and even if there were, A's attack raises sovereignty issues. So, the application of the law enforcement model leaves such problems unresolved, and especially the problem of the apparent moral justifiability of many instances of covert political action, and therefore of covert political cyberaction, even though their unlawfulness and seeming irrelevance to, at least, conventional law enforcement activity.

Although admittedly the distinction is not clear cut, let us nevertheless distinguish between two species of covert political cyberaction—namely, covert political cyberattacks and cyberespionage. As mentioned above, cyberattacks do not include purely defensive measures, such as firewalls and password protection. Again, as mentioned above, cyberattacks, if successful, are harmful (directly or indirectly) in one or more of the following ways: (i) physical or psychological harm to human beings per se; (ii) physical destruction; (iii) cyberharm, for example, data destruction; and (iv) institutional harm.

As already stated, covert political cyberattacks are, in many cases—arguably, the paradigm cases—covert, unlawful, and harmful actions short of war, undertaken by one nation-state against another nation-state (or non-state political actor) for political purposes—quintessentially, the purpose of national security. Since such actions are typically unlawful, an immediate response might be as follows: (i) one's own government ought not to authorize covert political cyberattacks and one's own security agencies ought to cease to carry out such attacks; (ii) foreign governments who authorize covert political cyberattacks and their security agencies who carry them out ought to be investigated, and if appropriate prosecuted and punished in accordance with (presumably) international law.[62] In short, the law enforcement model ought to be relied on to deal with this problem.

Unfortunately, as argued above, in the case of covert political cyberactions, this law enforcement approach is not practicable, given the attribution problem and the current state of the international criminal justice system—including not simply the absence of morally justified, international law, but also the absence of effective enforcement mechanisms. This is not to say that it is not worth striving to bring into existence a more effective, morally justified, international criminal justice system in respect of cyberattacks in general and covert political cyberattacks in particular; quite the contrary, as in fact we suggest below. However, to reiterate, it is to say that full-blown renovation and application of the law enforcement model with respect to covert political cyberactions is not practicable at this stage in the development of the international order. Moreover, this suggestion fails to address the putative legitimacy of covert political action as a means of securing national security.

So, the question to be addressed is: can our own covert political cyberattacks be morally justified in an overall context in which other nation-states are routinely directing such attacks on us and on one another? In short, can covert political cyberattacks be morally justified in what is in effect a state of nature—a cyberstate of nature?[63]

The existence of this cyberstate of nature notwithstanding, covert political cyberattacks do need to be morally justified; we are not advocating a so-called realist view of the international order.[64] In particular, they need to be

---

[62] However, covert political action directed by A against a foreign state (B) may well not be unlawful in A even if the same kind of action would be unlawful in A if conducted by B against A. Moreover, some forms of covert political action might not be unlawful in international law. Espionage, including cyberespionage, has these features.

[63] If the use of both terms *cyber* and *nature* to describe a single state is not a contradiction in terms.

[64] Assuming realism is the view that conflict between nation-states is somehow outside morality.

justified, at least in the first instance, by recourse to some morally weighty political purpose. For example, it was not morally justifiable for Russia to launch a covert cyberattack on Estonia's ICT infrastructure merely because it judged it to be in its political interest to do so, much less to launch covert cyberattacks on the Ukraine in the context of its wrongful invasion of Ukraine. On the other hand, if the US finds itself under frequent and ongoing covert cyberattack from, say, China, and these attacks threaten to destroy or seriously disrupt key US ICT infrastructure, then the US may well be morally justified on self-defence grounds in responding in kind.

So, on the one hand, we confront a cyberstate of nature, and on the other, we are not absolved from the need to provide moral justifications for our own covert political cyberattacks. We suggest that a number of familiar moral principles remain in play, even though in a somewhat different form. The principles in question exist in both the criminal law (and are, therefore, in part constitutive of the law enforcement model) and just war theory, even though in somewhat different forms. First, there is the principle of self-defence. In criminal justice contexts, this paradigmatically involves personal self-defence. A single person (Defender) is attacked by another person (Attacker) and it is the life of Defender that he is morally entitled to defend, and defend by use of lethal force, according to the principle of self-defence. In the case of war, by contrast, it might be the territorial integrity or political independence of one nation-state that is under attack by an armed force from another nation-state, and it is this that the nation-state being attacked is morally entitled to defend by waging war, according to the principle of self-defence. However, in the case of cyberattacks (other than those species of cyberattack constituting cognitive warfare—see 6.2 below) what is to be defended is typically somewhat different from both these cases. It is, for instance, national infrastructure (e.g., an electrical power grid), or the integrity of an institution such as an electoral process that is under attack, but not necessarily human lives, territorial integrity or political independence—and this is what the nation-state in question is entitled to defend, according to the principle of self-defence. Certainly, the defence of national infrastructure and the integrity of institutions is a national security issue. On the other hand, cyberattacks by authoritarian states on foreign companies, which are carried out to acquire commercial-in-confidence information, and thereby provide their own domestic companies with an economic advantage (i.e., industrial espionage), might not constitute a national security issue.

258   CYBERSECURITY, ETHICS, COLLECTIVE RESPONSIBILITY

Secondly, there is the principle of necessity. This principle is relative to an end (e.g., necessary as a means to preserve one's life, necessary as a means to win the war, and/or necessary as a means to secure national security). In the paradigmatic criminal justice context Defender is attacked by Attacker, and Defender will be killed unless he kills Attacker. Accordingly, it is necessary for Defender to kill Attacker, since killing Attacker is the only means by which Defender can preserve his life. Or alternatively, it is necessary if it is not the only means, but the only means that Defender knows will be effective and will not put Defender at risk of serious injury. Accordingly, there is an implicit harm minimization principle according to which Defender should not kill Attacker, if there is another effective means of self-preservation (e.g., flight or injuring Attacker without killing him), which would not involve killing Attacker (i.e., would do less harm). In the case of a defensive war, it is territorial integrity or political independence that will be compromised or lost unless the attacked state responds by taking up arms. Accordingly, it is necessary to wage war since this is the only means by which territorial integrity or political independence can be preserved. Moreover, there is an implicit harm minimization principle in play according to which war should not be waged if less harmful means, such as internationally imposed economic sanctions, would be effective. However, the application of the necessity principle in respect of cyberattacks is somewhat different again. As we saw above, in the case of a cyberattack it might be the defence of a national infrastructure asset that is in question or that of data security, including commercial data. Thus, the principle of necessity might licence a tit-for-tat response in the form of a cyberattack on the attacking state's critical infrastructure, but not recourse to all-out war—although there is the problem of escalation. Moreover, consistent with the principle of necessity, the defensive response might be more harmful than the attack to which it is responding.[65] Further, in the case of cyberattacks, the potential alternative means are somewhat different. Thus, the tit-for-tat cyberattack on the attacking state's critical infrastructure might not be morally justified if, for example, various technical cybersecurity defensive measures, such as firewalls, would be effective. So, here again, there is an implicit harm minimization principle in play.

Thirdly, there is the principle of proportionality. As with principle of necessity, the application of this principle in relation to cyberattacks differs from

---

[65] Michael Gross also makes this point in response to the application of a legal principle of equivalence. See Michael Gross, "Proportionate Self-Defense in Unarmed Conflict," in *Soft War*, 225.

its application in criminal justice contexts, and for that matter, in conventional warfare, by virtue of the following considerations. For one thing, there are differing ends in play, and these are somewhat incommensurable: preservation of lives versus integrity of a democratic institution; political independence versus security of commercial-in-confidence data; security of personal data versus maintenance of critical infrastructure. For another, the potentially available set of means to realize each of these different ends is correspondingly diverse, even if there is some overlap and some analogies: lethal force versus internationally imposed economic sanctions versus offensive cyberattacks.

Fourthly, there is the principle of discrimination (sometimes referred to as the principle of difference), which amounts to some version of the general moral principle prohibiting the intentional harming of innocent third parties. This general principle can be applied in theatres of war, in law enforcement, in personal self-defence, and indeed in other analogous settings. We return to this issue below. Here we simply note that, while there is overlap, nevertheless collateral damage is typically different in cyberconflict than in personal self-defence scenarios or in conventional wars (e.g., corrupted data on computers versus physical injury—including on a very large scale in the case of war). Moreover, as we saw above, given the high degree of connectedness of users of the internet, and the dual-use character of much cyber-based infrastructure (see Chapter 7), it is extraordinarily difficult to avoid or contain collateral damage and harm from cyberattacks in the manner that the damage and harm caused by kinetic attacks using most forms of conventional kinetic weaponry can be contained. For instance, if State A launches a cyberattack on State B's computer-based electrical grid in response to State B's cyberattack on State A's computer-based electrical grid, innocent citizens of B will inevitably suffer. Indeed, such a cyberattack is scarcely possible without the *intention* to damage or harm innocent citizens. In this respect, cyberattacks seem more like biological attacks (see Chapter 7, Section 7.1). Nor is this problem resolved by recourse to culpable noncombatants.[66] If a computer-based electricity grid is rendered ineffective by a cyberattack undertaken in self-defence, and therefore, performed so as to cause serious damage and harm to the aggressor, then it is likely that most of those who suffer, including by losing their lives, will be vulnerable, innocent persons (e.g., the

---

[66] Miller, "Civilian Immunity, Forcing the Choice and Collective Responsibility."

sick, the impoverished, the very old and the very young).[67] Accordingly, compliance with the principle of discrimination will need to be considerably relaxed and/or rethought in cyberconflict in general, and in covert political cyberaction in particular, on pain of either suffering unacceptable levels of damage and harm from cyberattacks (i.e., threats to national security), without taking any effective self-defensive measures or else resorting to kinetic war as one's self defensive measure.

As is typically the case in states of nature, there is a principle of reciprocity in play in respect of covert political actions, including covert political cyberactions. We note that the principle of reciprocity is not normally taken to be constitutive of just war theory nor is it typically invoked by proponents of the law enforcement model.[68] Moreover, as mentioned above, the various moral principles constitutive of just war theory and of the law enforcement model take on a different form, or at least must be differently applied, in contexts of covert political action. And, of course, they are differently applied in war than in criminal justice contexts in peacetime.

Here we need to distinguish two different, although related, tasks. One task would be to identify or fashion a principle (or principles) of reciprocity with respect to covert political cyber action in the light of the moral good of the national security of any given single liberal democratic nation-state (and do so in conjunction with a continuing commitment to the other moral principles mentioned above—namely, self-defence, necessity, proportionality, and discrimination). A second related task would be to identify or fashion a principle (or principles) of reciprocity with respect to covert political cyber action in the light of the *collective* good of global cybersecurity, (and do so in conjunction with a continuing commitment to the principles of self-defence, necessity, proportionality, and discrimination). This second task is, in effect, an attempt to address the collective action problem posed by covert political cyber action for all nation-states or, at least, for liberal democratic nation-states operating in a cyber state of nature comprised of both liberal democratic states and authoritarian states

---

[67] Michael Gross's suggestion in "Proportionate Self-Defense in Unarmed Conflict" of proportionate self-defence in tandem with "direct attacks on civilian targets subject to the conditions of participatory liability" (226) does not resolve the issue of the innocents (i.e., those who do not have participatory liability).

[68] Although it is apparently a principle of international law. See Mark Osiel, *The End of Reciprocity: Terror, Torture and the Law of War* (Cambridge: Cambridge University Press, 2009). See also Gross, "Proportionate Self-Defense in Unarmed Conflict" in respect of the legal principle of equivalence.

(i.e., the task of moving out of this cyber state of nature or, more realistically, of mitigating it).

Accordingly, we suggest the following as a strategy: (1) the principle of reciprocity, at least in its retrospective form (of which more below), applies to covert political cyberattacks and its effect is to render some such attacks morally permissible, notwithstanding that these same attacks may well not be permitted under *some* of the more stringent conditions (e.g., the last resort condition, imposed by just war theory supposing it were to be applied to them) and certainly not under the Law Enforcement model (at least as it typically applies in criminal justice contexts in liberal democracies during peacetime); (2) a principle of discrimination applicable to covert political cyberattacks involving the use of soft (i.e., non-lethal) cyberweapons is in some respects less stringent than the one constitutive of just war theory, since innocents can be deliberately targeted by soft cyberweapons in some circumstances. This principle is also less stringent than the one which is generally applicable under the Law Enforcement model. Let us turn first to the principle of reciprocity.

We suggest that the type of principle of reciprocity relevant to our discussion here is what we might refer to as a particularist principle. There are, of course, multiple principles of reciprocity. Thus, the principle of universalizability (e.g., "only perform actions under maxims that could be universally adopted") and (relatedly) the 'Golden Rule' ("Do unto others as you would have others do unto you") are sometimes referred to as reciprocal in character. However, these latter principles are essentially universal or generalist principles that advocate, and have as a *constitutive* aim to generate, morally acceptable forms of behaviour that are to be common to everyone under most circumstances (e.g., everyone should tell the truth because lying is not universalizable since no one wants to be lied to). Therefore, these principles only advocate reciprocity in a weak, because relatively constrained, sense. By contrast, the principles of reciprocity that we have in mind are of the 'an eye for an eye, a tooth for a tooth', 'quid pro quo', or 'tit-for-tat' variety. These principles advocate reciprocity in a strong sense that allows one's actual behaviour to differ radically, depending on how one is treated by another, and it allows one's behaviour to be relatively morally unconstrained by content-based principles. For instance, a person is only obliged to tell the truth to those who tell the truth to him. According to these particularist principles of reciprocity, one should act towards each other person on a case-by-case basis, and in particular, based on whether the actions that the other person performs are harmful or beneficial to oneself. Thus, A should harm B if B harms A, and A should benefit B if B benefits A. Moreover, the reciprocal harms or benefits should be of the same type as the initiating ones, or, at least, they should be proportionate. So, reciprocity is a two-way relational (the reciprocity relation) particularist principle, rather than a generalist or universalist principle advocating morally acceptable forms of behaviour that are to be common to everyone under most circumstances.

Notwithstanding the above, we note that particularist principles of reciprocity can have ends beyond themselves, while retaining their particularist character.[69] They can do so because these ends are not constitutive of the reciprocity relation but are rather external to this relation. Moreover, these ends can be morally worthy ones. Thus a tit-for-tat principle might be deployed to establish an optimific equilibrium without ceasing, thereby, to be particularist in character.[70] We further note that the application of particularist principles of reciprocity can be scaled up in the sense that multiple agents in a social setting could adopt them in their interactions with one another. If so, these principles are likely to have become conventionalized or even legalized. Moreover, principles of reciprocity governing interactions between nation-states can be conventionalized or embodied in international law (e.g., the international law that a benefit given by A to citizens of B should be given by B to citizens of A). Further, the moral permissibility of complying with a convention or law of reciprocity depends not only on the moral features of the initial particularist principle of reciprocity which underpins it, but also on the moral consequences of the conventionalization or legalization of that principle. Hence a particularist principle of tit-for-tat revenge might be morally permissible for any single individual considered on his or her own (at least in theory), while widespread adoption of the principle, supposing it was to be conventionalized or even legalized, might be morally disastrous, given the great magnitude of harm consequent upon general compliance.

On one rendering of particularist principles of reciprocity they are essentially retrospective in form. These take their inspiration from the ancient prescription, "an eye for an eye and a tooth for a tooth."[71] As is well

[69] Miller "War, Reciprocity and the Moral Equality of Combatants."
[70] Miller, "Cyber-attacks and 'Dirty Hands.'"
[71] See also "Rule 9" in Schmitt, *Tallinn Manual*, 36–41.

known, this version is problematic; in any case, it is not the version we have in mind. Rather we propose, firstly, a version of a retrospective principle in which there is a restriction with respect to the purposes it is to serve. More specifically, a morally acceptable version of this retrospective principle would justify nation-state A engaging in covert political cyberattacks against nation-state (or non-state actor) B, in circumstances in which B had engaged or was engaging, in unjustifiable cyberattacks on A—but only if A's attacks were in the service of A's morally justifiable political purposes and national security.

So far so good. However, there is another version of the reciprocity principle that is salient. This version is prospective in form. It is a tit-for-tat principle in the service of bringing about a morally desirable future state of affairs situation.[72] The state of affairs in question is an equilibrium state among nation-states (i.e., a collective good—see Chapter 5, Section 5.1); more specifically, a morally justifiable equilibrium under the rule of international law. Moreover, this collective good is a state of affairs with respect to which there is a collective (joint) moral responsibility (see Chapter 5, Section 5.2), on the part of nation-states. This is not tit-for-tat in the service of the very general purpose of doing whatever is in one's political self-interest, legitimate or otherwise (in the manner of rational choice theories); nor is it tit-for-tat measures short of war in the service of the narrow purpose of averting a future large-scale lethal attack, which would constitute war (as might be justified under some extension of just war theory). Of course, in this equilibrium state of affairs, there would be no covert political cyberattacks, or at least they would be few and far between. So, this principle does not justify dirty hands actions in the manner of its sister retrospective principle; rather it has as its purpose to eliminate, or at least greatly reduce, dirty hands actions, and in this case covert political cyberattacks, thereby moving the international order out of its current cyberstate of nature and into a cybersocial contract (so to speak). However, the equilibrium that is its raison d'être is at best a

---

[72] So, this is not the same as the tit-for-tat principle deployed in rational choice theory, understood in terms of rational self-interested actors. Randall Dipert, "Ethics of Cyberwarfare," *Journal of Military Ethics* 9, no. 4 (2010): 384–410, suggests in passing the application of rational choice theory and its tit-for-tat principle to cyberconflict. In Miller's view, rational choice theory is useful up to a point as a descriptive theory, but not as a normative theory. Moreover, the practical reasoning required to move to the social contract presupposes joint action at some point among at least some of the main actors. Elsewhere, Miller has argued against the adequacy of rational-choice-based modes of practical reasoning in joint action. Seumas Miller, "Rationalising Conventions," *Synthese* 84 (1990): 23–41.

long-term goal; it is unlikely to be achieved anytime soon.[73] Or, perhaps it can be achieved in respect of, say, cyberattacks on critical infrastructure, but not in respect of cyberespionage or other less harmful forms of covert political cyber action.

We suggest that these two contrasting principles of reciprocity, one retrospective the other prospective, one relatively permissive (though less permissive than the prescription, "an eye for an eye and a tooth for a tooth") the other much less so, may well both be applicable to covert political cyberattacks.[74] If so, then there are moral justifications for covert political cyberattacks, other than that of self-defence.

The retrospective principle of reciprocity justifies the pursuit of one's morally legitimate political interests by means of dirty hands actions, including covert political cyberattacks, given the other side is pursuing their political interests by such means. So, it is relatively permissive and might encourage reciprocal attacks, which would not otherwise be justified. On the other hand, at other times, it may have a deterrent effect and discourage initiating attacks. At any rate, its application is unlikely to lead to a large reduction, let alone the elimination, of covert political actions in general or of covert political cyberattacks in particular.

The contrasting prospective principle of reciprocity justifies tit-for-tat covert political cyberattacks in the cyberstate of nature if they are undertaken in the pursuit of the cybersocial contract—that is, a future morally justifiable equilibrium state under the rule of international law in which dirty hands actions are eliminated or greatly reduced. So, it is far more restrictive than its sister principle, although it does permit present covert political cyberattacks if they deter future ones and are likely to lead to the cybersocial contract.

Having discussed various normative principles that ought to govern what is in our view a, if not the, fundamentally important species of cyberconflict—namely, covert political cyberaction—let us now turn in the following section to a related form of cyberconflict, cognitive warfare (prior to a consideration of hard cyberweapons and so-called autonomous weapons, in particular, in section 6.3).

[73] What we are calling the prospective reciprocity principle has a retrospective aspect insofar as its application is only triggered by a past (or present) attack.

[74] They may also conflict with one another. However, we do not see one dominating the other in all cases of conflict.

6.2 Cognitive Warfare

6.2.1 Characterizing Cognitive Warfare

Cognitive warfare has been defined in various ways. Here are a couple of influential definitions of what is meant by this term: "Cognitive Warfare is a strategy that focuses on altering how a target population thinks—and through that how it acts";[75] "the weaponization of public opinion, by an external entity, for the purpose of (1) influencing public and governmental policy and (2) destabilizing public institutions."[76]

Accordingly, cognitive warfare is a recent development that has emerged from prior related nonkinetic forms of warfare, such as psyops[77] and Information Warfare. In doing so it has relied heavily on new communication and information technologies—notably, AI. Key features of cognitive warfare include its targeting of entire populations (as opposed to, for instance, merely military ones in wartime), its focus on changing a population's behaviour by changing its way of thinking rather than merely by the provision of discrete bits of false information about specific issues (e.g., denying the extent of casualties in a kinetic war), its reliance on increasingly sophisticated psychological techniques of manipulation, and its aim of destabilizing institutions—notably, political institutions, such as democratically elected governments, but also epistemic institutions, such as news media organizations and universities. Importantly, cognitive warfare has been able to harness the new channels of public communication, such as social media, upon which populations have become increasingly reliant. Moreover, in some contrast with traditional ideological contestation (e.g., the ideological conflict between the Soviet Union and the West during the Cold War, in which each of the protagonists have a system or quasi-system of ideas (or ideology) to try to promote), cognitive warfare also has a very strong initial focus on sowing division and undermining cooperation in its target population by emphasizing existing differences and promoting polarizing views

---

[75] Backes and Swab, "Cognitive Warfare: The Russian Threat to Election Integrity in the Baltic States." A version of the material in this section appeared in Seumas Miller, "Cognitive Warfare: An Ethical Analysis" *Ethics and Information Technology* 25, no. 46 (2023): 1–10.

[76] Bernal et al., "Cognitive Warfare," 10.

[77] Michael Robillard, "Counter-Terrorism and PSYOP," in *Counter-Terrorism: The Ethical Issues*, ed. Seumas Miller, Adam Henschke, and Jonas Feltes (Cheltenham: Edward Elgar, 2021), 143–55.

(e.g., promoting both extreme left-wing *and* extreme right-wing views). In short, cognitive warfare makes heavy use of computational propaganda (see Chapter 4).

Cognitive warfare is likely to be more successful in the context of the already destabilizing effects of war, economic depression, pandemics, and other disasters or in a context of a preexisting polarized society (e.g., the UK in the context of Brexit, the US in the aftermath of the Global Financial Crisis, or the Middle East in the context of the Israel-Arab conflict). Hence Russia and China seized upon the opportunity of the COVID pandemic to increase their operations in cognitive warfare (e.g., to promote various conspiracy theories in the US population). Again, Russia infamously utilized Cambridge Analytica to sow discord in the US Presidential elections. Moreover, terrorist groups, such as Al-Qaeda and the Islamic State, have utilized cognitive warfare techniques to recruit disaffected youths in various liberal democratic and authoritarian states to their cause, and importantly to sow discord by getting their enemies to overreact, as in the case of the 9/11 attack of the Twin Towers, which proved to be a spectacular success for Al-Qaeda in terms of its visibility and prestige among disaffected and radicalized Muslims.

It is important to understand that cognitive warfare is taking place in preexisting social, institutional, and technological contexts in which there have already been destabilizing effects arising from the proliferation, on a massive scale, of disinformation, misinformation, conspiracy theories, propaganda, hate speech, and so on, much of which has not been done in the service of an explicit political purpose—though it may have served such a purpose inadvertently.

We also need to distinguish between, on the one hand, computational propaganda (e.g. disinformation, hate speech, and propaganda/ideology/quasi-ideology/groupthink) the content of which is explicitly or implicitly expressive of the political ideology of the communicator (e.g., extremist jihadist ideology communicated by members of Islamic State, right-wing Russian nationalism communicated by Russian state officials, the ideology of the Chinese Communist Party communicated by Chinese state officials), and on the other hand computational propaganda, the content of which is not thus expressive (e.g., antivaxxer conspiracy theories or right wing US nationalist quasi ideology communicated by *Russian* state officials to US audiences to sow discord in the US).

6.2.2 Cognitive Warfare: Defensive Measures

The challenges posed by the advent of cognitive warfare are considerable, not least for liberal democracies committed to ethical or moral values and principles, such as freedom of communication, democratic processes, the rule of law, evidence-based truth telling, and so on. Thus, while there is a need to curtail disinformation, there is a requirement that this be done without undermining freedom of communication. Again, there is a need to combat states engaged in cognitive warfare, but it is problematic for a liberal democratic state to do so by spreading its own self-serving disinformation or by seeking to manipulate citizens of authoritarian states. A further issue pertains to responsibility. Given the nature of cognitive warfare, there is a need for a variety of institutions, other than merely governments and security agencies, to shoulder responsibilities for combating cognitive warfare (e.g., to shoulder responsibilities for building resilience to disinformation, ideology, and the use of manipulative techniques). What precisely are these responsibilities and to which institutions ought they be allocated? Speaking generally, we suggest that there is a collective responsibility (understood as joint responsibility—see Chapter 4, Section 4.3.5, and Chapter 5, Section 5.3) on the part of multiple institutions (or, at least, the members thereof), including government, security agencies, media organizations and institutions of learning such as schools and universities.

In Chapter 4, we proposed a raft of countermeasures to combat computational propaganda. These included the following ones repeated here for convenience:

- Government to enact legislation to hold mass social media platforms, such as Facebook and Twitter, legally liable for illegal content, such as incitement and hate speech, on their platforms.
- Mandatory licensing of *mass* social media social platforms (e.g., monopolist or oligopolist platforms), to be introduced, with the licences to be held on the condition that the content on their platforms is compliant with legal requirements, their compliance or noncompliance to be determined and adjudicated by an independent statutory authority established by government (e.g., the Australian Office of e-Safety Commissioner).
- Lawful content that, nevertheless, fails to meet minimum epistemic and moral standards (e.g., is demonstrably false, *and* which is significantly

artificially—for example, by means of bots—or otherwise illegitimately *amplified*) is to be liable to removal by social media platforms, but only in accordance with the (publicly transparent) adjudications of the above-mentioned independent statutory authority. The minimum epistemic and moral standards in question to be established by the independent statutory authority, following on a process of public debate, expert input, and so on.

- Account holders with mass social media platforms can retain their anonymity as far as the public are concerned (with some exceptions—see next bullet point), but nevertheless are to be legally required to register with the independent statutory authority, which will then issue a unique identifier but only after verifying the identity of the account holder (e.g., by means of his or her passport, driver's licence and the like). This will enable them to be identified and prosecuted if they use their accounts to engage in unlawful online activity.
- Communicators of politically significant content (including, but not restricted to, content with national security implications) on mass media channels of public communication who have very large audiences (e.g., greater than one hundred thousand followers) will be legally required to be publicly identified, other things being equal.

These measures are all relevant to cognitive warfare. However, they are not sufficient to combat a hostile state engaged in cognitive warfare—and, for that matter, probably not sufficient, without some redesign of epistemic institutions, to combat computational propaganda in other settings (see Chapter 4). What more needs to be said about measures in liberal democracies needed to combat a hostile state engaged in cognitive warfare, such as in the case of Russia's computational propaganda campaign directed at the Ukraine, and China's directed at Taiwan?

Here we need to distinguish microlevel interpersonal speech, (e.g., John Brown speaking to Mary Smith on a street corner) from macrolevel speech utilizing mass media channels of communication. Moreover, we also need to distinguish two forms of such macrolevel speech. Firstly, there is *macrolevel socially directed speech* to a very large audience through mass media channels of *public* communication. Examples of this would be CNN news broadcasts and former US President Donald Trump's Twitter communications. Such communications reach audiences numbered in the millions, and they emanate from a single known source that is known to the members of the audience. Moreover, importantly, these communications are public in the sense that all of the above information is a matter of *mutual knowledge* to the communicators and members of the audience.[78] Thus, each individual communicator and audience member knows who the source is, what the communicative content is, and that everyone else in the audience knows this, and so on.

Secondly, there is macrolevel, *profile-based, individually targeted (microtargeted)* speech to millions via mass media channels of *ostensibly private* communication. This macrolevel speech might involve the use of bots to send millions of emails to selected individuals who are not necessarily aware that the same communications are being sent to millions of recipients and being sent (at least initially) from a single source, although automatically contextualized to what is known of the recipient. This form of macrolevel speech is favoured by computational propagandists, such as Cambridge Analytica, as we saw in Chapter 4.[79]

Clearly, as argued in Chapter 4, there is no moral right to engage in macrolevel, *profile-based, microtargeted* speech to millions via mass media channels of *ostensibly private* communication. Indeed, quite the reverse; there is a moral obligation on the part of governments to combat such speech, including by recourse to the means we suggested in Chapter 4 and briefly summarized above. However, it will also turn out that there is no moral right on the part of foreigners to engage in macrolevel socially directed speech to the domestic citizenry, and this has implications for banning, for instance, Russian mass media channels, such as RT. Accordingly, we are providing the justification for a policy advocated by David Sloss—namely, the banning of RT and like mass media outlets.[80] Before doing so, we need to get clearer on the notion of socially directed speech, a form of public communication.

Socially directed speech is speech in which the speaker speaks to the rest of the community qua member of that community and does so publicly in our

[78] The concept of mutual, or common, knowledge has been analysed extensively in the philosophical literature. See, for instance, N. V. Smith, ed., *Mutual Knowledge* (Cambridge, MA: Academic Press, 1982)

[79] There are other (i.e., other than the two distinguished here), more subtle forms of macrolevel communication that utilize mass media channels of public communication for propaganda, such as the so-called content farms favoured by China. These can consist of websites appealing to, for instance, a religious group known to have a large following in Taiwan, which is China's main propaganda target. These sites offer a wealth of useful, information to the religious adherents in question. However, Chinese ideology and selected facts are always embedded in the content of these websites. See Hung and Hung, "How China's Cognitive Warfare Works," 7.

[80] Sloss, *Tyrants on Twitter*.

above-described sense of public communication in terms of mutual knowledge. Here the community is to be loosely understood as a social group. So, it could be a small local community or a large national, or even international, community, or an academic, business, or political community (to name but a few instances of social groups in our loose sense of that term). Examples of socially directed speech include the UK Prime Minister making a national address, Dr Anthony Fauci appearing on CNN to encourage the US population to be vaccinated, and the mother of a black man slain by police pleading that the demonstrations in response be nonviolent.

What of a supposed moral right to engage in socially directed speech to millions via mass media channels of public communication (i.e., to engage in macrolevel socially directed speech)? There is, at least in principle, a moral right of citizen A qua member of A's political community to speak to the rest of A's political community. This is a liberty right in that if one person is exercising it at one time, then others may not be able to at that time, and indeed it may be that not everyone can exercise this right, even over a reasonably lengthy period; there are just too many citizens for this to be possible.[81] More specifically, in modern mass societies, the exercise of this liberty right requires access to mass media channels of public communication. But whereas mass media channels enable mass audiences, and everyone can be a member of a mass audience, they do not enable mass speakers to those mass audiences. It is not possible, even in principle, for *everyone*, or even most of the population, to reach a mass audience. Only a few can be mass communicators. There are too many citizens and too few channels of public communication for everyone to be a mass communicator. Accordingly, here as elsewhere, there is a need for a fair procedure to govern this liberty right, a fair procedure that might be difficult to find. However, in the case of a foreign state actor seeking to communicate to a domestic audience other than its own, there is no need to identify such a fair procedure since such a foreign actor does not possess the liberty right in question. Thus, Russian state actors (and Russians citizens more generally), do not have a moral right (specifically, a liberty right) to engage in macrolevel communication on politically significant matters to US citizens.

Naturally, foreign actors do not have a right to engage in *socially directed* communications in particular to members of a domestic audience other

---

[81] Leif Wenar, "Rights," Stanford Encyclopedia of Philosophy, 24 February 2020, https://plato.stanf ord.edu/entries/rights/.

than their own. After all, they cannot engage in socially directed action as is it defined above, given they are not members of the relevant community. However, it might be suggested that, notwithstanding that foreign actors do not have the above-described liberty rights (or, obviously, the right to engage in socially directed communications), nevertheless, foreign state actors have a *conditional* moral right to use channels of mass communication to publicly communicate to members of a domestic audience other than their own. The exercise of such a macrolevel moral right of foreign state actors (e.g., Russian state actors), supposing it exists, would be conditional on members of the domestic audience in question (e.g., US citizens) being prepared to grant them this right (i.e., to allow them to use the channels of mass communication in question and, presumably, to agree at least in principle to their communications). Here we need to invoke the concept of a joint right once again.

Consistent with the above, there is a joint moral right of members of a political community qua members of *that* community to give an audience to speakers who do not have a right to *socially directed* speech to them via mass media channels of public communication. Thus, US citizens have a joint right to (in effect) listen to Russian state actors on RT. However, this joint right carries with it the joint right *not* to do so. Thus, US citizens have a joint moral right to ban foreign state actors from using mass media channels of public communication, including social media, to publicly communicate politically significant messages to *them* (i.e., to US citizens. As is the case with other joint rights of members of the citizenry, this joint right can be exercised on behalf of the citizenry by their democratically elected representatives. In short, a liberal democratic government, such as the US government, has a moral right to ban foreign state actor from using mass media channels of communication to publicly communicate politically significant messages to the citizens of the liberal democracy in question and may have a moral obligation to do so if, for instance, the communications in question consist of computational propaganda. Indeed, if the foreign state in question is engaged in cognitive warfare, then there is a clear moral obligation to institute such bans. Accordingly, we agree with Sloss that Russian and China state actors' accounts with Facebook, Twitter and other big tech should be revoked, given that these actors have engaged in cognitive warfare with liberal democratic states, and specifically have engaged in computational propaganda campaigns aimed at undermining key institutions in liberal democratic states, such as the US and Taiwan.

It is important to note that this above-mentioned joint moral right with respect to macrolevel, socially directed, politically significant speech is consistent with the *microlevel interpersonal right* of each member of a community to listen to foreign state actors via channels of communication that are not mass media channels of public communication. Thus, the bans mentioned above would not apply to microlevel communications by Russian citizens based in Russia to US citizens based in the US. On the other hand, this microlevel interpersonal right is not an absolute right. As with most, if not all, moral rights it can be overridden under certain conditions. However, it is essentially the fundamental natural moral right of human beings to engage in free speech, and as such there is a strong presumption against infringing it, a presumption that can only be overridden by specific weighty moral considerations and not, for instance, by blanket appeals to national security.

6.2.3 Cognitive Warfare: Offensive Measures

Thus far we have concerned ourselves with defensive measures against cognitive warfare. It is now time to turn to a consideration of offensive measures. Naturally, as we saw in relation to countermeasures to cyberattacks more generally, in an overall context of self-defence, nonkinetic offensive measures against attackers are justified (supposing they are likely to be effective) by a principle of reciprocity (in its retrospective and/or prospective form).

Let us assume that the offensive measures in question are nonkinetic. If so, and if these are directed at culpable attackers, then it might be thought that there are few, if any, restrictions, other than the likelihood of effectiveness, and perhaps of compliance with a principle of reciprocity.[82] If certain members of an enemy state are spreading disinformation, propaganda, ideology and hate speech, and doing so by recourse to computational propaganda and other manipulative means, then the defender is morally entitled to do likewise, at least if the target audience consists of the culpable members of

---

[82] It is unclear whether a third-party state (C) has any obligation to use offensive cognitive warfare measures to intervene to defend members of a state (A) being subjected to unjustified cognitive warfare by members of a hostile state (B), by analogy with the obligation that C might have to use lethal force against B, if B was waging an unjust kinetic war against A. There is, presumably, an expectation that an individual or state can stand up for themselves verbally (so to speak), even if they cannot be expected to stand up for themselves physically. On the other hand, there may be issues of great imbalances of communicative reach by virtue of, for instance, B's possession of far more sophisticated mass communication technologies.

the enemy state in question. Perhaps so. However, two immediate problems arise at this point.

Firstly, these nonkinetic measures may have lethal or other kinetic effects characteristic of kinetic wars. Consider, for instance, the dissemination of disinformation, propaganda/ideology/quasi-ideology/groupthink and hate speech designed with a view to inciting violence (see Chapter 4). More generally, the use of cognitive warfare techniques cannot be insulated from their kinetic effects, and certainly not from their intended kinetic effects. Afterall, the whole point of engaging in cognitive warfare is ultimately to change behaviour.

Secondly, many of these nonkinetic measures will not be effective if they only target culpable attackers. Consider, for instance, propaganda comprising (in part) disinformation that is aimed at weakening the enemy's war effort (in the overall context of a kinetic war); the obvious target is the civilian population. Moreover, the application of the culpable/nonculpable distinction to cognitive warfare is problematic, and certainly does not mirror the relatively clear combatant/noncombatant distinction relied upon by just war theorists and others in relation to the use of lethal force in kinetic wars.

The application of the culpable/nonculpable distinction in cognitive warfare is problematic since, for instance, many civilian members of an authoritarian state the security agencies of which are engaging in cognitive warfare might support the cognitive war in the weak sense that they verbally endorse it to their friends and family, but are otherwise without influence and offer no material support. Moreover, in doing so they might themselves be unknowing victims of the disinformation and manipulative propaganda of the authoritarian state in question. Given that they are victims in this sense, perhaps they are not culpable. But, if so, how are they to be distinguished in practice from fellow citizens who differ only in that they are fully aware of the techniques of disinformation and manipulative propaganda being deployed by their security agencies and verbally endorse the use of these techniques? Members of the latter group are culpable (or more culpable than members of the former group) but, nevertheless, unable in practice to be distinguished from members of the former group.

Let us distinguish cognitive warfare conducted in the context of a kinetic war from cognitive warfare conducted in peacetime (i.e., conducted in circumstances in which there is no kinetic war). Thus, since the invasion of Ukraine by Russia in February 2022, Ukraine and Russia are engaged in a cognitive war in the context of a kinetic war. By contrast, Russia has waged a cognitive war of sorts against the US (e.g., by virtue of its efforts to inter-fere in the US Presidential elections, and sow discord more generally), but is not doing so in the context of a kinetic war being waged by Russia against the US. Arguably, in the context of the latter kind of case (i.e., a morally jus-tified, we assume, cognitive war being waged in peacetime by a liberal dem-ocratic state), it is not necessary, and may be counterproductive at least in the medium to long term, to resort to harmful offensive cognitive warfare measures that target nonculpable (or, at least, much less culpable) members of the hostile state in question. Rather the following threefold combination of measures is likely to be sufficient: (1) essentially *defensive* cognitive meas-ures (e.g., implementing the measures mentioned above to combat compu-tational propaganda, including banning the hostile state's propaganda on the channels of public communication in the defending state); (2) developing counternarratives to the hostile state's disinformation and manipulative propaganda but counternarratives that are not essentially false or manip-ulative, and therefore not *harmful* offensive measures; and disseminating these counternarratives in an ongoing, systematic manner to the hostile state's population; (3) deploying harmful offensive measures that target *cul-pable* members of the enemy state, as appropriate (e.g., using profile-based, microtargeting techniques to disseminate disinformation and manipulative propaganda to culpable actors in the hostile state, such as members of secu-rity agencies).

What of cognitive warfare undertaken in the context of a kinetic war (or perhaps the threat of a kinetic war)? Given that there is much more at stake in a kinetic war than in a purely cognitive war and given what is at stake is in the here and now, a loosening of the restriction to avoid using harmful offensive measures against nonculpable members of the belligerent state is justified. (As above, we assume the perspective of a liberal democratic state deter-mining its morally justified response to the morally unjustified use of cogni-tive warfare by a hostile state, albeit this time in the context of a kinetic war being justly waged by the liberal democratic state against the hostile, indeed belligerent, state.) In the context of a kinetic war the general principles of necessity and proportionality have a clear application. However, in the con-text of a kinetic war, the culpable/nonculpable distinction as it applies to the use of the methods of cognitive warfare has much less purchase. Of course, the culpable/nonculpable distinction is akin to the closely related moral and legal principle, the principle of discrimination, which has application to ki-netic wars. According to the principle of discrimination, noncombatants cannot be intentionally targeted, although it is allowable for them to be un-intentionally killed in military operations if those operations are compliant with the principle of military necessity and if the number killed is not dis-proportionate by the lights of the principle of proportionality. However, as we saw above, the principle of discrimination (or related principles) has much less purchase if the intended harm to noncombatants, or innocent (i.e., nonculpable) civilians otherwise demarcated, is not death or serious physical injury, as it might not be in the case of a cyberattack. Accordingly, intention-ally harming nonculpable citizens by means of a cyberattack that disabled their electronic communications systems, or by disseminating disinforma-tion to them, might be morally justified under some circumstances (e.g., if it did not indirectly cause death or serious physical injury).

The justification in question would rely on the following general considerations: (1) the nature of the harm done by the use of the (inher-ently morally wrongful, let us assume) cognitive warfare technique in ques-tion (e.g., creating false beliefs in nonculpable citizens that results in the undermining of their well-founded (initial) confidence in the ability of their security forces to win a kinetic war); (2) the use of the cognitive warfare tech-nique in question is effective, and there is no more effective, less harmful (all things considered) means available to achieve the morally weighty military or political end it serves;[83] and (3) the use of a morally wrongful means taken in conjunction with the harm done by it was not disproportionate relative to the moral weight attached to the military or political end ultimately achieved by this means (e.g., the morally weighty end of facilitating victory in the just kinetic war in question greatly outweighed the harm done by the cognitive warfare technique).

A final point pertains to deaths or serious injury to nonculpable citizens that might result from the use of techniques of cognitive warfare in the con-text of waging a just kinetic war. If these deaths or serious injuries were not intended, then the use of the cognitive techniques in question might well be morally justified by recourse to the principles of necessity and propor-tionality. Here there would be parity of reasoning with the morally justified, unintended killing of nonculpable citizens (or, at least, noncombatants) by combatants using lethal force in accordance with the principles of neces-sity, proportionality, and discrimination. If, on the other hand, the deaths

---

[83] Or a means that is as effective but less harmful, or almost as effective but much less harmful, and so on.

of, or serious injuries to, the nonculpable citizens were intended, then they would likely violate the principle of discrimination. However, in these latter cases involving intended deaths or injuries, there are likely to be moral complications arising from two factors. Firstly, there is an indirect (causal) relationship between the use of these cognitive techniques and the resulting deaths or serious injuries in question. Secondly, those who directly cause the serious death or injuries must themselves bear some (and perhaps full) moral responsibility for these death or injuries, even though they were acting based on beliefs and other attitudes to some extent induced by those who targeted them with the cognitive warfare techniques intended to produce such acts. Arguably, in these sorts of case there is joint moral responsibility; the users of the techniques of cognitive warfare and their targets are jointly morally responsible for the resulting deaths or injuries to the nonculpable citizens. The use of techniques of cognitive warfare successfully to incite violence against nonculpable citizens would be an example of this.

Having discussed the use of nonlethal techniques of cognitive warfare at some length it is now time to turn to a highly controversial lethal form of cyberweaponry—namely, autonomous weapons.

6.3 Autonomous Weapons

Autonomous robots are able to perform many tasks for more efficiently than humans (e.g., tasks performed in factory assembly lines, autopilots, and driverless cars). Moreover, they can perform tasks dangerous for humans to perform (e.g., defuse bombs).[84] However, autonomous robots can also be weaponized and in a manner such that the robots control their targets (and, possibly, the selection of their weapons). Further, by virtue of developments in artificial intelligence, the robots have superior calculative and memory capacity. In addition, robots are quite literally without fear in battle; they don't have emotions and care nothing for life over death.

New and emerging (so-called) autonomous robotic weapons can replace some military roles performed by humans and enhance others. Consider, for example, the Samsung stationary robot that functions as a sentry in the demilitarized zone between North and South Korea. Once programmed and activated, it has the capability to track, identify, and fire its machine

---

[84] Miller, *Shooting to Kill*, Chapter 10.

guns at human targets without the further intervention of a human operator. Predator drones are used in Afghanistan and the tribal areas of Pakistan
to kill suspected terrorists. While the ones currently in use are not autonomous weapons, they could be given this capability in which case, once
programmed and activated, they could track, identify, and destroy human
and other targets without the further intervention of a human operator.
Moreover, more advanced autonomous weapons systems, including robotic
ones, are in the pipeline.

In this section, we explore the moral implications of autonomous robotic
weapons by addressing the following questions. Firstly, in what sense are such
weapons really autonomous? Secondly, do such weapons necessarily compromise the moral responsibility of their human designers, programmers
and/or operators, and if so, in what manner and to what extent? Finally,
should autonomous weapons be prohibited?

6.3.1   What are Autonomous Weapons?

Autonomous weapons are weapons system that, once programmed and activated by a human operator, can—and, if used, do in fact—identify, track,
and deliver lethal force without further intervention by a human operator. By
programmed, it is meant, at least, that the individual target or type of target
has been selected and programmed into the weapons system. By *activated*
is meant, at least, that the process culminating in the already programmed
weapon delivering lethal force has been initiated. This weaponry includes
weapons used in nontargeted killing, such as autonomous antiaircraft
weapons systems used against multiple attacking aircraft; more futuristi-
cally, weapons used against swarm technology (e.g., multiple lethal minia-
ture attack drones operating as a swarm so as to inhibit effective defensive
measures); and weapons used in targeted killing (e.g., a predator drone with
face-recognition technology and no human operator to confirm a match).

We need to distinguish between so-called human in-the-loop, human
on-the-loop, and human out-of-the-loop weaponry. It is only human out-
of-the-loop weapons that are autonomous in the required sense. In the case
of human-in-the-loop weapons, the final delivery of lethal force (e.g., by a
predator drone) cannot be done without the decision to do so by the human
operator. In the case of human on-the-loop weapons, the final delivery of lethal force can be done without the decision to do so by the human operator;

however, the human operator can override the weapon system's triggering mechanism. In the case of human out-of-the-loop weapons, the human operator cannot override the weapon system's triggering mechanism; so, once the weapon system is programmed and activated there is, and cannot be, any further human intervention.

The lethal use of a human-in-the-loop weapon is a standard case of killing by a human combatant, and as such is presumably, at least in principle, morally permissible. Moreover, other things being equal, the combatant is morally responsible for the killing. The lethal use of a human-on-the-loop weapon is also in principle morally permissible. Moreover, the human operator is, perhaps jointly with others (such as his or her commander—see discussion below on collective responsibility as joint responsibility) morally responsible, at least in principle, for the use of lethal force and its foreseeable consequences. However, these two propositions concerning human on-the-loop weaponry rely on the following assumptions:

(1) The weapon system is programmed and activated by its human operator and either;

(2) (a) On each occasion of use the final delivery of lethal force can be overridden by the human operator and (b) this operator has sufficient time and sufficient information to make a morally informed, reasonably reliable judgement whether or not to deliver lethal force or;

(3) (a) On each occasion of use the final delivery of lethal force can be overridden by the human operator and (b) there is no moral requirement for a morally informed, reasonably reliable judgement on each and every occasion of the final delivery of force.

A scenario illustrating (3)(b) might be an antiaircraft weapons system being used on a naval vessel under attack from a squadron of manned aircraft in a theatre of war at sea in which there are no civilians present.

There are various other possible such scenarios. Consider one in which there is a single attacker on a single occasion in which there is insufficient time for a reasonably reliable, morally informed judgement on the part of the defender. Such scenarios might involve a kamikaze pilot or suicide bomber. If such weapons were to be morally permissible the following conditions at least would need to be met: (i) prior clear criteria for identification/delivery of lethal force to be part of the design of the weapon and used only in narrowly circumscribed circumstances; (ii) prior morally informed judgement regarding criteria and circumstances, and (iii) ability of operator to override system. Here there is also the implicit assumption that the weapon system can be 'switched off' entirely (as well as being able to be overridden on any single occasion considered on its own), as is not the case with, for instance, biological agents released by a bioweapon.

What of human out-of-the-loop weapons (i.e., autonomous weapons)? As mentioned above, these are weapons systems that once programmed and activated can identify, track, and deliver lethal force without further intervention by a human operator. They might be used for nontargeted killing in which case there is no uniquely identified individual target, such as in the above-described cases of incoming aircraft and swarm technology. Alternatively, they might be used for targeted killing. An example of this would be a predator drone with face-recognition technology and no human operator to confirm a match. However, the crucial point to be made here is that there is no human on-the-loop to intervene once the weapons system has been programmed and activated. Three questions now arise. Firstly, are these weapons systems autonomous in the full-blown sense of moral autonomy in common use in respect of many, if not most, freely performed, morally informed human actions? (We say "morally informed" since taking someone's life is a morally significant action, and therefore the person taking this life ought to be making a morally informed decision.) Secondly, are humans fully morally responsible for the killings done by autonomous weapons or is there a so-called responsibility gap? Thirdly, should such weapons be prohibited?

6.3.2 Moral Autonomy

Two sets of distinctions need to be kept in mind regarding the notion of an autonomous agent, whether human, Martian, or otherwise.[85] The first distinction is between rationality and morality. An autonomous agent is a rational agent. However, arguably, being rational is not a sufficient condition for autonomy. Rather, an autonomous agent needs also to be a moral agent.

The second distinction pertains to sources of potential domination. An autonomous agent is one whose decisions are not externally imposed; he or

---

[85] Stanley Benn, *A Theory of Freedom* (Cambridge: Cambridge University Press, 1988) and Seumas Miller, "Individual Autonomy and Sociality," in *Socialising Metaphysics: Nature of Social Reality*, ed. F. Schmitt (Lanham: Rowman & Littlefield, 2003), 269–300.

she is not dominated by external forces or other persons. However, an autonomous person is also possessed of self-mastery; he or she is not dominated by internal forces (e.g., addiction).

Autonomous agents are able to choose their ultimate ends: those ends that are not simply the means to further ends. Perhaps one's own personal happiness is an ultimate end chosen by many in individualistic social groups, although human beings can choose different ultimate ends (e.g., high social status, great political power, justice for the poor and downtrodden, the survival of future generations threatened by climate change, and so on). If a creature did not choose its ultimate ends, then those ends must surely have been brought about either by the intervention of some other creature, or by some inanimate causal process. Either way, the autonomy of the creature in question is compromised. Evidently, robots (or, at least, currently existing robots) cannot choose their ultimate ends since these are programmed or otherwised designed into them.

A further point about autonomous agents (in the sense of autonomous agents who are moral agents) pertains to moral emotions (e.g., caring deeply about what is worth caring about). Someone can be rational, up to a point, without necessarily being moral. Consider, for example, a highly intelligent psychopath. Such a person may well pursue their goals efficiently and effectively and make sophisticated evidence-based judgements in doing so. So evidently, psychopaths can be highly rational. However, psychopaths do not care about other people and are happy to do them great harm if it suits their own purposes. Moreover, psychopaths, even if they recognize the constraints of morality and pay lip service to them, do not feel the moral force of moral principles and ends. In short, psychopaths can be rational and yet are not moral agents. So, rationality and morality seem to be different, albeit related, concepts. More generally, moral agents are rational agents who are sensitive to moral properties in the sense that they recognize moral properties as such, feel the force of them, and respond appropriately to them. While robots are sensitive to physical properties (e.g., heat and light), and can comply with rules, including rules that are proxies for moral principles, they are not sensitive to moral properties qua moral properties. Accordingly, robots (or, at least, currently existing robots) are not moral agents.

Considering the above, we can now see that autonomous human beings are ones who decide for themselves what is important and valuable to them and possess the capacity to make reason-based choices on the basis of recognizing, assessing and responding to relevant considerations, including nonmoral facts, moral principles, and ultimate moral ends. When we call an act autonomous, we mean that it is something done by such a person, on the basis of such a response. As we have seen, robots are not autonomous beings in this sense.

There is a presumption that all human adults, at least, have achieved the status of autonomous beings (in the above sense). This presumption is defeasible. We may be able to show that a person is so deficient in various constitutive features of autonomy, such as rationality or self-mastery, that they should not be counted as autonomous, and that others might be justified in making decisions on their behalf. But absent such defeat, we all possess the status of autonomous human beings. By contrast, there is no such presumption to be defeated in the case of robots.

6.3.3 Moral Responsibility and Autonomous Weapons

Let us now return to *so-called* autonomous weapons: human out-of-the-loop weapons. We have seen that so-called autonomous robots, and therefore autonomous weapons, are not autonomous in the same sense that human beings are, since they do not choose their ultimate ends and are not sensitive to moral properties. However, the question that now arises concerns the moral responsibility for killings done by autonomous weapons. Specifically, do they involve a responsibility gap such that their human programmers and operators are not morally responsible or, at least, not fully morally responsible for the killings done using these weapons?[86]

Consider the following scenario, which we contend is analogous to the use of human out-of-the-loop weaponry. There is a villain who has trained his dogs to kill on his command and an innocent victim on the run from the villain. The villain gives the scent of the victim to the killer dogs by way of an item of the victim's clothing and then commands the dogs to kill. The killer dogs pursue the victim deep into the forest and now the villain is unable to intervene; the villain is 'out of the loop'. The killer dogs kill the victim. The villain is legally and morally responsible for murder. However, the killer dogs are not, even though they may need to be destroyed on the grounds of the risk they pose to human life. So, the villain is morally responsible for

---

[86] Ronald Arkin, "The Case for Ethical Autonomy in Unmanned Systems," *Journal of Military Ethics* 9 (2010): 332–41). Arkin argues in favour of the use of such weapons.

murdering the victim, notwithstanding the indirect nature of the causal chain from the villain to the dead victim. The chain is indirect since it crucially depends on the killer dogs doing the actual physical killing. Moreover, the villain would also have been legally and morally responsible for the killing if the scent was generic, and therefore carried by a whole class of potential victims, and if the dogs had killed one of these persons. In this second version of the scenario, the villain does not intend to kill a uniquely identifiable individual,[87] but rather one (or perhaps multiple) members of a class of individuals.

By analogy, human out-of-the-loop weapons—so-called killer-robots— are not morally responsible for any killings they cause.[88] Consider the case of a human in-the-loop or human-on-the-loop weapon. Assume that the programmer/activator of the weapon and the operator of the weapon at the point of delivery are two different human agents. If so, then other things being equal, they are jointly (that is, collectively) morally responsible for the killing done by the weapon, whether it be of a uniquely identified individual or an individual qua member of a class.[89] No one thinks the weapon is morally, or other than causally, responsible for the killing. Now assume this weapon is converted to a human out-of-the-loop weapon by the human programmer-activator. Surely this human programmer-activator now has full individual moral responsibility for the killing, as the villain does in (both versions of) our killer dog scenario. To be sure, there is no human intervention in the causal process after programming-activation. But the weapon has not been magically transformed from an entity only with causal responsibility to one which now has moral responsibility for the killing.

It might be argued that the analogy does not work because killer dogs are unlike killer-robots in the relevant respects. Certainly, dogs are minded creatures whereas computers are not; dogs have some degree of consciousness and can experience, for example, pain. However, this difference would not favour ascribing moral responsibility to computers rather than dogs. If anything, the reverse is true. Clearly, computers (or, at least, currently

---

[87] It is not a targeted killing.

[88] See R. Sparrow, "Killer Robots," *Journal of Applied Philosophy* 24 (2007): 63–77. For criticisms see Uwe Steinhoff, "Killing Them Safely: Extreme Asymmetry and Its Discontents," in *Killing by Remote Control: The Ethics of an Unmanned Military*, ed. B. J. Strawser (Oxford: Oxford University Press, 2013), 179–210. See also Christian Enemark, *Moralities of Drone Violence* (Edinburgh: Edinburgh University Press, 2023), 167–72.

[89] Moreover, each is fully morally responsible. Not all cases of collective moral responsibility involve a distribution of the quantum (so to speak) of responsibility.

existing computers) do not have consciousness, cannot experience pain or pleasure, do not care about anyone or anything (including themselves), and as we saw above, do not choose their ultimate ends—and more specifically, cannot recognize moral properties, like courage, moral innocence, moral responsibility, sympathy, or justice. Therefore, they cannot act for the sake of principles or ends understood as moral in character, such as the principle of discrimination. Given this apparent nonreducibility of moral concepts and properties to nonmoral ones, and specifically physical ones,[90] computers or, at least, the current generation of computers, can at best only be programmed to comply with some nonmoral proxy for moral requirements. For example, "Do not intentionally kill morally innocent human beings" might be rendered as "Do not fire at bipeds if they are not carrying a weapon or they are not wearing a uniform of the following description." However, here as elsewhere, the problem for such nonmoral proxies for moral properties is that when they diverge from moral properties, as seemingly they inevitably will in some circumstances, the wrong person will be killed or not killed, as the case may be (e.g., the innocent civilian wearing camouflage clothing to escape detection by combatants on either side and carrying a weapon for personal protection is killed, while the female terrorist concealing a bomb under her dress is not).

Notwithstanding the above, some have insisted that robots are minded agents; after all, it is argued, they can detect and respond to features of their environment and in many cases, they have impressive storage/retrieval and calculative capacities. However, this argument relies essentially on two moves that are highly controversial. Firstly, rational human thought—notably, rational decisions and judgements—are downgraded to the status of mere causally connected states or causal roles (e.g., via functionalist theories of mental states). Secondly, and simultaneously, the workings of computers are upgraded to the status of mental states (e.g., via the same functionalist theories of mental states). For reasons of space, we cannot here pursue this issue further. Rather, we simply note that this simultaneous downgrade/upgrade faces prodigious problems when it comes to the ascription of (human-like) autonomous agency. For one thing, human autonomous agency involves the capacity for moral reflection (e.g., the generation of novel moral ideas) and it is unclear how a being

---

[90] The physical properties in question would not only be detectable in the environment but also would be able to be subjected to various formal processes of quantification, and so on.

that is insensitive to moral properties could engage in such reflection. For another, to reiterate, computers do not choose their own ultimate ends. At best, they can select between different means to the ends programmed into them. Accordingly, they cannot be autonomous agents. So, while killer robots are morally problematic, this is not for the reason that they are autonomous agents in their own right— but this brings us to our third and final question.

6.3.4 Prohibition of Autonomous Weapons

Our final question concerns the prohibition of autonomous weapons in the sense of human out-of-the-loop weapons. This question should be seen in the light of our conclusions that such weapons are not morally sensitive agents, and their use does not involve a responsibility gap. Rather, there are multiple human actors implicated in the use of autonomous weapons: there is collective moral responsibility in the sense of joint individual moral responsibility.[91] The members of the design team are collectively, which is to say jointly, morally responsible for providing the means to harm (i.e., the weapon) and the political and military leaders and those who follow their orders are collectively (jointly) responsible for these weapons being used against a certain group or individual. Those who follow these orders include intelligence personnel who are responsible for providing the means to identify targets, and the operators who are responsible for its use on a given occasion, since they programmed/activated the weapons system. Moreover, all the above individuals are collectively—in the sense of jointly—morally responsible for the deaths resulting from the use of the weapon, but they are responsible to varying degrees and in different ways. For instance, some provided the means (i.e., designed or manufactured the weapon), others gave the order to kill a given individual, and still others pulled the trigger. These varying degrees and varying ways are reflected in the different but overlapping collective end content of their cooperative or joint activity. Thus, the designers and manufacturers of this weapon have as a collective end that the weapon be used and that, if the weapon is used, then some combatants in some war will be killed; after all, the designers and manufacturers are producing the weapon for precisely this end. Accordingly, other things

---

[91] Miller, "Collective Moral Responsibility."

being equal, they have a share, jointly with others, in the collective moral responsibility for the outcome when it is so used. Again, a military leader, in issuing orders to his subordinates who are responsible for operating the weapon, that they use the weapon to kill enemy combatants in this theatre of war at this time, has, in conjunction with his subordinates, the collective end that the enemy combatants be killed. Accordingly, other things being equal, the leader and his subordinates are collectively moral responsible for this outcome.

It is important to note that each contributor to such a joint lethal action is individually morally responsible for his/her own individual action contribution (e.g., an individual weapons operator who chose to deliver lethal force on some occasion or perhaps, in the case of an on-the-loop weapon, not to override the delivery of lethal force by the weapon on this occasion). This is consistent with there being collective (joint) moral responsibility for the outcome (e.g., the death of an enemy combatant and the death of innocent civilians).

It is also important to note the problem of accountability that arises for morally unacceptable outcomes involving 'many hands' or joint action and indirect causal chains. Consider, for example, an out-of-the-loop weapon system that kills an innocent civilian rather than a terrorist because of mistaken identity and the absence of an override function when the mistaken identity is discovered at the last minute. The response to this accountability problem should be to incorporate institutional accountability into the weapon's design. Thus, in our example the weapons designers ought to be held jointly institutionally, and therefore jointly morally responsible for failing to include an override function, which is a failure to ensure the safety of the weapon system. Likewise, the intelligence personnel ought to be held jointly institutionally, and therefore, jointly morally, responsible for the mistaken identity. Analogous points can be made with respect to the political and military leaders and the operators.

As we have seen, human-*out*-of-the-loop weapons can be designed to have an override function and an on/off switch controlled by a human operator. Moreover, in the light of our above example and like cases, in general, autonomous weapons ought to be provided with an override function and on/off switch. Indeed, to fail to do so would be tantamount to an abnegation of moral responsibility. However, against this it might be argued that there are *some* situations in which there ought not to be a human on the loop (or in the loop).

Let us consider some candidate situations involving human-*out*-of-the-loop weapons that might be thought not to require a human in or on the loop.

(1) Situations in which the selection of targets and delivery of force cannot in practice be overridden on all occasions and in which there is no requirement for a context dependent, morally informed judgement on all occasions (e.g., there is insufficient time to make the decision to repulse an imminent attack from incoming manned aircraft and there is no need to do so since the aircraft in a theatre of war are clearly identifiable as enemy aircraft).

(2) Situations in which there is a need only for a computer-based mechanical application of a clear procedure (e.g., to deliver lethal force), under precisely specified input conditions (e.g., when an object being tracked is identified as an enemy submarine by virtue of its design and other features) in which there is no prospect of collateral damage (e.g. in open seas in the Arctic).

However, even in these cases it is difficult to see why there would be an objection to having a human on the loop (as distinct from in the loop) especially since there might still be a need for a human on the loop to accommodate the problems arising from false information or unusual contingencies. For instance, the enemy aircraft or submarines in question might turn out to be ones captured and operated by members of one's own forces. Alternatively, one's own aircraft and submarines might now be under the control of the enemy (e.g., via a sophisticated process of computer hacking), and therefore should be fired upon.

A further argument in favour of autonomous weapons concerns human emotion. It is argued that machines in conditions of war are superior to humans by virtue of not having emotions since stress/emotions lead to error. Against this it can be pointed out that human emotions inform moral judgement, indeed specifically moral emotions, such as caring for others, are in part constituted by moral judgement, and moral judgement is called for in war. For instance, the duty of care with respect to innocent civilians relies on the emotion of caring and is in part constitutive of it; but caring is not a property possessed by robots. Moreover, human stress/emotions can be controlled, and if controlled, utilized, to a considerable extent. Thus, combatants should not be combatants if not appropriately selected and trained, and the influence of stressors can be reduced (e.g., by requiring some decisions to be made by personnel at some distance from the action).

The upshot of this discussion is that human out-of-the-loop weapons are neither necessary nor desirable. Rather, autonomous weapons should always have a human on-the-loop (if not in-the-loop). Moreover, not to do so would be an abnegation of responsibility. Accordingly, autonomous weapons in the sense of human out-of-the-loop weapons should be prohibited.

7

Individual and Collective Responsibility for Cybersecurity

Webs of Prevention

Most cybersecurity issues require individual users, organizations, and security agencies to take responsibility for various aspects of cybersecurity. This is especially the case given that any weak link in the security chain has the potential not only to impact the person responsible for that link (e.g., the person's personal computer), but may impact others if, for instance, the person in question's machine is utilized for attacks on others within one's organization, or indeed other linked organizations, or some remote and unrelated machine through a DDoS attack. In short, there is a need for a so-called web of prevention if the security of not only particular individuals and organizations is to be maintained, but also that of critical infrastructure, such as power grids, upon which multiple individuals and organizations rely. Clearly, such a web of prevention does not currently exist, although fragments of it certainly do (e.g., at the level of particular organizations).

This chapter explains the concept of a web of prevention and how it might apply to preventing and countering cyberthreats. In doing so, it must be kept in mind that state actors, in particular, are not simply engaged in cooperating with one another in order to prevent cyberattacks from malevolent non-state actors, but are also, quite often, themselves engaged in cyberconflict (i.e., mounting cyberattacks against one another—see Chapter 6). It might be suggested that, inevitably, there needs to be webs (plural) of preventions. Evidently, this is so. Accordingly, although we now live in an interconnected global cyberdomain, and therefore it might seem to make sense to speak of a single global web of prevention, unfortunately, the reality of serious and ongoing cyberconflict between the most powerful state actors, including the US and China, means that there is unlikely to be a single global web of prevention. Rather, there can only be, at least for the foreseeable future, multiple interconnected, but in some respects, inconsistent (by virtue of this

Cybersecurity, Ethics, and Collective Responsibility. Seumas Miller and Terry Bossomaier, Oxford University Press.
© Oxford University Press 2024. DOI: 10.1093/oso/9780190058135.003.0008

cyberconflict between state actors inter alia) webs (plural) of prevention. In this important respect, the concept of a web of prevention to prevent and counter cyberthreats is unlike, for instance, the concept of a web of prevention to prevent and counter biological threats (see 7.1 below). However, in this chapter for ease of exposition we will sometimes refer to a single web of prevention against cyberthreats.

There is a second important difference between a web (or webs) of prevention against cyberattacks and a web of prevention to prevent and counter biological threats. In the latter case, it might reasonably be held that it is morally unacceptable to use offensive biological weapons, even in response to a biological weapons attack. By contrast, it is evident that the use of offensive cyberweaponry may well be entirely morally justified, as is the case with conventional weapons. Accordingly, a web of prevention against cyberattacks needs to include an offensive cybercapability, both as a deterrent (and, therefore, as a preventative measure), and as a weapon to be used in self-defence. As a weapon used in self-defence, it is in part a preventative measure: it prevents future enemy attacks. However, it is not entirely preventative in that, if used, it is also a destructive weapon of attack. We will assume in this work that cyberweapons, if used only in justified self-defence or in compliance with our prospective principle of reciprocity (see Chapter 6, Section 6.1.2), are part of a web of prevention, even if this is a somewhat strained use of the term *prevention*.

In this chapter, we identify a number of features of a technologically and institutionally based web of prevention against cyberthreats. Such a web institutionalizes the individual and collective moral responsibilities, the discharging of which is required to protect individuals, organizations, critical infrastructure, and in the case of cognitive warfare, political and other institutions. In short, a web of prevention is required to protect the nation-state and its constitutive individuals and collectives against cyberattacks. We note that whereas our principal concern in this work is with cybersecurity, as opposed to cybersafety, in practice a web of prevention needs to undertake both tasks and do so in a coordinated manner. An important reason for this is the close relationship between safety and security. Accidents (a safety issue) occur in part because of negligence. However, harm caused by *culpable* negligence is, we suggest, a security issue.[1] Moreover, in the case of dual-use cybertechnology (technology used for both good and evil—see 7.2 below),

---

[1] Miller, *Dual Use Science and Technology, Ethics and Weapons of Mass Destruction*, 9.

safety and security issues are very closely connected; indeed, necessarily intertwined. Dual-use cybertechnology, by definition, not only poses a very significant safety risk it can also be used by malevolent actors to cause great harm (i.e., dual-use cybertechnology poses both a security and a safety risk).

Since dual-use cybertechnology provides great benefits as well as grave safety and security risks, and since it is pervasive, it creates a particular challenge in the design of a web of prevention against cyberthreats. Naturally, an adequate web of prevention against cyberthreats needs to ensure that in providing protection against cyberthreats it does not unnecessarily or disproportionately undermine the benefits that cybertechnology in general (including dual-use cybertechnology) provides. However, in the case of dual-use cybertechnology, given its dual-use character, this problem of balance is acute.

Moreover, the precise nature of the problem has not been assisted by a lack of clarity in relation to the definition of dual-use technology.[2] In a general sense, as the earlier chapters in this book have demonstrated, cybertechnology can be used to provide great benefits to humankind, but also to cause great harm, including the creation of weapons that cause great harm, as in the case of predator drones (see Chapter 6, Section 6.3). However, there are more restricted (and more helpful) senses of dual-use science and technology than this. In one very restricted sense, certain forms of cybertechnology can cause great harm by virtue of being capable of being used not only for peaceful purposes but also as weapons, even though, as a dual-use cybertechnology, it was not specifically designed as a weapon. Examples of this are self-replicating programs that install themselves in computers for beneficial purposes but can also function as viruses (see section 7.2.2.1 below). In a second, less restricted sense of dual-use technology, the technology can be used as a weapon. An example of this is encryption used in ransomware attacks (see section 7.2.2.3 below). Naturally, a web of prevention against cyberthreats needs to be designed in such a manner as to afford protection against purpose-built cyberweapons as well as against threats emanating from misused dual-use cybertechnology. We focus on the issue of dual-use cybertechnology in detail in 7.2 below.

However, we begin this chapter with a description of the concept of a technologically and institutionally based web of prevention and its relationship to the concept of collective moral responsibility. In doing so, we

---

[2] Ibid., Chapter 2.

introduce some theoretical notions that underpin this concept of a web of prevention—namely, multilayered structures of joint action, joint institutional mechanisms, and chains of collective institutional responsibility. There follows the section on dual-use cybertechnology. This completes this chapter and sets the stage for the conclusion to this book, which sets forth a set of ethical guidelines in relation to some of the key features of a web of prevention against cyberthreats.

7.1 Technologically and Institutionally Based Webs of Prevention

As its name implies, technologically and institutionally based webs of prevention against threats consist of an integrated mix of technological and institutional components. For instance, in the case of a web of prevention against the COVID-19 pandemic, the components would include vaccines (biotechnological), QR code or Bluetooth (technological) tracking for detecting and informing close contacts, and regulations (institutional) imposing testing and quarantine requirements. These components, among others, need to be combined and integrated in a manner that efficiently and effectively realizes the overall purpose that is the raison d'être of the web of prevention (e.g., eliminating or substantially curtailing the COVID-19 pandemic). Importantly, this overall purpose is a collective moral responsibility (in a complex sense, to be elaborated below). Accordingly, the web of prevention is not only constituted by technological and institutional components, it is justified by, and indeed is the embodiment of, a collective moral responsibility to discharge an overarching moral purpose (e.g., to save lives and livelihoods on a very large scale).

The concept of a web of prevention was introduced some decades ago in relation to biological threats.[3] More recently, Novossiolova et al. had this to say concerning such a web of prevention:

---

[3] See, for instance, Rappert and McLeish, *A Web of Prevention*; Seumas Miller, "Moral Responsibility, Collective Action Problems and the Dual Use Dilemma in Science and Technology," in *On the Dual Uses of Science and Ethics*, ed. Brian Rappert and Michael Selgelid (ANU Press, 2013); Graham Pearson, "The Web of Prevention," in *Preventing Biological Threats: What You Can Do. A Guide to Biological Security Issues and How to Address Them*, ed. S. Whitby, T. Novossiolova, G. Walther, and M. Dando (Bradford, UK: Bradford Disarmament Research Centre, University of Bradford, 2015); Miller, *Dual Use Science and Technology, Ethics and Weapons of Mass Destruction*; Jonas Feltes, *CBRN Threats, Counter-Terrorism and Collective Moral Responsibility* (Delft, Netherlands: TU Delft, 2021).

It is evident, not least from the ongoing COVID-19 pandemic that biological threats regardless of their origins constitute a major security concern.
Biological threats are complex and multifaceted and hence, their effective
prevention and countering require multiple lines of collaborative action and
sustained cross-sectorial coordination. It is helpful to think of this required
approach as an integrated and comprehensive web of prevention in which
the efforts aimed at preventing the accidental release of biological agents or
toxins, including naturally occurring disease and the efforts aimed to prevent the deliberate release of biological agents and toxins and the misuse
of life sciences are complementary and reinforce each other . . . against the
backdrop of the dynamic international security context, on the one hand,
and the rapidly evolving life science landscape, on the other. In doing so, it
underscores the utility of the concept of a web of prevention for elucidating
the need for continuous interaction between the international biosafety
and international biosecurity regimes, in order to ensure that the life sciences are used only for peaceful purposes.[4]

We suggest that just as in the case of biological threats there is a need to design and implement a web of prevention against cyberthreats. Naturally,
cyberthreats and biological threats are different, and therefore an adequate
web of prevention against biological threats would not suffice as an adequate
web of prevention against cyberthreats. First, as mentioned above, ongoing
cyberconflict between the most powerful nation-states eliminates the option of a single global web of prevention against cyberthreats.[5] Second, offensive use of cyberweapons is potentially morally justified as a preventative
measure, whereas this is, arguably, not so in the case of biological weapons.
Third, cyberthreats consist of human artefacts or, in the case of computational propaganda, communicative actions; they are not biological phenomena. Nevertheless, there are important analogies to be drawn between
biological and cyberthreats, and therefore general lessons to be learnt in the
design and implementation of a web of prevention against cyberthreats from

[4] Tatyana A. Novossioloval, Simon Whitby, Malcolm Dando, and Graham S. Pearson, "The Vital
Importance of a Web of Prevention for Effective Biosafety and Biosecurity in the Twenty-First
Century," *One Health Outlook* 3, no. 17 (2021): 3–17. https://doi.org/10.1186/s42522-021-00049-4.
[5] The existence of state-based biological warfare programs means that this distinction between
cyberthreats and biological threats should not be overstated. However, there does seem to be a realistic chance of eliminating biological warfare between state actors, whereas this is not a realistic
option in the case of cyberwarfare.

the work already done on a web of prevention against biological threats—or so we suggest.

There are several features of biological threats and the web of prevention against biological threats, that are relevant to preventing cyberthreats. First, as is the case with biological threats (see above quotation), cyberattacks are a "major security concern" that is "both complex and multifaceted" (e.g., ransomware, disinformation, computational propaganda, malevolent nonstate actors, malevolent state actors). Importantly, as is the case with a dangerous biological agent such as the SARS-CoV-2 virus that causes COVID-19, computer viruses can potentially, and very rapidly, infect any computerized device that is connected to the internet, or indeed any computer network—including not simply human operated computers but many computerized devices, such as fridges and other household appliances, not operated by human beings. This interconnectedness of computers and computerized devices via the internet, and in particular the explosion of computerized devices connected to the internet but not operated by human beings (i.e., the so-called Internet of Things, IoT), brings with it considerable benefits. For instance, smart fridges can request supplies via the resident's smartphone when supplies run low and when the fridge determines that the resident is close to a suitable store, reducing unnecessary shopping travel; or, driverless trucks in mining operations can reduce costs and safety risks, and so on. However, the advent of the IoT has also brought with it huge security risks analogous, by virtue of this large-scale interconnectedness, to the risks associated with pathogens such as SARS-CoV-2. Consider, for instance, that a terrorist might hack into the driverless truck's computer system and divert it to drive into a crowd of people (see 7.2 below) or that an authoritarian leader might order his cyberoffence team to hack into the controllers of the floodgates of the largest dam in a neighbouring democracy.

Second, strategies of cyberthreat prevention and countering are analogous to those to be used in response to biological threats. Thus: "effective prevention and countering requires multiple lines of collaborative action" (e.g., technological and regulatory) and "sustained cross-sectorial coordination" (e.g., criminal justice and national security agencies), business organizations, and epistemic institutions (e.g., universities and news media organizations). Third, the required approach is that of "an integrated and comprehensive web of prevention" (i.e., a technologically and institutionally based web of prevention). Fourth, this web of prevention needs to address cybersecurity concerns against the backdrop of a "dynamic international security context" (e.g., the rise of a powerful authoritarian China), a resurgent belligerent Russia (e.g., in relation to the Ukraine), international terrorism, a global pandemic, under-regulated global big tech companies (e.g., Facebook and TikTok), the destabilizing effects of disinformation, and more generally computational propaganda and a rapidly evolving cybertechnology landscape (e.g., IoT, cloud computing, machine learning (ML), facial-recognition technology). Fifth, the safety and security concerns in relation to cyberthreats are interconnected as they are in relation to biological threats (e.g., the use of dual-use technology in both the biological sciences and in cybertechnology).[6] Accordingly, the web of prevention needs to address both safety and security concerns in a manner that involves "continuous interaction" between cybersafety and cybersecurity regimes. Sixth, as is the case with the web of prevention against biological threats, the web of prevention against cyberthreats needs to operate at several levels (e.g., individual, social, organizational, national, and global). Moreover, at the national and global levels there is a need to address multiple somewhat disparate, yet interconnected, issues including, for instance, the regulation of social media in relation to computational propaganda and the prevention of cyberconflict in the form of cyberattacks on critical infrastructure.

While the detailed design and justification of a web of prevention against cyberthreats would take us beyond the remit of this book, we note that we have already, in effect, identified and justified a number of the features of such a web of prevention in earlier chapters: a prohibition on 'out-of-the-loop' autonomous weapons, analogous to the Biological Weapons Convention prohibiting the use of biological weapons; regulations with respect to privacy, such as the European Union's General Data Protection Regulation, analogous to biosecurity regulations regarding transport and storage of pathogens; and the development of ethical guidelines to give direction to policy regarding various forms of cyberconflict. Moreover, in the Conclusion we provide a consolidated list of these features of a web of prevention against cyberthreats and related threats in relation to which cybertechnology has a preventative role (e.g., pandemics). Before doing so, as mentioned above, we need to discuss some key concepts that underpin the notion of a web of prevention—namely, collective responsibility and institutional responsibility, multilayered structures of joint action, joint institutional mechanisms, and chains of collective institutional responsibility. One

---

[6] Miller, *Dual Use Science and Technology, Ethics and Weapons of Mass Destruction*.

reason it is important to do this is because, ultimately, cybersecurity depends on the ability of human beings, acting individually and jointly, to take responsibility for cybersecurity, albeit developing and utilizing cybertechnology and institutional arrangements to do so. In addition, we need to discuss, as mentioned above, the concept of dual-use cybertechnology and the challenges it poses for a web of prevention against cyberthreats.

7.1.1   Collective Responsibility, Institutional Responsibility, and Webs of Prevention

A web of prevention in the sense used here is essentially an institutional arrangement that involves cooperative action on the part of multiple individuals, private and public sector organizations, security agencies, and governments. Moreover, to reiterate, in the case of a web of prevention against cyberthreats, it is an institutional arrangement in which cybertechnology is deployed to prevent and counter cyberthreats. Naturally, as discussed in Chapter 6, countering cyberthreats in the interest of national security (or, for that matter, global cybersecurity) may require the use, or threat of the use, of offensive (soft and hard) cyberweapons and cognitive warfare weapons—or, for that matter, of conventional weapons or other noncyber means (e.g., economic sanctions).[7]

Further, a web of prevention involves, as mentioned above, institutionally embodying a collective moral responsibility that, in the case of interest to us here, is the collective moral responsibility to prevent and counter cyberthreats. This collective moral responsibility is embodied in institutional roles defined in terms of institutional rights and duties (i.e., role occupants with defined institutional responsibilities, and associated social norms and conventions).[8] In some cases, these roles are preexisting ones to which additional responsibilities are attached (e.g., managers who now have a responsibility to institute cybersecurity protocols, such as virus protection measures, and employees who now have a responsibility to ensure their own compliance with these protocols). In other cases, new roles are created for cybersecurity purposes (e.g., in cyberemergency response teams [CERTs]).

---

[7] We note that modern conventional weapons typically utilize cybertechnology without being cyberweapons per se.

[8] On the related concepts of social norms and conventions, see Miller, *Social Action*, Chapters 3 and 4, respectively.

Moreover, over time, the discharging of these individual and collective institutional and moral responsibilities establishes and maintains cybersecurity social norms.

The notion of institutional responsibilities includes,[9] inter alia, legal responsibilities and regulatory responsibilities.[10] Moreover it presupposes some notion of an institution.[11] Such organizations (or systems thereof) are complex social forms that reproduce themselves and include governments, police organizations, universities, hospitals, business corporations, markets, and legal systems. A normative theory of institutions specifies what the purpose or function of types of institution *ought to be*, as opposed to what in fact it is. One normative theory of social institutions, Miller's normative teleological theory,[12] is based on an individualist theory of joint action.[13] Put simply, by this account the organizations or systems of organizations, including webs of prevention, are ones that provide collective goods by means of joint activity. The collective goods in question include the fulfilment of aggregated moral rights, such as needs-based rights for security (police organizations), material wellbeing (businesses operating in markets), education (universities), governance (governments), and so on.[14] In the case of interest to us here, the collective good produced by the web of prevention in question is, of course, cybersecurity.

In the light of the above, we can distinguish three possible ways of understanding institutional responsibility. Firstly, there is the responsibility *to institutions*. This is the responsibility (possibly moral responsibility) that an individual, group, or institution might have *to an institution(s)*. This responsibility to an institution might consist in establishing, redesigning, maintaining, or simply refraining from undermining the institution in question (e.g., the responsibility of members of the community not to engage in harmful hacking activities against financial, welfare, and other institutions, and the responsibility of government to private and public sector institutions and the wider community to establish an e-Safety Commission). Here the property *institutional* does not necessarily qualify the notion of

---

[9] John Ladd, "Philosophical Remarks on Professional Responsibility in Organizations," *International Journal of Applied Philosophy* 1, no. 2 (1982): 58–70.

[10] R. A. Duff, *Answering for Crime: Responsibility and Liability in the Criminal Law* (Oxford: Hart Publishing, 2007)

[11] John R. Searle, *Making the Social World* (Oxford: Oxford University Press, 2011).

[12] Miller, *The Moral Foundations of Social Institutions*.

[13] Ibid. Ch. 1.

[14] So collective goods in this sense are not public goods in the economists' sense of public goods (i.e., nonexcludable, nonrival goods).

responsibility; rather it is part of the *content* of the responsibility. Secondly, there is responsibility *of institutions*. For example, financial institutions are ascribed *legal* responsibilities with respect to protecting their customers and investors personal financial information from hackers seeking to engage in identity theft; this responsibility is discharged in part by installing appropriate protective cybertechnology. Thirdly, there is the responsibility *of institutional role occupants*. This is the institutional responsibility of the human beings who occupy institutional roles: responsibility qua institutional role occupant (e.g., the institutional responsibility of police officers not to engage in unauthorized access to sensitive information, and the institutional responsibility of the Chief Security Information Officer in an organization to notify the relevant cyberauthority of cyberattacks).

Evidently, individual role occupants are *individually* institutionally responsible for at least some of their actions and omissions. For example, an employee of an organization might be individually institutionally responsible for seeing to it that they protect their computer password. This is responsibility in its forward-looking sense. Moreover, if the employee fails to protect her password (e.g., she writes it on a sticker on her keyboard), then she is individually responsible for not having protected it. This is responsibility in its backward-looking sense. Moreover, this failure attaches to a person qua institutional role occupant, and the person might reasonably be subjected to disciplinary measures by their employer.

On the other hand, a number of institutional role occupants might be *collectively* institutionally responsible for some outcome.[15] The paradigmatic cases here are ones of joint action; actions involving cooperation between institutional actors to achieve some outcome. For instance, all the employees in the organization in question might be collectively responsible for cybersecurity insofar as maintaining cybersecurity requires each employee to ensure their passwords are protected, to report any cyberattacks, and so on. For it may well be that the cybersecurity of the organization is only as good as the weakest link in the chain. This is, of course, not to say that some employees of the organization do not have specialist cybersecurity responsibilities by virtue of the roles they occupy and the expertise they have (e.g., members of the cybersecurity team).

[15] Dennis F. Thompson, "Moral Responsibility and Public Officials: The Problem of Many Hands," *American Political Science Review* 74, no. 4 (1980): 259–73; Larry May, *Sharing Responsibility* (Chicago: University of Chicago Press, 1992); Michael Zimmerman, "Sharing Responsibility," *American Philosophical Quarterly* 22, no. 2 (1985): 115–22.

Collective responsibility of the kind in question here is the responsibility that attaches to the participants of a joint action for the performance of the joint action, and for the realization of the collective end of the joint action. Our account elaborated in Chapter 5, Section 5.3, conceptualizes collective moral responsibility for joint action as *joint responsibility*.[16] On this view of collective responsibility as joint responsibility, collective responsibility is ascribed to individual human beings only, even though jointly.[17] Moreover, institutional actors can be ascribed collective institutional responsibility when they act jointly in accordance with their institutional roles. Consider the cybersecurity team. Each member of the group is individually institutionally responsible for their contributory action, and for the aimed-at outcome (the collective end) of the set of actions. However, each member is individually responsible for that outcome, *jointly with the others*; therefore, the conception is relational in character. Thus, in this cybersecurity example, each member of the team is institutionally responsible jointly with the others for responding efficiently and effectively to cyberthreats, because each performed their contributory action in the service of that collective end (i.e., preventing or mitigating cyberattacks). Accordingly, the members of the cybersecurity team are collectively institutionally responsible for preventing or mitigating cyberattacks, in both the forward-looking and the backward-looking senses of responsibility.

Note that there can be cases where there the morally significant collective end of a joint action is realized, yet one or more individuals fails to successfully perform their individual action, as well as cases where the morally significant collective end of a joint action is not realized, yet one or more individuals successfully perform their individual action.[18] In the former kind of case, assuming the individual (or minority) has the collective end (and presumably, therefore, did not intentionally fail to perform their

---

[16] Gregory Mellema, "Collective Responsibility and Qualifying Actions," in *Midwest Studies in Philosophy*, vol. XXX, ed. Peter A. French, 2006; Miller, "Collective Moral Responsibility: An Individualist Account."

[17] Accordingly, there is no need to hold that collective responsibility attaches to collective entities per se, as done by collectivist theorists such as Margaret Gilbert, *On Social Facts* (Princeton, NJ: Princeton University Press, 1989) and (in a somewhat different vein) Christian List and Philip Pettit, *Group Agency* (Oxford: Oxford University Press, 2011). For criticisms of these collectivist accounts, see Seumas Miller and Pekka Makela, "The Collectivist Approach to Collective Moral Responsibility," *Metaphilosophy* 36, no. 5 (2005): 634–51 and Andras Szigeti, "Are Individualist Accounts of Collective Moral Responsibility Morally Deficient?," in *Institutions, Emotions and Group Agents: Contribution to Social Ontology*, ed. A. Konzelmann-Ziv and H. B. Schmid (Dordrecht: Springer, 2014), 329–42.

[18] Miller, *Shooting to Kill*, Chapter 5, Section 5.4.

contributory action), the individual shares in the collective moral responsibility for the realization of the end, despite their individual failure in relation to their contributory action. In the latter kind of case, again assuming the individual has the collective end, the individual shares in the collective moral responsibility for the failure to realize the end, despite their individual success in relation to their contributory action. It is consistent with this that, if an individual (or minority) *culpably* failed to realize their individual end, yet knew that the collective end would nevertheless be realized, then that individual does *not* share in the collective moral responsibility of the successful outcome, since the individual did not in fact have the collective end. It is also consistent with the above that, if an individual (or minority) *culpably* failed to realize their individual end with the knowledge that, as a consequence of this culpable failure of theirs, the collective end would not be realized, then the individual (i) does not have the collective end and (ii) is individually morally responsible for the collective failure to realize the collective end. So, there might be no collective moral responsibility for the failure.

A final point. Whereas collective responsibility implies joint responsibility for the realization, or failure to realize, the *collective end* of the joint action in question, it does not necessarily imply that each participant is individually responsible for the failure (or success) of some other participant to perform this other participant's *contributory action*.

7.1.2 Multilayered Structures of Joint Action

While the concept of collective responsibility as joint responsibility is relatively easy to understand and apply in the case of small teams of actors with narrowly specified tasks and goals (collective ends), matters are more complex when it comes to large organizations.

Naturally, collective institutional responsibility is a key feature of organizational activity. However, the problem is how to understand this notion of collective institutional responsibility at the level of large organizations. Organizational action typically consists in, what Miller has elsewhere termed, *a multilayered structure of joint actions.*[19] One relevant illustration of the notion of a layered structure of joint actions is a cybersecurity

---

[19] Miller, *Social Action*, Chapter 5; Miller, *Moral Foundations of Social Institutions*, Chapters 1 and 2.

department comprised of three (let us assume for purposes of simplification) cyberteams: a cyberthreat intelligence team (TI), an incident response team (IR), and an engineering team (EN). Suppose, at an organizational level, a number of joint actions are severally necessary and jointly sufficient to achieve some collective end (e.g., to prevent or mitigate malware attacks).[20] Thus, the epistemic action of the TI team gives early warning to the IR team (which can act to prevent or mitigate a cyberattack), and if necessary to the EN team (which can enable it to patch a defect in the system that the cyber-attack is exploiting). Assume that the action of TI is, in fact, a joint action, as are the actions of the IR and EN teams. Moreover, assume also that the actions of the TI, IR, and EN teams are severally necessary and jointly suf-ficient to achieve the collective end of preventing or mitigating the ongoing cyberattack (e.g., a virus). Taken together, these actions constitute a fourth joint action comprised of the three joint actions of TI, IR, and EN teams, respectively.

At the first level, there are individual actions directed to three distinct col-lective ends: the collective ends of (respectively) collecting and disseminating cyberthreat intelligence, responding to the cyberattack, and removing the cybersystem vulnerability. Thus, at this level there are three distinct joint actions (of TI, IR, and EN teams, respectively). However, taken together these three joint actions constitute a single second-level joint action. The col-lective end of this single second-level joint action is to mitigate the effects of the ongoing cyberattack; and from the perspective of this second-level joint action, and its collective end, the three first-level *joint* actions are three second-level *individual* actions that are constitutive of the single second-level joint action. We note that, typically in organizations, not just the nature, but also the quantum of the individual contributions made to the collective end will differ from one team member to another.

Obviously, given the crucial role of institutions and institutional actions in the prevention of cyberattacks, it is important for the purposes in this chapter that organizations that are institutions can be understood in purely individualist terms and by recourse to the core notion of joint action. Hence, the significance of the technical notion of a multilayered structure of joint action.

---

[20] Here, there is simplification for the sake of clarity. For what is said here is not strictly correct, at least in the case of many actions performed by members of organizations. Rather, typically some threshold set of actions is necessary to achieve the end. Moreover, the boundaries of this set are vague.

7.1.3 Joint Institutional Mechanisms

Collective institutional responsibility is involved not only in joint (institutional) actions, but also in the related phenomena of *joint institutional mechanisms*. These important mechanisms are often embedded in organizations, although this is not necessarily the case. Some of these mechanisms are also technologies (e.g., blockchains; see 7.1.5 below). Others are not technologies or, at least, only contingently so (e.g., voting systems). Consider the ubiquitous joint institutional mechanism of voting. There is voting for political office, voting in the cabinet of a parliamentary democracy, voting by shareholders of corporations, voting in committees (including voting in relation to cybersecurity measures), and so on. Consider, for instance, shareholder voting. Voting rights belong to shareholders, and each share gives the shareholder one vote. Shareholder A exercises their institutional right (if not duty) by casting their vote in an election, and A does so only if shareholders B, C, D, and so on also vote, and only if there is something or someone (e.g., candidates to be directors on the board of directors) to vote for. Thus, in addition to the actions of voting, there are the actions of the candidates X, Y, Z, and so on to be directors. That they stand as candidates is (in part) constitutive of the input to the voting mechanism; after all, voters vote *for candidates*. So, there are interlocking and differentiated actions (the inputs). Furthermore, there is some result of the operation of the mechanism: some candidate, say, Smith, is voted in by virtue of having secured the required number of votes (the output). What of the mechanism itself? A key constitutive feature of this voting mechanism is as follows: to receive the required number of votes *is* to be successful in the election.[21] Importantly, that Smith is voted in is not something necessarily aimed at by all the participants; specifically, those who voted against Smith were not aiming at getting Smith elected. Since we are assuming Smith did in fact receive the required number of votes, it follows that those who voted for him have realized the collective end of their joint action. Importantly, there is also a collective end of *all* the voters and *all* the candidates (or at least all those voting and standing for election in good faith). This is the collective end that those who get the required number of votes—whoever they happen to be—are, thereby, members of the board of directors. This is a collective end of all bona fide participants in the joint

---

[21] There are, of course, any number of alternative voting systems in democracies. However, this does not materially affect the analysis here.

institutional mechanism and reflects the commitment of the participants to the key constitutive feature of the institutional mechanism (i.e., that a candidate with the required number of votes is and is entitled to be a board member). Accordingly, participants in this joint institutional mechanism perform the individual actions of casting a vote and/or standing as a candidate and have as a collective end that the those who get the required number of votes—whoever they are—are thereby members of the board. So voting is a species of joint action, and more specifically, a joint institutional mechanism.

Notice that in the case of the shareholder voting mechanism, if some one person has the majority of the voting power, as Mark Zuckerberg does in Meta, then the democratic feature of voting mechanisms is undermined—and so, too, will be the (restricted) weighted feature of some voting mechanisms. Accordingly, it is important to establish what the institutional purpose or purposes (i.e., collective ends, in our parlance) of a joint institutional mechanism (e.g., a voting mechanism) are. Under some conditions, some of these institutional purposes might be thwarted. For instance, arguably, it is institutionally incoherent, because it is largely pointless, to design a voting mechanism for a public institution—including a corporation—in such a manner that one person could always ensure that their vote carried the day. Note that this is not to say that any voting mechanism designed in a way that some had more votes than others is necessarily incoherent. That is, it is not that *all* shareholder voting mechanisms are necessarily institutionally incoherent. After all, having a greater say is not the same thing as always having the decisive say. But it is to say that voting mechanisms in public institutions—institutions whose raison d'être is a collective good—should not be designed in such a manner that an individual person can always ensure that their vote carries the day.[22]

7.1.4 Chains of Institutional Responsibility

Collective institutional responsibility is also involved in an institutional phenomenon that is an extension of joint institutional mechanisms in the above sense—namely, in what we refer to as *chains of institutional responsibility*.[23] Consider a criminal investigation team, including interviewers, cyberforensic

[22] Miller, *Moral Foundations of Social Institutions*, Chapter 10.
[23] Seumas Miller, "Police Detectives, Criminal Investigations and Collective Moral Responsibility," *Criminal Justice Ethics* 33, no. 1 (2014): 21–39.

officers, and so on, investigating a major cybercrime. Let us assume that the team is engaging in a joint institutional action—namely, that of determining who is responsible, which is often a difficult undertaking given the problem of attribution (see Chapter 6, Section 6.1). Moreover, they do so having as a collective end to determine the *factual* guilt or innocence of this and other suspects. At some point, these police investigators complete this process and provide a brief of evidence to the prosecutors according to which, and based on all the evidence, the members of an organized cybercrime group (CCG) are the offenders. So far so good, but the criminal justice processes do not terminate in the work of the investigators. For there is now the matter of the trial—that is, the determination by the members of a jury of the legal guilt or innocence of the members of CCG. Let us assume that the members of the jury perform the joint action of deliberating on the *legal* guilt or innocence of CCG, and jointly reach the verdict of guilty. The question that now arises concerns the institutional relationship between the joint institutional action of the investigators and the joint institutional action of the members of the jury. It is here that the notion of a chain of institutional responsibility is illuminating.

Let us assume in what follows that the collective end of the criminal justice process comprised of both the criminal investigators *and* the members of the jury (as well as others, but here we simplify) is that the factually guilty be found legally guilty (and the factually innocent not be found legally guilty). Note that from the perspective of this larger institutional process, the collective end of the investigators (i.e., that of determining the factual guilt or innocence of a suspect) is merely *proximate*, whereas that of the members of the jury is *ultimate*.

In chains of institutional responsibility, all the participants aim (or should be aiming) at the further (ultimate) end in addition to undertaking their own roles—and, therefore, aiming at the end definitive of their own role. Moreover, all the participants (at least, in principle) share in the *collective responsibility* for achieving that further end, or for failing to do so. In our cybercrime example, presumably the end in question is for the factually guilty to be found legally guilty (and the factually innocent not to be found legally guilty) and this is an end (i.e., a collective end) that is realized by the investigators working jointly with the members of the jury (and the other relevant institutional actors).[24] It is not an end that the investigators could

---

[24] Assuming there are only two possible verdicts (guilty and not guilty), which is not the case in some jurisdictions (e.g., in Scotland).

achieve on their own; they can only arrive at knowledge of factual guilt.[25] But equally it is not an end that the members of the jury could realize on their own, for they rely on the knowledge provided by the investigators.[26]

7.1.5 Blockchains

Blockchains are an important new species of institutionally embedded cybertechnology and, indeed, of institutional mechanism, that can be used to replace the role of a single, central authority (e.g., a bank in relation to certain financial transactions or a land title office in relation to buying and selling land) with a consensus mechanism that uses cryptographic techniques. Responsibility for ensuring the integrity of any given record of a transaction (e.g., that a payment was in fact made for a service provided or that Jones rather than Smith is the owner of a piece of land) is distributed among all the participants in the blockchain including, importantly, all those who had previously performed transactions recorded in the blockchain. As such, blockchains are a species of joint institutional mechanism that have as a collective end the integrity of the records of all the transactions in some domain. Accordingly, everyone reliably knows (at least in principle) what transactions have taken place, and therefore reliably knows who owns what. More specifically, each participant reliably knows what transactions he or she has engaged in and what he or she owns. However, since the blockchain mechanism involves a diachronic process in which any participant who performs a given transaction with another participant relies (in a certain sense) on previous transactions—and, therefore, on previous participants—it also constitutes a chain of responsibility in something like our above-described sense of that notion. Accordingly, as we saw was the case with joint institutional mechanisms and chains of collective responsibility, blockchains embody collective responsibility and chains of responsibility with respect to the preservation of the integrity of records of transactions. If, as seems to be the case, blockchains provide a more robust form of security for many kinds of

---

[25] The concept of knowledge is philosophically contested and the subject of a vast literature. But see, for instance, Paul Moser, *Knowledge and Evidence* (Cambridge: Cambridge University Press, 1989). Let us assume here that it is justified true belief which, of course, does not imply certainty, but perhaps, only evidence that establishes a high probability that one's true belief is true.

[26] Chains of institutional and moral responsibility consist of a process in which the completion of one stage institutionally triggers the commencement of the next (e.g., arrest is followed either by the suspect being charged or released within a specified timeframe).

transactions conducted in cyberspace than does a single, central authority (e.g., banks and land title offices), then they are to be welcomed as an important tool in the service of cybersecurity.

Notice that in the case of some financial transactions (e.g., electronic transfers of funds and so-called cryptocurrencies, such as Bitcoin), the record of the transaction is in part constitutive (albeit defeasibly) of the transaction itself.[27] For instance, the transfer of $1 million from A's bank account to B's bank account is in part constituted by A's bank account displaying a deficit of $1 million and B's a credit of $1 million. This (joint) record of the transaction is only defeasibly constitutive since, for instance, the receiving bank needs to authorize the transaction for it to count as a transaction. Of course, in the case of Bitcoin, there are no banks to ensure the integrity of transactions; rather, it relies on blockchain technology, and more specifically, cryptographic techniques.

We stated above that blockchains are a consensus mechanism that relies in part on the use of cryptographic techniques, specifically secure hashes (see Glossary) and public/private key cryptography (PPK) (see Chapter 2, Section 2.3), to preserve the integrity of the record of any given transaction in some domain.

The first point is that each transaction is usually cryptographically signed using the private key of the PPK of the authority for the transaction. Anybody can check the transactions using the public key. This procedure is more secure, but conceptually little different from having scanned copies of conventionally signed and witnessed documents. These transaction signatures should not be confused with the cryptographic linking of blocks. Each block consists of one or more transactions, or at least records thereof. Moreover, the domain in question constitutes a series of interconnected blocks: a blockchain. Accordingly, the second point concerns blocks. Specifically, for a block to be fixed in the blockchain it requires a so-called proof of work that establishes who gets to sign off on the block, and possibly get paid for doing so. The linking mechanism uses a secure hash (see Glossary) of the previous block inside the current block. Hence, if the previous block were changed, the hash would change and conflict with the hash of the current block. Thus, if anybody tries to change the previous block (or any earlier block) all subsequent

---

[27] Bitcoin is reliant on protocols developed by a pseudonymous person, or persons, named Satoshi Nakamoto. See Don Tapscott and Alex Tapscott, *Blockchain Revolution: How the Technology Behind Bitcoin Is Changing Money, Business and the World* (London: Penguin, 2016), 5.

hashes will be wrong and the chain will be broken. And there is this further point in relation to who has the right to add the next block to the chain. The notion of a proof of work was introduced using a first-to-finish mechanism involving a mathematical puzzle that is hard to solve but easy to verify. Tapscott and Tapscott explain the process as follows: "Participants agree that whoever solves the puzzle first gets to create the next block. Miners have to spend resources (computing hardware and electricity) to solve the puzzle by finding the right hash, a kind of unique fingerprint for a text or a data file."[28]

A simple explanation of the concept of a proof of work, in layman's terms by way of an analogy, is as follows. Imagine that our blockchain consists of blocks of printed documents. We want to create a new block from a new batch of documents. They currently weigh 822 grams, but each block in our chain must weigh exactly, to the nearest milligram, one kilogram. In addition to the documents, there is a piece of card, like a certificate, from the previous block. It weighs 33.1g. So, our block is 144.9g short. The miners who want to complete the block do not know the weight. They must cut a piece of card to add to the block. If it weighs 144.9g (i.e., brings the block up to exactly 1 kg), their card is accepted, and the block is complete. If their piece of card is not the right weight, they do not find out if it was too heavy or too light, they simply must start again and try another piece. Now we could go a step further and identify the miners. Each miner has their own unique pattern on their card. Thus, the piece of card stored with the batch of documents would identify the miner who completed that particular block. This is rather like a digital signature with a PPK. Anybody can use the pattern to check on the author, but the pattern is very difficult to construct by anybody else. This piece of card is now the starter for the next block. Suppose now that somebody wants to swap a document in some block back in the chain. That replacement batch of documents now has a different weight. Thus, the next block will need to be mined again. The piece of card added will have a different weight than before, which means the next block must be remined, and so on.

The proof of work is a mechanism for distributed authorization of blocks without a central authority. Now if everybody involved in the blockchain had to agree to each new block, the blockchain is going to slow down as it grows. To avoid this, Ethereum (at the time of writing the second largest blockchain after Bitcoin) changed to a new approach, *proof-of-stake*:[29] "Proof-of-stake

---

[28] Tapscott and Tapscott, *Blockchain Revolution*, 31.
[29] https://ethereum.org/en/developers/docs/consensus-mechanisms/pos/. Accessed 21/7/2023.

(PoS) underlies Ethereum's consensus mechanism. Ethereum switched to its proof-of-stake mechanism in 2022 because it is more secure, less energy-intensive, and better for implementing new scaling solutions compared to the previous proof-of-work architecture".

In short, the set of past and present transactions—or, more accurately set of records of these transactions—in the domain in question, constitutes a series of interconnected blocks, each consisting of one or more transactions, or records thereof: a blockchain. Once a given block is fixed in the chain, it not only confirms its constitutive transactions (in the manner that a bank authorizes a transaction), but it exists as a permanent and almost completely irreversible record of the transaction(s) in question.[30]

Moreover, all the transactions in some domain, such as all the car sales or all the land sales in a particular jurisdiction, are transparent in various respects to all those who participate in these transactions (as opposed to being known to an intermediary such as a bank or land title office) as well as, of course, being known to those who might be given access to the records of the transactions by the participants. This transparency among all the participants in the blockchain is with respect to those features of all the transactions required to assure themselves of the integrity of the record of transactions, and as we saw in the case of the bank authorizing financial transactions, the record of the transactions in the blockchain is only *defeasibly* constitutive of the transaction—and therefore assurance is required. The use of block hashes, PPK, and proof of work or stake in blockchains are integrity-preserving mechanisms that provides assurance. Blockchains assure in part by enabling transparency with respect to its integrity-preserving qualities. Public keys are, by definition, transparent. What of private keys? The miner who gets to complete each block signs the block with her private key, since she requires evidence that it was indeed she who solved the puzzle and is according to the agreed institutional rules entitled to a reward, which is a piece of bitcoin in the bitcoin blockchain.

Moreover, what is also transparent is that PPK is being used and that it is an extremely robust cryptographic method (or, at least, has been prior

---

[30] It is *practically* impossible because rewriting a transaction would entail changing the past. See Tapscott and Tapscott, *Blockchain Revolution*, 7. However, a more recent cryptocurrency, Ether (within the Ethereum project), did in fact suffer a massive fraud. So even the practical impossibility claim is overstated.

to quantum computing).[31] This transparency, thereby, reduces the need for individual trust in the person with whom one is transacting. Secondly, the use of PPK can, if required, enable the participants in the transactions to be anonymous, and in addition, enable features of the transactions that are not required to be transparent to all to remain unknown to all. For instance, in relation to land transactions, everyone knows that a particular identified parcel of land was sold by Jones to Smith. However, the actual price paid might not be disclosed, as would probably have to be disclosed to the single, central authority, if that were the integrity preserving system in use. By contrast, in the case of Bitcoin, a feature of transactions that everyone knows might be the amount paid by one account holder to another (e.g., that 10 bitcoins were paid by X to Y). What everyone does not know (i.e., what is anonymous) is: (1) who X and Y in fact are, and (2) what good or service was paid for by X. Here X and Y are cryptowallets, the owners of which are not known, at least in principle, to anyone other than the owners themselves. Governments want to change this to combat illegal activity. In the case of Bitcoin, information about the goods or services paid for is not stored. For other systems, such as Ethereum, the exchange may be part of a smart contract,[32] and therefore contain more information about the transaction.[33]

7.1.6 Institutional and Moral Responsibility

Evidently some institutional actions—actions performed by the human occupants of institutional roles in their capacity as institutional actors—are not morally significant, and some morally significant actions are not institutional. On the other hand, many institutional actions are morally significant (e.g., actions that prevent ransomware attacks that would otherwise threaten the lives of patients in a hospital or morally culpable omissions that fail to prevent such attacks).

An important feature of the relationship between moral responsibility and institutional responsibility is that, as mentioned above, institutional arrangements assign moral responsibilities to persons on the basis of the

---

[31] D. Castelvecchi, "Are Quantum Computers about to Break Online Privacy?," *Nature* (6 January 2023): 221–22. https://www.nature.com/articles/d41586-023-00017-0. Accessed 31/10/2023.

[32] Smart contracts are computer programs that secure and execute settlements of recorded agreements between parties. Tapscott and Tapscott, *Blockchain Revolution*, 101.

[33] Castelvecchi, "Are Quantum Computers about to Break Online Privacy."

collective good that the institution exists to serve. Moreover, in many cases they assign moral responsibilities to persons that those persons did not previously have, and indeed in some cases such as the establishment of a new type of institution (e.g., cybersecurity agencies), that no person previously had.[34] In the case of the institutional roles of cybersecurity teams, cybercrime investigators, or cyberforensics specialists, the moral basis appears to be cybersecurity, or, more specifically the aggregate cybersecurity of the individuals and organizations within the jurisdiction in question. In the case of cybersecurity, the security of individuals and organizations within a given jurisdiction is likely to overlap with, and connect to, the security of individuals and organizations in other jurisdictions. This is in part because organizations, such as multinational corporations, are transjurisdictional entities, but also, and importantly for our purposes here, because the cyberdomain (e.g., the internet) is transjurisdictional.

Regarding the aggregate security of individuals and organizations, each member of a community has an individual human right to, say, some minimum level of security, if they need it. However, it is only when a certain threshold of aggregate need exists that the establishment of an institution takes place. For example, a police organization with its constitutive institutional role occupants—police officers—is not established because a single person's right to security is not being realized. When such a threshold of aggregate need exists, what is required is collective or joint action on the part of many persons. Accordingly, a cooperative enterprise or institution is established that has as a collective end, the provision of security to the needy many by means of the joint activity of the police officers who are members of the police institution. However, in the case of cybersecurity, as opposed to personal bodily security, the security of one person or organization is interdependent with the security of other persons and organizations. Moreover, this interdependence operates at the organizational, national, and global levels.

The (collective) duty to assist may, then, in certain cases imply the duty to establish and support institutions to achieve the object of the duty (e.g., cybersecurity teams, cybercrime units, cyber defensive and offensive arms of military organizations). Once such institutions, with their specialized role occupants, are in place, it may be that we generally have no further duty to

---

[34] Miller, *Moral Foundations of Social Institutions*. For a contrary view, see Bernard Gert, *Common Morality* (Oxford: Oxford University Press, 2007).

assist within the area of the institutions' operations. Indeed, it may be that generally we should not even *try* to assist, given our relative lack of expertise and the likelihood that we will get in the way of the role occupants. Moreover, these specialized role occupants have duties that they did not have before, and indeed that no one had before the establishment of the institutional role with its specific duties. For example, computer forensics specialists may have an institutional, and indeed now a *moral* duty to access computer hard drives in a manner that is not morally required of ordinary citizens, in part because most ordinary citizens do not have the required technical expertise (although some do), and indeed that was never morally required of anyone prior to the establishment of computer forensics units.

Once institutions and their constitutive roles have been established on some adequate moral basis, such as the duty to assist those facing threats to their security, then those who undertake these roles necessarily put themselves under obligations of various kinds—obligations that attach to, and are in part constitutive of, those roles. In the case of cybersecurity, in the light of the undeniable extraordinary scale and seriousness of cyberthreats, there is unquestionably an adequate moral basis for the establishment of relevant institutions, roles, and constitutive rights and obligations. Naturally, in the case of cybersecurity institutions and roles, as with all institutions and roles, there is a need for side constraints, including human rights constraints, on these roles and on the tasks constitutive of these roles. For instance, as we have elaborated in Chapters 2 and 4, for instance, there are privacy constraints on cybersecurity measures. Moreover, as also elaborated in Chapters 2 and 4 and elsewhere in this book, the application of cybertechnology in the service of cybersecurity can generate moral risks, such as function creep and power imbalances between the state and its citizenry.

In light of the above discussion of institutional purposes (and the associated normative-teleological account of social institutions), it follows that in order to understand the specific content of institutional role morality we need to examine the purposes—to meet aggregate cybersecurity needs, in the cases of interest to us—that the various institutions and their constitutive roles have been formed to serve, and the way in which roles must be constructed in order to achieve those purposes. Of course, one typically comes to have an institutional role through voluntary action, but the morality that comes with that role is not itself ultimately grounded in the individual's choice, but rather in the larger purposes (collective ends) of the role, or so the argument goes.

In this chapter, thus far, we have elaborated the concept of a web of prevention and suggested that it is central to the task of preventing and countering cyberthreats. In doing so, we have provided analyses of key theoretical notions underpinning the concept of a web of prevention (e.g., multilayered structures of joint action, joint institutional mechanisms, and chains of collective institutional responsibility). These notions pertain to webs of prevention as institutional arrangements embodying collective moral responsibilities to establish and maintain collective goods of which cybersecurity is a paradigm case. It is now time to turn to a key problem confronting the design of a web of prevention for preventing and countering cyberthreats—namely, dual-use cybertechnology.

7.2 Dual-Use Cybertechnology

The expression *dual use* refers to scientific research or technology that can be used for both beneficial (good) and harmful (bad) purposes.[35] However, this general sense of *dual use* is too broad since it has the effect that almost everything could count as dual use. For instance, machetes are used for farming, but they were also used in the Rwandan genocide in 1994 as tools of murder. Therefore, we require a narrower notion of dual use. Most of the current debate has focused on research and technologies with implications not simply for weapons, but for weapons of mass destruction (WMDs) in particular (i.e., where the harmful consequences of malevolent use would be on an extremely large scale). That said, defining *dual use* simply in terms of WMDs yields too narrow a notion because it excludes, for instance, gain of function (GOF) research in the biological sciences,[36] which is research on vaccines for viruses that leads to knowledge of how to increase the virulence or transmissibility of those viruses, and thereby how to create a 'superbug' that is more

---

[35] See, for example, Miller and Selgelid, "Ethical and Philosophical Consideration of the Dual Use Dilemma in the Biological Sciences"; Koos van der Bruggen, Seumas Miller, and Michael Selgelid, *Report on Biosecurity and Dual Use Research* (The Hague: Dutch Research Council, 2011), 1–122; Miller, "Moral Responsibility, Collective Action Problems and the Dual Use Dilemma in Science and Technology"; Jonathan Tucker, ed., *Innovation, Dual Use, and Security* (Boston: MIT Press, 2012). Some material, as opposed to technologies, (e.g., toxins, might be dual use if, for instance, they are not naturally occurring but were manmade). However, for the sake of simplicity we will not refer to dual-use materials unless this is required in the particular case under discussion.

[36] National Science Advisory Board for Biosecurity Framework for Conducting Risk and Benefit Assessments of Gain-of-Function Research (Washington, DC: NSABB, 2015); Michael Selgelid, "Gain of Function Research: Ethical Analysis," *Science and Engineering Ethics* 22, no. 4 (2016): 923–64.

dangerous than the original virus. Accordingly, we need to provide a serviceable notion of *dual use* that avoids these two problems,[37] but on the assumption that any definition will involve a degree of stipulation.

On this (somewhat stipulative) definition, new and emerging science or technology is dual use if:

1. It can be used for both large-scale, significantly beneficial and large-scale, seriously harmful purposes—where either the harmful purposes involve the use of a weaponized version of the science or technology as means, and often WMDs in particular, or the serious, large-scale harm aimed at does not necessarily involve weapons or weaponization but does involve serious moral rights violations;

2. The serious, large-scale harm in question is able to be caused by a single act of using the technology—as opposed to multiple acts that in aggregate cause great harm;

3. A beneficial outcome is intended by the original researchers/developers (e.g., those who invented the technology in question);

4. The actual or potential harmful outcome is typically reasonably foreseeable by the original researchers/developers, and if it eventuates is either intended by secondary malevolent users, or at least their secondary use involves culpable negligence.

The intended great harm is typically (but not necessarily) delivered by a weapons system of some sort (e.g., chemical, nuclear, or biological weapons). Cybertechnology is apparently not different in this respect since there are so-called cyberweapons (e.g., the Stuxnet virus used to shut down Iranian nuclear facilities).[38] Moreover, the intended harm might be caused by something other than a weapons system. For instance, a homicidal lunatic might dump an extremely dangerous man-made toxin (e.g., a pesticide)—a toxin that also has beneficial uses—into a city's supply of clean water with the intention of killing many the city's residents. In this situation, the R&D that enables the production of the toxin might well be regarded as dual-use in character. However, the toxin is not per se a weapon.

---

[37] An earlier version of some material in 7.2 is in Miller, Dual Use Science and Technology, Ethics and Weapons of Mass Destruction Ch. 7.

[38] Michael Kelley, "The Stuxnet Virus Was Far More Dangerous than Previously Thought," *Business Insider*, 21 November 2013. https://www.businessinsider.com/stuxnet-was-far-more-dangerous-than-previous-thought-2013-11 ; Stamatis Karnouskos. "Stuxnet Worm Impact on Industrial Cyber-Physical System Security," in *IECON 2011-37th Annual Conference on IEEE Industrial Electronics Society*, 4490–94 (IEEE, 2011), 1.

Furthermore, the harm in question is not merely epistemic harm in the sense of harm consisting merely of believing what is false or of being in a state of ignorance. Of course, epistemic harm may lead to nonepistemic harm. For instance, ignorance of the toxic nature of some liquid may result in a child or even the members of a whole community drinking it and suffering death, consequently. But that is another matter; for death is not in and of itself an epistemic harm.[39]

Let us now apply the above definition of dual-use technology to cybertechnology with a view to determine which types of cybertechnology, if any, are dual-use in character. In doing so we modify the above definition somewhat. Importantly, as will be argued below, according to this (modified) definition, cybertechnology used to effect mass destruction and in which the weapons used are controlled by computers (including the selection of targets and, perhaps, the selection of the weapons themselves) constitutes dual-use technology, as do various forms of computer viruses and ransomware.

Consider the following two cases of truck terrorism:

1. A terrorist hacks into a truck's computer system and diverts it into a crowd of people. The internet and computer interaction with machinery are essentially enabling technologies for the terrorist hacker's remote control of the weapon (i.e., the truck).
2. The terrorist writes a computer worm that hunts for trucks connected to the IoT and causes those it infects to drive into a crowd of people. The internet and computer interaction with machinery are essentially enabling technologies for the computer worm's control of the weapons (i.e., the trucks).

The second example involves computer autonomy (as opposed to human autonomy—see Chapter 6, Section 6.2.1) in the selection of trucks and targets, and therefore the cybertechnology is conceptually integral to weapons of destruction in a way the essential enabling technology of the internet, let alone roads, is not. In our terrorist hacker example, the weapon is the truck, and it is selected by and under the control of the human hacker—although the

---

[39] This is so even if the death in question was of, say, a brilliant scientist whose most recent discovery died with him. Moreover, refraining from informing someone of a fact that they have a moral right to know may well result in great harm but it might not do so. So rights violations are typically but not necessarily harmful. Only technology that can be used for large-scale seriously harmful purpose can be dual use, even if the uses are rights violations.

terrorist hacker only controls the weapon remotely and indirectly, and in doing so relies on the essentially enabling technology of the computer interaction with machinery and the internet.[40] By contrast, in our computer worm example, while the weapons are trucks, they are selected by and under the control of the computer worm: the cybertechnology. Accordingly, the cybertechnology consisting of the computer worm utilizing the essentially enabling technology of the internet and computer interaction with machinery is conceptually integral in a strong sense to the weapons of destruction (i.e., the trucks).

The upshot of this discussion is that cybertechnology, while epistemic in character, can nevertheless be conceptually integral, in a strong sense, to weaponry. Arguably, therefore, the epistemic character of cybertechnology does not necessarily prevent it from being dual-use technology. Moreover, as is illustrated by our computer worm example, cybertechnology could potentially be used to kill very large numbers of people as a result of the release of a single virus. We conclude that some forms of cybertechnology used to effect mass destruction, such as our illustrative computer worm, may well constitute dual-use technology. Indeed, in section 7.2.2 below, we argue that computer worms (and related computer viruses) and autonomous robots are in fact species of dual-use cybertechnology, at least in certain configurations. However, before arguing that these *are* instances of dual-use cybertechnology, we need to argue for the proposition that the internet and certain other cybertechnologies are *not* species of dual-use technology, despite the tendency to believe that they are. Critical cybertechnological infrastructure such as the internet, for instance, is often referred to as dual use.

7.2.1 Identifying Dual-Use Cybertechnology

Infrastructure, such as dams, telephone cables, and powerlines, if deliberately destroyed or severely damaged for a prolonged period by weapons in

---

[40] As Scott Vella has pointed out, it also relies on a means to exploit the hack, which would constitute dual-use cybertechnology. These are tools which facilitate the exploitation of vulnerabilities, for instance, a bug in the truck's code that allows the hacker to gain control of it. However, these are the exact same tools that software security testers use to detect these vulnerabilities and evaluate their severity, and thus the software engineering teams can fix it. Hence, it has benefits as well as harmful uses, can be used to effect large scale hacks (i.e., on power grids, and so on), are intended as software testing tools, and can reasonably be predicted to be misused by malicious actors. As a result, these tools are also conceptually integral in a strong sense to the weapon of destruction (e.g., the truck).

the context of war, may lead to widespread suffering, even death. However, it would not follow that such critical infrastructure was dual use in our sense. Of course, such infrastructure may well be dual use in the quite different sense that it is used by both civilians and the military.[41] Moreover, its destruction may harm both civilians and the military. So, the population at large is vulnerable to great harm by virtue of its dependence on critical infrastructure. However, the infrastructure in and of itself is not a weapon or other vehicle being used to harm; rather it is the thing being damaged or destroyed, from which harm to the population results.

The internet is critical infrastructure; indeed, critical global infrastructure—and, as we saw above, potentially an essential enabling technology for weapons of mass destruction. A good deal of interpersonal, organizational, local, national, and international communications and data transfer are now dependent on the internet. Accordingly, central national and global institutions are dependent on the internet. For example, the global financial system depends on the internet. However, this dependence makes these institutions, and therefore the societies in part constituted by these institutions extraordinarily vulnerable should this critical infrastructure, or important parts of it, be severely damaged for a prolonged period by, say, terrorists. Moreover, the internet is used by civilians and military alike. So, the population at large, indeed multiple populations, are vulnerable to great harm by virtue of their dependence on the internet. Nevertheless, as is the case with other types of critical infrastructure, the internet per se is not dual-use technology in our sense.[42]

Developments in information and communications technology (ICT) not only enable the provision of critical infrastructure; they also enable the efficient collection, storage, analysis, communication, and dissemination of information on an unprecedented scale. Consider, for example, social media. Also consider big data.[43] Big data simply means all or, at least, a very

---

[41] In addition, it might be claimed to be dual use in the sense that any communication or transport infrastructure can be used for good or harm. However, infrastructure is typically a couple of removes from the device that causes the good or harm. For this reason, we decline to describe infrastructure as dual use, albeit this is essentially a stipulation on our part.

[42] It might be claimed that this has counterintuitive results in that, for instance, it follows that a nuclear power station is not dual use, even though it may produce the fuel for nuclear bombs as a by-product. Presumably, at some level of description a nuclear power station is not dual use (i.e., if its by-products of fuel are not per se useable in nuclear weapons but in need of further enrichment. However, the internet is a good deal more neutral in this respect than a nuclear power station. Specifically, the internet per se does not necessarily have 'fuel' (e.g., extremist propaganda, as a by-product).

[43] V. Mayer-Schonberger and K. Cukier, *Big Data: A Revolution That Will Transform How We Live, Work and Think* (London: John Murray, 2013).

significant fraction, of the very large amount of data in some domain; for example, all the financial transactions in a global capital market in a twenty-four-hour period.

Facebook and Twitter enable the immediate communication of information to vast audiences, and this has had a revolutionary effect on, for instance, political campaigns in the US, like those of Barack Obama, Hillary Clinton, Donald Trump, and Joe Biden. Again, the collection, storage and analysis of big data creates an extraordinary treasure-trove for those seeking to benefit humankind (e.g., for demographers projecting future population numbers, climate scientists trying to determine the rate of global warming, or other aspects of climate change).

Of course, social media and big data are also able to be used for harmful purposes. Terrorists use social media to recruit, incite, and provide access to training manuals, such as how to make an improvised explosive device (IED). Authoritarian governments use big data to monitor their citizens intrusively and thereby violate their civil liberties.

Nevertheless, neither social media nor big data are dual-use technologies in our sense, for the ultimate weapons-based harm done by terrorists who use social media—namely, the murdering of innocent people, is not directly done either by the data or information relied on in social media communications (whether it be true or false, believed or not believed), or by the knowledge or belief of this data or information, or by mere communication of the data or communication (leading to knowledge or belief on the part of others). This is because epistemic states (e.g., knowledge or belief) and epistemic harm, (e.g., false beliefs), are not weapons per se, and certainly not weapons that directly cause death and destruction. Nor is the weapon-based harm directly done by the essentially communicative acts performed by terrorists on social media. Social media is not per se a weapon as is, for example, a nuclear warhead; nor is social media weaponized as is, for example, an aerosolized pathogen in a container fitted to a weapons delivery system. Again, the ultimate weapons-based harm done by authoritarian governments who collect and analyse data about their citizens—namely, the forcible incarceration, torture, and/or murder of their citizens—is not directly done by the essentially epistemic acts performed by those who collect and analyse this data. Naturally, the collection and analysis of some of this data (e.g., personal information of citizens), may constitute a violation of the privacy rights of the citizenry and may, as such, be morally wrong. But dual-use technology, as we are using the term, typically involves weapons-based harm, and if it does not then, nevertheless, it directly causes[44] large-scale harm of a considerably more serious kind than mere violation of individual informational privacy rights (but see 7.2.2.5 below on facial-recognition technology). That said, communications by terrorists on social media, and the use of cybertechnology in a manner that violates privacy rights, are morally problematic, and indeed at least potentially pose security risks. The former because they can influence persons to cause death and destruction; the latter because violations of privacy and autonomy can rise to a level where they seriously harm an individual or group (e.g., by undermining their autonomy, or undermining liberal democratic institutions). As such, they need to be addressed in the web of prevention against cyberthreats. After all, this web of prevention is not only, or even primarily, concerned with threats emanating from dual-use cybertechnology.

For a similar reason, technology that enables cybertheft or cyberespionage is not as such dual-use technology in our sense. Cyberespionage is cybertheft of material that is reasonably regarded as confidential from a national security perspective and is stolen to realize some political or military purpose. Theft does not necessarily involve weapons-based harm; so, it does not meet this important criterion. Moreover, although dual-use harm is not necessarily weapons based, it is necessarily very serious harm on a very large scale. However, theft of property, if it involves harm (as opposed to simply being a violation of property rights)[45] it does so, all things being equal, at the lower end on the scale of harms, even though the ultimate consequences of theft of property, if it is ongoing and affects the property of large numbers of individuals, may well be extremely harmful.[46] However, despite the ultimate serious harm potentially done by cybertheft or cyberespionage, technology used to perform such actions is not dual-use technology in our favoured sense, all things being equal. For cybertheft is theft of intellectual property, and the possession by another person of one's intellectual property is essentially an epistemic condition; as such, it does not constitute a serious harm to oneself. Rather, it is what the person can do because of their new-found knowledge that is potentially profoundly harmful. Accordingly, the fact that

---

[44] Or, at least, is not at too great a remove from the harm done. Here there is a degree of vagueness in respect of the causal distance between the use of the putative dual-use technology and the harm done.

[45] The theft of property from a very rich person, if never detected, might not in fact harm this person.

[46] Naturally, other things might not be equal. Theft of a person's means of livelihood may put their life at risk.

cybertechnology is vulnerable to acts of cybertheft and cyberespionage does not make it dual-use technology.

Thus far in this section we have identified various harmful uses of certain forms of cybertechnology and argued that, nevertheless, the technology in question is not dual-use technology in our sense. The time has now come to discuss some salient species of cybertechnology that are strong candidates to be regarded as dual-use technology. We consider: (i) computer viruses;[47] (ii) autonomous robots; a third, fourth, and fifth putative species of dual-use technology—namely, (iii) encryption (e.g., used in ransomware); (iv) blockchain (or at least its use in cryptocurrencies, such as bitcoin); and (v) facial-recognition technology. Of these, computer viruses and autonomous robots seem to be clear instances of dual-use technology, whereas the encryption, blockchain and facial-recognition technology might seem not to be.

7.2.2 Dual-Use Cybertechnology: Viruses, Autonomous Robots, Encryption, Blockchain, and Facial-Recognition Technology

7.2.2.1 Computer Viruses
Computer viruses are akin to pathogens. They are potentially extraordinarily destructive weapons; indeed, they are potentially WMDs. However, like their biological counterparts, computer viruses are not necessarily harmful, nor do they necessarily hide themselves. They are essentially self-replicating programs that install themselves in computers without necessarily having the consent of the computer user. Moreover, the software technology underpinning computer viruses is extraordinarily beneficial. It is essentially the

---

[47] There are other forms of computer malware than viruses, which are candidates for being dual use in character. For instance, software/penetration-testing tools, also known as hacker tools, that enable a hacker to gain control of another system (including trucks, as in the example above) are dual use cybertechnology. Another type of putative dual-use cybertechnology is control software. For instance, devices (which could be critical infrastructure) install updates and patches provided to them by control software, but this software can thus also be used to roll out malware. Thanks to Scott Vella for this point. System updates can also be the delivery vector. The recent the Kaseya hack is indicative. https://news.sophos.com/en-us/2021/07/04/independence-day-revil-uses-supply-chain-exploit-to-attack-hundreds-of-businesses/. Accessed 6/7/2021. This attack reached a large number of corporations and effectively shut down a whole 800-supermarket chain in Sweden. The victims were all using Kaseya's VSA (Virtual Server Administration). In essence this meant that the system software managing the victim machine was updated remotely via control and patchupload from Kaseya. It was just such an uploaded patch which carried the payload.

technology that enables the construction of software agents that can collect, transmit, encrypt information.

For the last half-century computers have run many programs simultaneously through the mechanism of time-sharing. Most of these programs were neither directly started by, nor communicate with, any human users.[48] They do things like manage the file system, control network traffic, and other housekeeping things. But in the last two or three decades another type of program has appeared—the computer virus, an example of computer malware. It may arrive in several ways, either via computer networks or files copied from portable media.

We can classify malware for our present discussion into three categories: local, device-oriented, and global. Local malware does things that will usually impact a single user or small group of users (e.g., ransomware that encrypts the hard disk). We consider this below. *Device-oriented* means that the malware is sent out to attack a controller of some physical device. If the device has the potential to cause widespread destruction, then this fits our definition of *dual use*. The destruction does not have to be human. It could be costly infrastructure, machinery, or even a virtual entity, such as a stock market. The most remarkable such piece of malware in recent years was the Stuxnet virus.

Computer worms, and many other forms of malware, are potentially dual-use technologies, despite their epistemic character, because they can be conceptually integral in a strong sense to weaponry. Moreover, in the case of Stuxnet inter alia this cybertechnology has been used as a weapon, indeed a weapon of war, and could easily have been used as a WMD (although probably not now, given enhanced cybersecurity systems). We conclude that computer viruses—qua self-replicating programs—are a species of dual-use technology.

7.2.2.2 Autonomous Robots

Autonomous robots can provide great benefits (e.g., as used on factory assembly lines or in defusing bombs). However, as we saw in Chapter 6, Section 6.3, autonomous robots, such as predator drones, can also be weaponized

---

[48] However, these are processes of the initial boot process, which is started by a user—or, an administrator; or, at the very least, is provisioned to be started by an automated process by an administrator—when the machine is booted. Furthermore, these programs are either part of the operating system or part of another program that the user has chosen to install. As such, the user starts these programs at the time of installing the operating system or program, or by purchasing a machine with preinstalled software. Thanks to Scott Vella for this point.

and cause great harm. Are autonomous robots, therefore, a species of dual-use technology?[49]

Autonomous robots are, we suggest, a species of dual-use technology, irrespective of whether they are human in-the-loop, on-the-loop, or out-of-the-loop. Here there are several considerations. Firstly, once weaponized, autonomous robots are conceptually integral in a strong sense to their weapons; that is, they utilize the essentially enabling technology of the internet and computer interaction with machinery (the weapon). Secondly, autonomous robots have the potential to be armed with WMDs (e.g., chemical or nuclear devices).

However, human out-of-the-loop autonomous weapons have a degree of *computer* autonomy that the human in-the-loop or on-the-loop autonomous weapons do not. In short, in the case of autonomous weapons, computer autonomy underpins (in part) the conceptual integration of the cybertechnology with the weapon—and, thereby, justifies the claim that autonomous robots are a species of dual-use technology. However, as was argued in Chapter 6, Section 6.2, the autonomy in question (computer autonomy) should not be confused with human autonomy or be taken to have extinguished human moral responsibility. Moreover, as argued in Chapter 6, Section 6.2, this form of dual-use cybertechnology—namely, autonomous (out-of-the-loop) predator drones and the like—should be prohibited.

7.2.2.3 Encryption and Ransomware

Encryption offers enormous benefits. The whole of e-commerce depends upon being able to feed credit card numbers safely into a website, relying on the TLS encryption protocol. Encryption of course appears throughout the ages in a military context. Turing's cracking of the Enigma-encrypted communications[50] saved many lives in the Second World War. Yet the encryption itself, or the breaking thereof, was one step removed from harm, which in this case was done by U-boats, or the torpedoes they launched.

However, it does not seem possible to use encryption as a weapon. On the other hand, encryption can be used to cause serious moral rights violations on a large scale. Moreover, when so used, arguably, it is conceptually integral to those rights violations. However, the question arises as to whether it

---

[49] Seumas Miller, "Collective Responsibility for Robopocolypse," in *Super Soldiers: The Ethical, Legal and Social Implications*, ed. Jai Galliott and M. Lotze (Cheltenham: Ashgate, 2015), 153–66.

[50] Alan M. Turing, Turing's treatise on enigma. Unpublished Manuscript, 1939. https://www.archives.gov/files/press/press-releases/2015/images/turing-enigma-treatise.pdf

is conceptually integral to the large-scale, serious harms caused, and it seems that it is not since there is typically a human being that intervenes between the encryption process and the harm caused (e.g., a nurse who gives the wrong dosage of a drug to the patient because the encrypted correct dosage information is not available to her because of a ransomware attack). Accordingly, encryption does not appear to be a dual-use technology. Let us consider the issue in more detail.

Take, for instance, the victim of ransomware, whose data on a hard disk is maliciously encrypted for financial or other gain. In a ransomware attack, the target computer becomes infected by a piece of malware. It may arrive as an email attachment, from a dubious website, a downloaded Trojan horse app, or by other means. Once installed it then sets about encrypting the hard disk, rendering the data unreadable to its owner. Decryption requires a key, for which a ransom is required.

But dual use requires harm to a significant number of people. Consider the case of a large hospital, such as the case of the ransomware attack on the NHS (Chapter 5, Section 5.1). Patient records, treatment procedures, and schedules are now kept online. A terrorist hacker could gain access to this system and encrypt the contents of patient records, or perhaps encrypt records of just selected unspecified patients. Many hundreds of patients could suffer serious harm, including death, as a result. However, in such cases encryption is not a weapon per se. However, the question arises as to whether encryption is, nevertheless, dual-use technology. This is doubtful. Rather encryption's role in ransomware, at least as described above, seems more akin to rewriting in code the information regarding the doses of life-preserving medicines to be given to patients and refusing to decode it. Accordingly, it seems that while the encryption is conceptually integral to the initial rights violation, it is not conceptually integral to the serious harm done. Certainly, encryption is conceptually integral to the data security rights violation (i.e., the prevention of access to the data). However, this rights violation in and of itself does not directly cause the harm done since this harm was done by the nurse in administering the wrong dosage to the patient.[51]

This distinction between dual-use technology in our favoured sense and a closely related technology such as encryption is not without practical

---

[51] The death of the patient is the much more serious rights violation. However, the person who is morally responsible for the death qua rights violation is the person who is responsible for the ransomware attack, and not the nurse.

significance. For unlike autonomous weapons, presumably cybertechnology that can be used to autonomously encrypt the data on a hard disk should not be subjected to the same stringent regulation and, in the case of autonomous out-of-the-loop weapons, prohibited (e.g., criminalized or otherwise banned from use).

7.2.2.4 Blockchain Technology and Cryptocurrencies

As we saw above, blockchains are a species of joint institutional mechanism having as a collective end the integrity of the records of all the transactions in some domain (e.g., a finance domain), and do so in part by virtue of a dependence on cryptography. Moreover, blockchains can be used to replace the role of a single, central authority, such as a bank in relation to certain financial transactions. Undoubtedly, blockchain technology has huge potential benefits. However, depending on its applications, blockchains have a downside, and potentially a very great one, if used in some finance domains.

According to Joseph Stiglitz, for instance, cryptocurrencies greatly facilitate illicit activities that undercut the very functioning of our society by undermining the basis of the financial system. Stiglitz said: "(Cryptocurrencies) are becoming significantly important in terms of undermining the basis of our financial system and transparency of our financial system. I'm of the view that now is the time that regulators all over the world to basically shut down cryptocurrencies."[52] If this is the case then blockchain, at least as used in cryptocurrencies, is a candidate for being dual-use technology.

The argument here is essentially that the financial transactions in cryptocurrencies and those transacting are conducted on platforms that are not transparent to authorities, or indeed potentially to anyone other than those engaged in unlawful transactions, such as money laundering. Hence, as we saw in Chapter 4, the close relationship that can exist between cryptocurrencies and the Dark Web. Accordingly, blockchain (e.g., as potentially used in cryptocurrencies) might be argued to be a dual-use technology. However, if, as argued above, encryption (e.g., as used in ransomware) is not a dual-use technology, then it seems neither are cryptocurrencies. Of course, the use of encryption in ransomware is different from its use in cryptocurrencies. However, both rely on cryptography and

---

[52] Park Ga-young, "Stiglitz Urges Regulators to Shut Down Cryptocurrencies," *Korea Herald*, 28 October 2021, https://www.koreaherald.com/view.php?ud=20211028000916

can be used to cause large-scale harm. Moreover, in both cases the relationship between the use of the technology and the harm done is indirect. In the case of cryptocurrencies, if Stiglitz's argument is accepted, they can be used to undermine the financial system (and, ultimately facilitate crimes and transnational criminal organizations that rely on money laundering) by affording protection to unlawful financial activity. Accordingly, perhaps cryptocurrencies should be prohibited. However, if encryption per se is not dual-use technology and cannot be weaponized, then it is difficult to see what would justify prohibiting blockchain technology as such, although of course it may well be that certain uses of it should be curtailed or even prohibited (e.g., cryptocurrencies).

7.2.2.5 Facial-Recognition Technology

As we saw in Chapter 4, Section 4.4, facial-recognition technology is used by law enforcement, border protection, and in national security contexts to enhance security. It clearly provides very considerable security benefits. However, as we also saw, it can be used by criminals to locate and track their victims and law enforcement personnel (e.g., undercover operatives). Of course, facial-recognition technology is very useful to the security agencies of authoritarian states engaged in the violation of the moral rights of their citizens. Moreover, as mentioned in Chapter 4, Section 4.4, criminals and other malevolent actors can apply ML techniques to existing facial images to create new images, and thereby thwart law enforcement.

Of course, as we saw above, some dual-use technologies are such that the technology could be weaponized and, specifically, that it could be conceptually integral in a strong sense to a weapon. However, other dual-use technologies are not able to be weaponized but can, nevertheless, be used to cause violations of moral rights. Our definition of dual-use technologies also requires that they be able to cause large-scale serious harm. As noted in Chapter 4, Section 4.4, there are now literally billions of facial images on the internet (e.g., on social media websites, or stored in government and private sector databases). Moreover, as the private firm Clearview AI has demonstrated, literally billions of these images can be scraped off the internet, stored in databases, and used for harmful (as well as beneficial) purposes. Evidently, the magnitude of the potential harm arising from the use of facial-recognition technology is very great. Moreover, the magnitude of the harm potentially caused by a single powerful actor, such as an authoritarian state is also very great. However, much of this harm is indirect and relies on intervening human agents making use of other technology and performing additional consequential actions, as in the case of the incarceration of innocent people who have been located and tracked using facial-recognition technology. The direct harm caused by facial recognition is primarily a violation of privacy (and, to that extent, of autonomy), even though if the perpetrator of this harm is an authoritarian nation-state the magnitude of the violations of privacy could be very great. However, it might be held that a violation of privacy, even involving a violation of autonomy to some extent, is not in and of itself sufficiently harmful to reach the threshold required for facial-recognition technology to count as a dual-use technology. Consider, for example, the contrast between the widespread use of facial technology to locate and track dissidents in an authoritarian state and an attack on a nuclear power station that leads to the death of thousands. On the other hand, it can be argued that since one's face is constitutive of one's personal identity, the violation of one's privacy/autonomy involved in the complete loss of control of the distribution of one's facial image and of the uses to which it is put is a serious moral wrong. It can also be argued that the use of an integrated society-wide surveillance system deploying facial-recognition technology (in conjunction with CCTV systems and related technology) to continuously track and monitor an individual over many years would be a very significant human rights violation in and of itself—and if this activity was scaled up to the point where it constituted something akin to a so-called surveillance society, then the threshold of harm required for facial-recognition technology to count as dual-use technology might well be thought to have been reached. However, the counterpoint might be that the facial-recognition technology used in such a system ought to be more appropriately regarded as a set of coordinated facial technology equipment (i.e., as multiple items of facial technology equipment). If so, it would be more akin to the coordinated use of a set of hand-held weapons by multiple actors, and as such not dual-use technology in our specialized sense. Nevertheless, the system as such might well be regarded as dual-use technology.

7.3 Webs of Prevention Against Cyberattacks

In this chapter, we have taken as our starting point the need for webs of prevention to prevent and counter cyberthreats, and we have provided an account of this concept and its relation to institutional and collective responsibility. Specifically, as already noted, we have adopted a theory developed by Miller that collective moral responsibilities are jointly held individual moral responsibilities, the discharging of which typically produces or maintains a collective good, such as security, to which participants have a joint moral right. Consequently, we eschew collectivist accounts of moral responsibilities. Collective entities per se, as opposed to the individual human beings who occupy their roles, do not have moral responsibilities—although in some cases, of course, they have legal and other institutional responsibilities. The significance of this is that the moral responsibility for cybersecurity rests squarely on the shoulders of individual human beings cooperating with one another, although relying on mediating institutional arrangements, especially webs of prevention, that they have designed and put in place. It goes without saying that cybertechnology plays a key role in these webs of prevention.

In addition, we have proffered analyses of key notions implicated in the concept of a technologically and institutionally based web of prevention against cyberthreats, including that of dual-use cybertechnology. Other notions are moral rights and principles constitutive of liberal democracy. It is the potential conflict between these rights and principles (e.g., between the collective good of cybersecurity and individual moral right to privacy/autonomy) that has generated many of the ethical problems addressed in this work. It is now time to outline a salient set of *ethical guidelines* developed in this and earlier chapters. These guidelines are intended not only to assist those interested in a general sense with figuring out solutions to the various ethical problems that have arisen in securing cyberspace, but potentially also to provide some direction to cybersecurity policymakers in relation to the measures constitutive of effective and *ethically sustainable* webs of prevention against cyberthreats in the various domains that we have been concerned with in this work.

Conclusion
Ethical Guidelines

The following ethical guidelines have been devised to give direction to regulation and other measures constitutive of effective and ethically sustainable *webs of prevention*. Webs of prevention are to be understood as institutionally embedded means of discharging *collective moral responsibilities* directed to the realization of the *collective good of cybersecurity*. These webs of prevention are fashioned to combat cyberthreats in the various thematic cybersecurity domains discussed in this work—namely, privacy/autonomy (Chapter 2), freedom of political communication (Chapter 3), criminal justice (Chapter 4), public health (Chapter 5), and cyber conflict (Chapter 6). Moreover, in elaborating these ethical guidelines and designing the associated webs of prevention particular attention needs to be paid to the dual-use character of many cybertechnologies and the sense in which they might be dual use. All dual-use technologies are able to provide great benefits and do great harm. However, some but not others are able to do great harm by virtue of being able to be weaponized. The dual-use cybertechnologies considered in detail in Chapter 7 were self-replicating programs (computer viruses), autonomous robots (autonomous weapons), encryption, blockchain, and facial-recognition technology, all of which can be used to provide great benefits to humankind, but in the hands of malevolent actors they can also do great harm. However, of these only viruses and autonomous robots were able to be weaponized and, thereby, cause large-scale, serious harm. By contrast, encryption, blockchain, and facial-recognition technology cannot be weaponized. Nevertheless, they are able to be used for moral rights violations (and, ultimately, large-scale, serious harm). That said, the harm ultimately caused by encryption or by blockchain was argued to be at some considerable remove from the use of the technology, suggesting that they ought not to be regarded as dual-use technologies (at least in terms of our favoured definition of dual-use technology). By contrast, the use of an entire system of facial-recognition

Cybersecurity, Ethics, and Collective Responsibility. Seumas Miller and Terry Bossomaier, Oxford University Press.
© Oxford University Press 2024. DOI: 10.1093/oso/9780190058135.003.0009

technology (as opposed to the discrete use of a single piece of facial-recognition equipment) seemed to warrant it being categorized as dual-use technology, or so we argued.

The significance of the distinction between dual-use technologies and other technologies, and between categories of dual-use technologies, lies in the different regulatory stances to be taken. Thus, other things being equal, dual-use technologies that are able to be weaponized ought to be subject to more stringent regulation and, in some cases if weaponized (e.g., autonomous, out-of-the-loop weapons), prohibition.

A. Privacy/Autonomy, Confidentiality, and Security (Chapter 2)

There are a range of technical cybersecurity measures that, other things being equal, ought to be deployed to protect data security, and therefore privacy/autonomy and confidentiality. (Naturally, other things might not be equal; for instance, the data in question might be stolen or be held by a criminal organization). These measures include passwords, antivirus software, firewalls, encryption (including encryption technologies, such as blockchains), applying patches when vulnerabilities are identified, and so on. However, as noted in Chapters 1 and 2, these technical measures are, typically, only as effective as the competence and conscientiousness of the human beings who apply them. Moreover, the responsibility to ensure these measures are applied is, typically, a collective moral responsibility in need of institutional embodiment in the form of a fragment of a web of prevention (e.g., awareness raising programs, compliance requirements, and cybersecurity agencies). Consider, in this connection, webs of prevention for the Internet of Things or IoT, webs that might only be as effective against determined hackers as their weakest link. Such a web of prevention would need to consist of cybersecurity measures being designed into the devices and the owners and users of these devices ensuring that they applied these measures. Clearly effective webs of prevention do not currently exist in the IoT.

In Chapter 2, privacy, confidentiality, autonomy/identity and ownership rights were distinguished. Privacy is a constitutive human good, and confidentiality is a constitutive institutional good, at least in the case of security agencies. Security is a collective good, and in the case of data security, a collective good comprised in part of aggregated individual rights to privacy.

Therefore, the sharp contrast often drawn between privacy rights and security is not sustainable.

In relation to bulk databases, the following general ethical principles ought to guide policy makers. First, there is a presumption against the collection, analysis, and use of bulk sensitive personal data without the consent of the persons concerned. Moreover, the consent in question needs to be genuine consent. Thus, if a company providing an essential service requires one to consent to their collection and use of one's personal information to have access to their service, then there must be alternative providers of this service who do not require consent to the collection and use of one's personal information.

Second, while the presumption against the collection, analysis, and use of bulk sensitive personal data can be overridden for some law enforcement purposes, the specific purposes and specific databases in question need to be justified in terms of principles of necessity and proportionality. Justification cannot simply consist in a general appeal to security.

Third, bulk database cross-linkages also need to be justified. It is unacceptable for data, including surveillance data, originally and justifiably gathered for one purpose (e.g., taxation or combating a pandemic) to be interlinked with data gathered for another purpose (e.g., counterterrorism), without appropriate justification.

Fourth, insofar as the use of bulk data created for law enforcement, health, or other purposes can be justified for the investigation of serious crimes, and privacy and other concerns mitigated, it is imperative that their use be subject to accountability mechanisms to guard against misuse.

Fifth, integration of the bulk databases of the personal and public information of citizens and the application of face recognition, phone metadata and the like to track the movements and activities of citizens has the potential to create a power imbalance between governments and citizens in favour of governments, or between corporations and consumers in favour of corporations. These power imbalances need to be in favour of citizens and consumers, respectively.

In relation to encryption, the following additional general ethical principles ought to guide policy makers.

First, consistent with the above, privacy is not an absolute right, and therefore contrary to some privacy advocates, the claim that there are no circumstances in which very strong encryption could be morally and, therefore, potentially legally, impermissible, is not sustainable.[1] On the other

---

[1] Bossomaier does not accept this claim.

hand, very strong encryption might be justified in some circumstances (e.g., if the devices in question belong to dissidents in an authoritarian state or if the threat posed by cybercriminals is so severe that citizens and businesses require devices equipped with very strong encryption).

Second, very strong encryption might be morally permissible if other means are sufficient for legitimate law enforcement purposes (e.g., the use of bulk metadata, hacking, and insertion of snooping devices is sufficient).

B. Political Communication: Freedom and Responsibility in Cyberspace (Chapter 3)

Social media platforms, such as Facebook, Twitter, YouTube and TikTok (a Chinese-owned company based in Beijing), are used by billions of communicators worldwide, as are search engines, such as Google. Their use has thereby enabled the moral right to political communication to be exercised on a very large scale. However, the advent of social media platforms has brought with it, firstly, a desire and a capacity on the part of authoritarian governments to censor legitimate political communications. Secondly, in part as the result of the use of cybertechnology (e.g., bots), big data analytics, and psychological manipulative techniques, the advent of social media platforms has led to an exponential increase in the spread of disinformation, misinformation, conspiracy theories, hate speech and propaganda/ideology/quasi-ideology/groupthink (i.e., an exponential increase in computational propaganda, on the part of a wide array of actors, including individual citizens, single-issue pressure groups, right-wing and left-wing extremist groups, terrorist groups, criminal organizations, and in some cases, such as Russia and China, governments). So, the moral right of freedom to communicate has frequently not been exercised responsibly. Moral obligations to seek and communicate truths, rather than falsehoods, have not been discharged, resulting in large-scale social, political, institutional, and ultimately physical harm. How is computational propaganda to be countered while respecting the right to political communication?

Our strategy emphasizes three related underlying conditions that facilitate computational propaganda: (1) the strength of epistemic norms (some such norms being a species of joint action) in a population targeted by computational propaganda; (2) the intellectual health of the epistemic institutions (understood as organizations providing a collective good through joint epistemic activity), such as schools, universities, and news/comment media companies (redesigned, if necessary, to ensure that they undertake their fundamental institutional roles (e.g., as a responsible free press) in that population, and; (3) their degree of embeddedness in, and influence on, the population that hosts them. In this context, we recommend a strategy for institutional redesign of the global technology companies and of public political communication on social media platforms. Here there are several guiding ethical principles.

First, there is the general principle that there is a very morally weighty public interest in liberal democracies in efficient, effective channels of public communication that are accessible to all (a collective good), a public forum for political communication that is accessible to all, and compliance with norms of evidence-based truth seeking. This public interest overrides private interests, commercial or otherwise. This principle evidently has the following consequences for big tech companies.

Insofar as the big tech companies are to remain market-based companies, they must respect the principles of free and fair competition; accordingly, they might need to be downsized to achieve this. Insofar as they are infrastructure providers of platforms, then they must be redesigned to ensure that they provide the required collective good(s). This may require them or, perhaps, some of them to be transformed into publicly owned enterprises, and at the very least it would require greater transparency in relation to, for instance, the algorithms they use. Moreover, the regulation of content cannot be left to the tech giants in the absence of their having the legal status of publishers. Perhaps the entire compliance task needs to be performed by an external, independent institution (if this is practicable), although, if it is practicable, it is a task that should be paid for by the tech companies themselves and/or their advertisers or others who use their platforms.

Further, there needs to be mandatory licensing of *mass* social media social platforms (e.g., monopolist or oligopolist platforms), with the licences to be held conditionally on the content on their platforms complying with legal requirements, their compliance or noncompliance to be determined and adjudicated by an independent statutory authority established by government (e.g., the Australian Office of e-Safety Commissioner). This external independent institution might need to have the fact-checking role in addition to monitoring content for the purpose of identifying illegal content.

The second principle pertains to lawful content which, nevertheless, fails to meet minimum epistemic and moral standards (e.g., is demonstrably false). This principle needs to be applied consistently with the principle of freedom of *interpersonal* (as opposed to organizational) communication. Accordingly, content that is demonstrably false, *and* which is significantly artificially or otherwise illegitimately *amplified* (e.g., by means of bots), is to be liable to removal by social media platforms, but only in accordance with the (publicly transparent) adjudications of the above-mentioned independent statutory authority. The minimum epistemic and moral standards in question to be established by the independent statutory authority following on a process of public debate, expert input, and so on.

The third principle pertains to the obligations of public communicators to respect privacy rights. Account holders with mass social media platforms can retain their anonymity as far as their public communications are concerned (with some exceptions—see next point), but nevertheless must be legally required to register with the independent statutory authority, which will then issue a unique identifier only after verifying the identity of the account holder (e.g., by means of his or her passport, driver's licence and the like). This will enable them to be identified and prosecuted if they use their accounts to engage in unlawful online activity.

However, communicators of politically significant content (including, but not restricted to, content with national security implications) on mass media channels of public communication who have very large audiences (e.g., greater than 100 thousand followers) will be legally required to be publicly identified (other things being equal).

C. Criminal Justice (Chapter 4)

Machine Learning

Machine learning (ML) is a powerful new technology that, however, has its limitations. For instance, the data sets it relies on may well contain false or biased data, and in relation to predictive policing and prediction of legal adjudications, there is reliance on the potentially false assumption that future criminal behaviour will repeat past criminal behaviour and that new legal cases will have similar features to past ones. Thus, profiling practices that rely on ML techniques that utilize such data sets can end up generating morally unjustified, racially based profiles of offenders, and thereby entrench existing racist attitudes among police officers and others.

The use of ML techniques must respect moral principles constitutive of liberal democracies, such as the principle that the state has no right to interfere with a citizen if the actions of the citizen have not otherwise reasonably raised suspicion of unlawful behaviour. For instance, possession of data indicating a crime hotspot would not in and of itself justify stopping and searching, or arresting a person merely because the person happened to be at that location. Accordingly, if such a person was to be stopped and searched or arrested, there would need to be additional evidential facts based on, presumably, real-time observation (e.g., visual evidence of carrying a gun).

Universal DNA Databases

The collection of genomic and biometric data in population-wide or other bulk databases—notably, DNA profiles and facial-image data—is rapidly expanding, and information obtained from DNA analysis can be integrated with facial-image data and other forms of personal data to generate a detailed picture of individual lives. Further, techniques for extracting information from DNA include ones that enable not simply the identification of an individual as an offender, or at least as present at a crime scene (as with traditional DNA techniques), but rather the extraction of detailed genetic information associated with a person's externally visible physical traits, ancestry, ethnicity, and inherited diseases.

DNA is a powerful weapon in combating crime—and serious crimes, such as murder, rape, and grievous bodily harm, in particular. However, the genome of a person is constitutive of that person's individual-specific (biological and personal) identity; as such, a person has stringent (but not absolute) privacy/autonomy rights to their DNA profile. That said, there is a complication. That same genome is *in part* constitutive of the individual-specific (biological) identity of the person's relatives. Accordingly, there is a species of joint right to control genomic data in play here, and not merely an exclusively individual right. A further stringent (but not absolute) moral right is the right not to self-incriminate, a form of self-defence. People who have committed a heinous crime retain the right not to, in effect, speak against themselves or otherwise intentionally facilitate their own conviction.

Considering these moral principles, we offer the following ethical guidelines. First, universal databases should not be permitted if they require compelling everyone to provide DNA. Rather, only the DNA of those convicted of serious crimes should be collected and retained permanently.

Second, a person reasonably suspected by law enforcement of committing a serious crime, or who is among a group of familial relatives, one or more of whom is suspected of committing a serious crime, has (respectively) an individual or joint (i.e., collective) moral responsibility and ought to have a (derived) legal responsibility to provide their DNA to law enforcement for exculpatory or inculpatory purposes (providing that adequate data security and disposal protocols, such as destruction of the data within a reasonable time-frame, are in place).

Third, the individual moral responsibility to provide one's DNA to law enforcement is overridden by the moral right not to self-incriminate.

Fourth, law enforcement should not have the legal right to access DNA databases collected for other purposes, except in two sorts of case. In the first kind of case, there is a particular already uniquely identified person who is reasonably suspected of having committed a serious crime, and access to their DNA data is granted under warrant. In the second kind of case, there is a particular already uniquely identified person who is *not* suspected of having committed a serious crime, but who is a member of a group of familial relatives, one or more of whom are reasonably suspected by law enforcement of having committed a serious crime, and access to the nonsuspect's DNA data is granted under warrant.

Fifth, persons intending to provide their DNA for another purpose (e.g., to a health provider or to a commercial provider to determine their ancestry) have a moral right, and should have a legal right, to be informed that their DNA data might be accessed by law enforcement in the above-described circumstances.

Facial-Recognition Technology

One's facial image is an image of a constitutive feature of one's identity (i.e., one's face), and given the tight connection between identity and autonomy, control of one's facial image is importantly connected to individual autonomy—specifically, a person has a moral right to control the use of images of his or her face (e.g., digital photos). However, unlike DNA profiles, facial images are easily and surreptitiously obtainable, and because of the widespread use of social media, ubiquitous. Moreover, social media is a rich source of data, exploited by companies such as Clearview AI. Accordingly, control of the uses of one's facial image is increasingly difficult, and therefore, given the possibility of using facial-recognition technology to identify and track a person, the threat to individual privacy and autonomy posed by facial-recognition technology is considerable.

In a liberal democracy, to reiterate, the state has no right to seek evidence of wrongdoing on the part of a particular citizen or to engage in selective monitoring of that citizen, if the actions of the citizen in question have not otherwise reasonably raised suspicion of unlawful behaviour, and if the citizen has not had a pattern of unlawful past behaviour that justify monitoring. Moreover, in a liberal democratic state there is a presumption against the state monitoring the citizenry. As stated above under Privacy/Autonomy, Confidentiality, and Security, this presumption can be overridden for specific purposes, but only if the monitoring in question is not disproportionate, is necessary or otherwise adequately justified and kept to a minimum and is subject to appropriate accountability mechanisms. Accordingly, the use of facial recognition by law enforcement should be prohibited unless it can meet these criteria. We note that, given the invasive nature of using facial-recognition technology to identify and track a person, the criterion of proportionality may well be difficult to meet.

D. Public Health, Pandemics, and Security (Chapter 5)

Health and Medical Information Commons

A health and medical information commons, involving the use of big data, AI, and cybertechnology, produces a *collective epistemic good* that is in the service of public health (e.g., combating pandemics) which is a collective good. If this personal data is necessary to yield the required quantum of knowledge to combat a pandemic, which is a serious threat to public health— and relatedly, national security (and, potentially, global security)—then, other things being equal (e.g., all necessary steps have been taken to ensure the data is secure[2]), there is a *collective moral responsibility* (understood as a joint moral responsibility) on the part of members of the relevant population to provide this data. Moreover, potentially this moral responsibility overrides

---

[2] Such as, for instance, homophonic encryption.

ownership rights. Privacy can be mitigated by anonymization and data security concerns (e.g., ransomware attacks) by various means, including the use of distributed models of data analysis. However, the degree to which privacy is necessarily reduced, taken in conjunction with the magnitude of the data security risk and the consequences thereof (e.g., ransomware attacks), are moral costs that needs to be overridden by the collective goods likely to be provided, if a health and information commons (or any such information commons or like database) is to be morally justified.

Contact Tracing and Phone Applications

If the use of phone applications for contact tracing inter alia is effective and necessary to combat a pandemic that is a serious threat to public—and relatedly, national security (and, potentially, global security)—then, other things being equal (e.g., the application is affordable), there is a collective moral responsibility on the part of members of the relevant population to use this application, and therefore an individual responsibility on the part of each member to do so (held jointly with the individual responsibility of other members). Moreover, potentially this moral responsibility overrides privacy rights and data security concerns. Though, as above, the moral weight to be attached to the benefits provided by such phone applications needs to outweigh the reduction in privacy, taken in conjunction with the magnitude of the data security risk (and the consequences thereof).

Vaccination

Cybertechnology is deeply implicated in combating pandemics, including the creation and use of vaccines—perhaps the most important tool in combating pandemics, like COVID-19. There are, as we have just been discussing, a variety of ethical issues that arise in relation to the use of cybertechnology in combating pandemics. However, there is a moral question that is prior to these other issues, and that has implications for our response to these issues—namely, the moral obligation, or lack thereof, to be vaccinated.

   If vaccination of the majority of the population is effective and necessary to combat a pandemic, which is a serious threat to public health—and relatedly, national security—then, other things being equal (e.g., the vaccine is safe), there is an enforceable collective moral responsibility on the part of members of the relevant population to be vaccinated, and therefore an enforceable individual responsibility on the part of each member to do so, absent special circumstances (e.g., a person with a serious adverse reaction to the vaccine). Enforcement would consist of enforced isolation if the vaccine were refused.

E. Cyber Conflict (Chapter 6)

Cyber Conflict and Covert Political Action

Cyberconflicts, including ones involving criminal organizations that threaten national security (either by their own self-directed actions or by their actions performed under the direction of states), frequently do not rise to the threshold of conflict reasonably characterized as war. Rather, they have been instances of conflicts short of war. These forms of cyberconflict are more appropriately regarded as instances of covert political action—a species of conflict short of war—or as an ancillary means of fighting a conventional war. However, just war theory is not an adequate normative theory of covert political action, since the latter is a species of dirty hands activity, even though it may be morally justified, in the service of the collective good of national security—and relatedly, the collective good of global cybersecurity. So, there is a pressing need to provide a normative theory of cyberconflict, understood as covert political action. That said, appropriately adjusted, the principles of necessity, proportionality, and discrimination are applicable to covert political action and to cyberconflict. In addition, there are two principles of reciprocity that are applicable. The first principle of reciprocity is a retrospective principle that justifies the pursuit of a nation-state's national security, as opposed to merely national interest, by means of covert political cyberattacks, given a hostile state is doing so against it—although the principles of necessity, proportionality, and a relaxed principle of discrimination remain in play. The second principle is prospective in form. It is a tit-for-tat principle in the service of bringing about in the future a more morally desirable state of affairs than exists in the present: a morally justifiable equilibrium under international law in which serious forms of cyberconflict (e.g., attacks on critical infrastructure) are eliminated or greatly reduced by virtue of the discharging of the collective responsibility of state actors to comply under the terms of the international laws in question.

Cognitive Warfare

Cognitive warfare is a species of cyberconflict waged by means of disinformation, propaganda, hate speech, and the use of psychological manipulation techniques. It seeks to sow discord in a polity and undermine political and other institutions. Thus, cognitive warfare involves the use of computational propaganda, and therefore the countermeasures to computational propaganda (see above) are also serviceable as counter measures to cognitive warfare. However, these measures are not sufficient. Here it is important to keep in mind that there is no moral right on the part of foreigners to use the channels of mass communication to communicate politically significant content to the domestic citizenry (e.g., no moral right of the members of RT to communicate to US citizens). Rather, US citizens have a joint right (potentially exercised via their democratically elected legislature) to ban such use by foreigners foreign of the channels of mass communication. This joint moral right with respect to macrolevel politically significant speech is consistent with the *microlevel interpersonal right* of each member of a community to listen to foreign state actors via channels of communication that are not mass media channels of public communication.

In addition to these essentially defensive measures, it may be necessary to have recourse to offensive cognitive warfare measures. But these measures give rise to problems for a liberal democratic state, unless only culpable members of the hostile foreign state in question can be targeted and unless these measures do not have lethal effects.

Unfortunately, however, these nonkinetic measures may have lethal or other kinetic effects that are characteristic of kinetic wars. Consider, for instance, the dissemination of disinformation, propaganda, and hate speech designed with a view to inciting violence. Moreover, many of these nonkinetic measures will not be effective if they only target culpable attackers. Consider, for instance, propaganda comprising (in part) in disinformation that is aimed at weakening the enemy's war effort, in the overall context of a kinetic war; the obvious target is the civilian population.

At this point, the general principles of necessity and proportionality have a clear application. Moreover, the culpable/nonculpable distinction as it applies to the use of the methods of cognitive warfare has much less purchase.

The justification for the use of the offensive methods of cognitive warfare by a liberal democratic state could rely on the following general considerations: (1) the nature and extent of the harm done by the use of the cognitive warfare technique in question (e.g., creating false beliefs in nonculpable citizens); (2) the use of the cognitive warfare technique in question is effective, and no more effective, less harmful technique is available (all things considered) to achieve the moral weighty military or political end it serves; (3) the use of a morally wrongful means, taken in conjunction with the harm done by it, is not disproportionate relative to the moral weight to be attached to the military or political end ultimately achieved by this means (e.g., the morally weighty end of facilitating victory in the just kinetic war in question) greatly outweighed the harm done.

Autonomous Weapons

The use of autonomous cyberweapons (i.e., out-of-the-loop weapons, as opposed to on-the-loop or in-the-loop) ought to be prohibited, whether in conventional wars or in covert political actions. That is, all weapons should be designed in such a manner that even if they are not directly and at all times operated by a human, they do have an override function and an on/off switch controlled by a human operator (i.e., there is a human operator in-the-loop even if not on-the-loop).

Glossary of Technical Terms

ACM	Association of Computing Machinery.
AES	Advanced Encryption Standard.
Affetiva	Company spun off from work by Rosalind Picard at MIT to pick up emotions in real time from microexpressions in video data.
alphanumeric	Characters that are either letters, upper- or lowercase, and the digits 0 to 9.
Android	Operating system developed by Google for smartphones, dominant outside iPhones.
ASCII	A code mapping letters, numbers, and special characters (such as ampersands and brackets) to binary numbers.
AWS	Amazon Web Services. One of the leading providers of cloud services.
Baidu	Large Chinese company specializing in activities such as ecommerce.
big data	Generally refers to very large datasets used for data mining and machine learning.
biometrics	Identification/authorization using some aspect of the human body, such as fingerprints or iris scans.
bit	the smallest unit of information, a binary, yes/no, on/off choice.
blockchain	A blockchain is a *distributed ledger*—a decentralized, distributed, digital ledger. It uses cryptography to secure a chain of blocks of information, with each new block including a hash of the previous block.
Bluetooth	A short-range (typically, several metres) communication protocol, for wirelessly connecting devices together (such as headphones) to a smartphone.
bot	An autonomous software agent.
Brave browser	One of several web browsers with enhanced privacy.
byte	Eight bits.
CERT	Cyber Emergency Response Team.
chatbot	A software agent for mediating a conversation, usually by text, on a website.
chatGPT	one of the most prominent examples of the new wave of interest in Generative AI.

Clearview AI	A company that has constructed huge databases of a billion or more images, scraped from social media and other sites. The ethics of this practice have been questioned.
Client-Side Scanning (CSS)	Inspection of documents of any kind (including images, and so on) before encryption, or after decryption, on a user device, such as a mobile phone.
Cloud Storage Computing	Storage or computation provided as a pay-by-use service, based on numerous servers distributed over the internet.
cookie	A data item stored by a website in a user's browser to record aspects of their interaction with the site.
Dark Web	A large, hidden part of the Web, accessed through the ToR browser and the GRAMS search engine.
DDoS	A Distributed Denial of Service, where the attack comes from many sources, typically an illicit botnet.
deep learning	A recent highly successful variant of machine learning, frequently using very large datasets and large sets of internal parameters.
Diffie-Hellman	The eponymous Diffie-Hellman key exchange is used for exchanging a private key over a public channel.
DMA	Digital Markets Act (EU).
DoS	A Denial of Service (DoS) is any type of attack where the attackers (hackers) attempt to prevent legitimate users from accessing the service. The attacker floods the service with too many requests to handle.
DNS	Domain Name Server.
DP3T Protocol	Decentralized Privacy-Preserving Proximity Tracing, protocol developed for COVID-19 contact tracing, used in the UK, and implemented in modified form jointly by Google and Apple.
DRM	Digital Rights Management.
DSA	Digital Services Act (EU).
DYN	A large domain name service.
e-shredding	Secure destruction of a file by writing over it with something else, such as zeros. Deleting a file usually only removes it from an index. It does not destroy the contents.
ECC	Elliptic Curve Cryptography.
Encrochat	A specialized phone, supposedly end-to-end encrypted, widely used by criminals.

end-to-end encryption	Encryption at the initial sending device (such as a phone) and decryption at the receiving device, with no intermediate decryption and re-encryption.
ethernet	A protocol for sending messages (frames) along cable and optical fibre, used widely throughout the internet.
Facetime	Encrypted video and messaging service offered by Apple on its devices.
firewall	Software that controls access to and from an external network.
FTP	File Transport Protocol.
gait analysis	Identification of a person through their walking style.
GDPR	General Data Protection Regulation (EU).
Generative AI	Artificial Intelligence software capable of synthesizing plausible, human-like text, and already showing promise for creating images and video. Central to the production of deep fakes.
GitHub	A worldwide, largescale, software repository.
hash function	Hash functions are trapdoor functions where it is easy to go one way but not the other, like breaking an egg. An ultra-simple (and abysmal) hash of a number might be the remainder after dividing by some number N. So, with N = 11 a hash value of 3 could have come from 3, 14, 25, and so on. These ambiguities are called collisions, where two numbers hash to the same value (i.e., they collide). In this primitive example, it's obvious that there can be no more than eleven hash values. All numbers hash to just one of these eleven values. The vast literature on hash functions endeavors to find algorithms with minimal collisions and other constraints. Hash functions are widely used in cryptography and computer science, generally. The current standard is SHA.
HIPAA	Health Insurance Portability and Accountability Act.
honeypot	Part of a website or other computer service used to attract hackers, where they can be observed without them being able to do any damage.
ICANN	Internet Corporation for Assigned Names and Numbers.
iCloud	The Apple cloud service.
IEEE	Institute of Electrical and Electronic Engineers.
IETF	Internet Engineering Task Force.
internet	The worldwide collection of networked computers and devices, each with a unique address, through which it may communicate with any other such device or computer.

internet domain	A collection of IP addresses grouped in some way, such as for universities (*.edu*) or countries (*.uk*).
IoT	Internet of Things.
IP address	An address of a computer with access to the internet, appearing as a series of numbers separated by dots, such as 167.33.255.2.
IPO	Initial Public Offering.
Kaseya hack	A ransomware attack on Kaseya's remote system update software, which brought down the Swedish Coop supermarket chain.
key	Used for encryption and decryption, like a password but usually much longer. It is not usually entered by a user, but accessed via a password, biometric or some sort of secure ID.
LASER Program	Los Angeles Strategic Extraction and Restoration Program.
machine learning (ML)	The use of computer software to learn from examples how to do a particular task, which may be purely epistemic or involve the control of devices.
malware	Portmanteau word for malicious software.
metadata	Data that describes data. So, for an email message metadata would include sender, recipient, subject, and so on.
MITM	Man in the Middle Attack. For example, suppose Alice and Bob have an encrypted channel in which each encrypts a message with the other's public key, which only the recipient can decrypt using his or her private key. Charlie, the man in the middle, attacks by posing as Alice when the channel is set up with his own public-private key pair. Bob's messages are received, decrypted by Charlie, and sent on to Alice, with Charlie's key, which Alice assumes to be Bob's.
NDNAD	National DNA Database (UK).
node, internet	A point of presence on the internet, with its own IP address.
NSA	National Security Agency.
onion routing	A way of securely routing information. Packets have an onion-like set of addresses and travel over many nodes. No node can see all the layers of the onion. Each node strips off the top layer to find the address of to where it should send the packet.
OpenWhispers	Software group founded by Moxie Marlinspike, developer of the Signal protocol for encrypted messaging.

patch	An update to a piece of software or operating system, for increased security or functionality, which does not require reinstallation of the software.
Petya	Petya is a family of encrypting ransomware, targeting Microsoft Windows-based systems.
phishing	Phishing is the attempt to obtain sensitive information such as user names, passwords, and credit card details (and money), often for malicious reasons, by disguising as a trustworthy entity in a digital communication, such as email.
port	A network access point on computer.
PPK	Denotes public and private keys, a cornerstone of cryptography. A document can be encrypted with either, but only decrypted with the complementary one. Thus, an email encrypted with a private key can be decrypted by anyone with the public key and vice versa.
QR Code	A two-dimensional pattern of black squares that can be decoded to a URL or other information. Decoding apps are usually available on smartphones.
ransomware	A form of malware (malicious software), which encrypts the contents of a computer disc. The attacker demands a ransom from the victim, promising to restore access to the data upon payment.
RC4	A stream cipher invented by Ron Rivest, now modified for greater security.
RSA	RSA algorithm developed by Rivest, Shamir, and Adleman; the foundational PPK system.
RTB	Real Time Bidding.
Signal	An end-to-end encrypted messaging and voice call application, part of the Open Whispers development of Moxie Marlinspike.
signature, digital	A cryptographic signature on a document, usually a comprising message digest of the document and its encryption with a private key. Anybody can check the signature with the public key and compare the message digest with the message digest of the supposed document.
social media	A generic term for computer platforms that facilitate communication between individuals and groups. Examples include Facebook, Twitter, Instagram, and TikTok.
spam	Mass unsolicited email, often with malintent.
SSD	Solid State Device: a form of computer storage based on solid state memory.
SSL	Secure Socket Layer (a ubiquitous part of a network protocol stack).

Stuxnet	Malware introduced into the Iranian Uranium enrichment centrifuges to disable them.
system log	A record kept by the operating system of all its activities, from file access to data comms.
Telegram	An end-to-end encrypted messaging service, with undisclosed cryptographic algorithms.
Tencent	Chinese internet technology company.
Titan	Hardware security chip in Android phones.
TLS	Transport Layer Security (part of a network protocol stack).
ToR	The Onion Router, literally. The ToR browser uses onion routing.
Trojan Horse	A type of malware that is often disguised as legitimate software. Trojans can be employed by cyberthieves and hackers trying to gain access to users' systems. Users are typically tricked by some form of social engineering into loading and executing Trojans on their systems.
troll	A user, possibly a bot, making critical, usually abusive, posts on social media.
Turing test	A conceptual test created by computing pioneer, Alan Turing, to determine if an entity is a human or a computer.
Vernam cipher	A cipher system, predating computers, using a code book to convert each character in a message by combining it with a corresponding character in the code book.
virus	A type of malware that, when executed, replicates itself by modifying other computer programs and inserting its own code. Like a biological virus, it uses software and services within the host to execute its code.
VPN	Virtual Private Network.
Wannacry	A worm that spreads by exploiting vulnerabilities in the Microsoft Windows operating system. Once installed, it encrypts files and demands a payment to decrypt them.
WhatsApp	A widely used end-to-end encrypted messaging system, originally from Open Whispers, but now owned by Meta (Facebook).
WMD	Weapons of Mass Destruction.
worm	A standalone malware computer program that replicates itself to spread to other linked computers.
WPA	Wi-Fi Protected Access.
XOR	Bitwise exclusive OR (true if two bits are different, false if they are the same).

Bibliography

Alazab, M. "Russia Is Using an Onslaught of Cyber Attacks to Undermine Ukraine's Defence Capabilities." *The Conversation*, 24 February 2022. https://theconversation.com/russia-is-using-an-onslaught-of-cyber-attacks-to-undermine-ukraines-defence-capabilities-177638.

Allendorfer, W. H., and S. C. Herring. "ISIS vs US Government: A War of On-line Video Propaganda." *First Monday*, 14 December 2015. https://firstmonday.org/ojs/index.php/fm/article/download/6336/5165.

Allhoff, F., A. Henschke, and B. J. Strawser, eds. *Binary Bullets: The Ethics of Cyberwarfare*. Oxford: Oxford University Press, 2016.

Anderson, David. Independent Reviewer of UK Terrorism Legislation, UK Government. "Report of the Bulk Powers Review." 2016. https://terrorismlegislationreviewer.independent.gov.uk/wp-content/uploads/2016/08/Bulk-Powers-Review-final-report.pdf.

Arkin, Ronald. "The Case for Ethical Autonomy in Unmanned Systems." *Journal of Military Ethics* 9 (2010): 332–41.

Ashworth, Andrew. "Self-Incrimination in European Human Rights Law—A Pregnant Pragmatism?" *Cardozo Law Review* 30 (2008): 751.

Backes, A., and A. Swab. "Cognitive Warfare: The Russian Threat to Election Integrity in the Baltic States." Paper presented at Belfer Center for Science and International Affairs, Harvard Kennedy School, Boston, MA, November 2019.

Balkin, Jack M. "Free Speech Is a Triangle." *Columbia Law Review* 118 (2018): 2040–41.

Bellaby, Ross. *The Ethics of Intelligence*. London: Routledge, 2014.

Benn, Stanley. *A Theory of Freedom*. Cambridge: Cambridge University Press, 1988.

Berger, J. M. "Defeating IS Propaganda. Sounds Good, But What Does It Really Mean?" The Hague: International Centre for Counter-Terrorism, 2017. https://icct.nl/publication/defeating-is-ideology-sounds-good-but-what-does-it-really-mean/.

Bernal, B., C. Carter, I. Singh, K. Cao, and O. Madreperla. "Cognitive Warfare." Brussels: NATO Report, 2020.

Bok, Sissela. *Lying: Moral Choice in Public and Private Life*. New York: Pantheon Books, 1978.

Bossomaier, Terry, Steven D'Alessandro, and Roger Bradbury. *Human Dimensions of Cybersecurity*. Boca Raton, FL: CRC Press, 2019.

Burke, Paul, Doaa El Nakhala, and Seumas Miller, eds. *Global Jihadist Terrorism: Terrorist Groups, Zones of Armed Conflict and National Counter-Terrorism Strategies*. Cheltenham: Edward Elgar, 2021.

Christen, Markus, Bert Gordijn, and Michele Loi, eds. *The Ethics of Cybersecurity*. Dordrecht: Springer, 2020.

Cocking, Dean, and Jeroen van den Hoven. *Evil On-line*. Hoboken: Wiley Blackwell, 2018.

Cook, James L., "Is There Anything Morally Special about Cyber War?" In *Cyber War: Law and Ethics for Virtual Conflicts*, edited by J. D. Ohlin, K. Govern, and C. Finkelstein, 20–33. Oxford: Oxford University Press, 2015.

Cook-Deegan, Robert, Mary Majumder, and Amy McGuire. "Introduction: Sharing Data in a Medical Information Commons." *Journal of Medicine, Law and Ethics* 47, no. 1 (2019): 7–11.

Cuellar, Mariano-Florentino and Aziz Z. Huq. "Review of Zuboff's *Age of Surveillance Capitalism*." *Harvard Law Review* 133 (2020).

D'Alessio, A. "Computational Propaganda: Challenges and Responses." *Academia Letters* Article 3468 (2021). https://doi.org/10.20935/AL3468.

Danks, Joseph and David Danks. "Beyond Machines." In *Binary Bullets: The Ethics of Cyberwarfare*, edited by F. Allhoff, A. Henschke and B. J. Strawser, 177–200. Oxford: Oxford University Press, 2016.

Degeling, Chris, Stacy Carter, Antoine van Oijen, Jeremy McAnulty, Vitali Sintchenko, Annette Braunack-Mayer, Trent Yarwood, Jane Johnson, and Gwendolyn Gilbert. "Community Perspectives on the Benefits and Risks of Technologically Enhanced Communicable Disease Surveillance Systems." *BMC Medical Ethics* 21, no. 31 (2020): 1–14.

Dennis, Ian. "Instrumental Protection, Human Right or Functional Necessity? Reassessing the Privilege against Self-Incrimination." *Cambridge Law Journal* 54, no. 2 (1995): 342–76.

Dipert, Randall R. "Ethics of Cyberwarfare." *Journal of Military Ethics* 9 (2010): 384–410.

Dirks, Emile, and James Leibold. "Genomic Surveillance: Inside China's DNA Dragnet." *ASPI Policy Brief*, Report no. 34 (2020).

Duff, R. A. *Answering for Crime: Responsibility and Liability in the Criminal Law*. Oxford: Hart Publishing, 2007.

Eberle, Christopher J. "Just War and Cyberwar." *Journal of Military Ethics* 12 (2013): 54–67.

Ellul, Jacques. *Propaganda: The Formation of Men's Attitudes*. Translated by Konrad Kellen and Jean Lerner. New York: Random House/Vintage, 1973.

Enderle, Georges. "Whose Ethos for Public Goods in the Global Economy?" *Business Ethics Quarterly* 10, no. 1 (2000): 131–44.

Enemark, Christian. *Moralities of Drone Violence*. Edinburgh: Edinburgh University Press, 2023.

Feltes, Jonas. *CBRN Threats, Counter-Terrorism and Collective Moral Responsibility*. PhD diss., Delt University of Technology, 2021.

Ferguson, Andrew G. *The Rise of Big Data Policing*. New York: New York University Press, 2017.

Ferguson, Andrew G. "Predictive Policing Theory." In *Cambridge Handbook of Policing in the United States*, edited by T. R. Lave and E. J. Miller, 491–510. Cambridge: Cambridge University Press, 2020.

Ford, Shannon. "*Jus Ad Vim* and the Just Use of Lethal Force-Short-of-War." In *Routledge Handbook of Ethics and War: Just War in the 21st Century*, edited by F. Allhoff, N. Evans, and A. Henschke, 63–75. Abingdon: Routledge, 2013.

Freadman, Richard, and Seumas Miller. *Rethinking Theory: A Critique of Contemporary Literary Theory and an Alternative Account*. Cambridge: Cambridge University Press, 1992.

Fried, Charles. "Privacy." *Yale Law Journal* 77, no. 3 (1969): 475–93.

Galliot, J., ed. *Force Short of War in Modern Conflict: Jus ad Vim*. Edinburgh: Edinburgh University Press, 2019.

Gerstein, Robert. "Privacy and Self-Incrimination." *Ethics* 80, no. 2 (1970): 87–95.

Gert, Bernard. *Common Morality*. Oxford: Oxford University Press, 2007.

Gilbert, Margaret. *On Social Facts*. Princeton, NJ: Princeton University Press, 1989.

Gore, Al. *The Assault on Reason*. New York: Penguin, 2007.

Gould-Davies, Nigel. "The Russia Report: Key Points and Implications." Hague: IISS, 2020. https://www.iiss.org/blogs/analysis/2020/07/isc-russia-report-key-points-and-implications.

Green, Stuart. *Lying, Cheating and Stealing* Oxford: Oxford University Press, 2006.

Green, Thomas H. *Lectures on the Principles of Political Obligation*. London: Longmans, Green and Co., 1895.

Gross, Michael, and Tamar Meisels, eds. *Soft War: The Ethics of Unarmed Conflict*. Cambridge: Cambridge University Press, 2017.

Harding, Luke. *The Snowden Files: The Inside Story of the World's Most Wanted Man*. London: Guardian Books, 2014.

Head, John G. *Public Goods and Public Welfare*. Durham: Duke University Press, 1974.

Henderson, Gerard. *George Pell: The Media Pile-On and Collective Guilt*. Melbourne: Connor Court, 2021.

Henschke, Adam. *Ethics in an Age of Surveillance: Virtual Identities and Personal Information*. New York: Cambridge University Press, 2017.

Henschke, Adam. "On Free Public Communication and Terrorism Online." In *Counter-Terrorism: The Ethical Issues*, edited by Seumas Miller, Adam Henschke, and Jonas Feltes. Cheltenham: Edward Elgar, 2021.

Henschke, A., A. Reed, S. Robbins, and Seumas Miller, eds. *Counter-Terrorism, Ethics, and Technology: Emerging Challenges at the Frontiers of Counter-Terrorism*. Dordrecht: Springer, 2021.

Hobbes, Thomas. *Leviathan*. Cambridge: Cambridge University Press, [1651] 1996.

Hollis, Duncan B. "Rethinking the Boundaries of Law and Cyberspace." In *Cyber War: Law and Ethics for Virtual Conflicts*, edited by Jens David Ohlin, Kevin Govern, and Claire Finkelstein, 86–93. Oxford: Oxford University Press, 2015.

Horwitz, Jeff. *Broken Code: Inside Facebook and the Fight to Expose Its Toxic Secrets*. New York: Penguin Doubleday, 2023.

Hung, Tzu Chieh, and Tzu Wei Hung. "How China's Cognitive Warfare Works." *Journal of Global Security Studies* 7, no. 4 (2020).

Ingram, Haroro, J. "A Brief History of Propaganda During Conflict." The Hague: International Centre for Counter-Terrorism, 2016. https://icct.nl/publication/a-brief-history-of-propaganda-during-conflict-a-lesson-for-counter-terrorism-strategic-com munications/.

Inness, Julie. *Privacy, Intimacy and Isolation*. New York: Oxford University Press, 1992.

Johnson, Loch. "The "Third Option" in American Foreign Policy." In *National Security Intelligence and Ethics*, edited by Seumas Miller, Mitt Regan, and Patrick F. Walsh. London: Routledge, 2021.

Jones, Peter, ed. *Group Rights*. Aldershot: Ashgate, 2009.

Kant, Immanuel. *Groundwork of the Metaphysics of Morals*. Translated by H. J. Paton. New York: Harper Collins, 1956.

Karnouskos, Stamatis. "Stuxnet Worm Impact on Industrial Cyber-Physical System Security." *IECON 2011—37th Annual Conference of the IEEE Industrial Electronics Society*. Melbourne (November 2011): 4490–94.

Kaye, David, and Michael Smith. "DNA Identification Databases: Legality, Legitimacy, and the Case for Population Wide Coverage." *Wisconsin Law Review* (2003): 413–459.

Kleinig, John, Peter Mameli, Seumas Miller, Douglas Salane, and Adina Schwartz. *Security and Privacy: Global Standards for Ethical Identity Management in Contemporary Liberal Democratic States*. Canberra: ANU Press, 2011.

Klimburg, Alexander. *The Darkening Web: The War for Cyberspace*. New York: Penguin Press, 2017.

Koepsell, David. *Who Owns You?: Science, Innovation and the Gene Patent Wars*. Hoboken, NJ: Wiley-Blackwell, 2015.

Kosseff, Jeff. "Defining Cybersecurity Law." *Iowa Law Review* 103 (2018): 1010.

Kraaijeveld, Steven. "Vaccinating for Whom? Distinguishing between Self-Protective, Paternalistic, Altruistic and Indirect Vaccination." *Public Health Ethics* 13, no. 2 (2020): 190–200.

Ladd, John. "Philosophical Remarks on Professional Responsibility in Organizations." *International Journal of Applied Philosophy* 1, no. 2 (1982): 58–70.

Lin, Herb, "Overview of Relevant IHL Rules and Principles That May Be Challenged by Cyberwar." Paper presented at the *Cyberwarfare, Ethics and International Humanitarian Law workshop*, Geneva, Switzerland, May 21–22, 2014.

List, Christian, and Philip Pettit. *Group Agency*. Oxford: Oxford University Press, 2011.

Lucas, George. "Just in Silico: Moral Restrictions on the use of Cyberwarfare." In *Routledge Handbook of Ethics and War: Just War in the 21st Century*, edited by F. Allhoff et al., 367–82. Abingdon: Routledge, 2013.

Lucas, George, *Ethics and Cyber Warfare*. New York: Oxford University Press, 2017.

Lynch, Michael. *The Internet of Us*. New York: Liveright, 2016.

Macnish, Kevin. *The Ethics of Surveillance*. London: Routledge, 2017.

Macnish, Kevin. "Government Surveillance and Why Defining Privacy Matters in a Post-Snowden World." *Journal of Applied Philosophy* 35, no. 2 (2018): 417–32.

Majumber, Mary, Juli Bollinger, Angela Villanueva, Patricia Deverka, and Barbara Koenig. "The Role of Participants in a Medical Information Commons." *Journal of Law, Medicine and Ethics* 47, no. 1 (2019): 51–61.

Mandiant Intelligence Centre. *APT1: Exposing One of China's Cyber Espionage Units* Washington, DC: Mandiant Intelligence Centre, 2013. http://intelreport.mandiant. com/Mandiant_APT1_Report.pdf.

Mann, Monique, and Marcus Smith. "Automated Facial Recognition Technology: Recent Developments and Approaches to Oversight." *UNSW Law Journal* 40, no. 1 (2017): 121–45.

Martin, James, and Chad Whelan. "Ransomware Through the Lens of State Crime." *State Crime* 12, no. 1 (2023): 1–25. https://www.scienceopen.com/document_file/c292f cee-3ae3-4f60-98cf-11a933700fba/ScienceOpen/SCJ_12_1_Martin%20and%20Whe lan.pdf

Mattei, Tobias A. "Privacy, Confidentiality, and Security of Health Care Information: Lessons from the Recent Wannacry Cyberattack." *World Neurosurgery* 104, no. 1 (2017): 972–74.

May, Larry. *Sharing Responsibility*. Chicago: University of Chicago Press, 1992.

May, Larry. "The Nature of War and the Idea of 'Cyber War.'" In *Cyber War: Law and Ethics for Virtual Conflicts*, edited by J. D. Ohlin, K. Govern, and C. Finkelstein, 3–15. Oxford: Oxford University Press, 2015.

Mayer-Schönberger, Viktor, and Kenneth Cukier. *Big Data: A Revolution that Will Transform How We Live, Work and Think*. London: John Murray, 2013.

McGuire, Amy, Jessica Roberts, Sean Aas, and Barbara Evans, "Who Owns the Data in a Medical Information Commons?" *Journal of Law, Medicine and Ethics* 47, no. 1 (2019): 62–69.

McKean, Margaret, and Elinor Ostrom. "Common Property Regimes in the Forest." *Unasylva* 180, no. 46 (1995): 3–15.

Mellema, Gregory. "Collective Responsibility and Qualifying Actions." In *Midwest Studies in Philosophy*, vol. XXX, edited by Peter A. French, 168–75. 2006.

Mill, John Stuart. *On Liberty*. London: Longman, Roberts and Green, 1869.

Miller, Seumas, "Truth-Telling and the Actual Language Relation." *Philosophical Studies* 49, no. 2 (1986): 281–94.

Miller, Seumas. "Ideology, Language and Thought." *Theoria* 74 (1989): 97–105.

Miller, Seumas. "Rationalising Conventions." *Synthese* 84 (1990): 23–41.

Miller, Seumas. "Joint Action." *Philosophical Papers* 21, no. 3 (1992): 275–97.

Miller, Seumas. "Collective Rights." *Public Affairs Quarterly* 1, no. 4 (1999): 331–46.

Miller, Seumas. "Academic Autonomy." In *Why Universities Matter*, edited by Tony Coady, 110–31. Sydney: Allen and Unwin, 2000.

Miller, Seumas. "Collective Rights and Minorities." *International Journal of Applied Philosophy* 14, no. 2 (2000): 241–57.

Miller, Seumas. *Social Action: A Teleological Account*. Cambridge: Cambridge University Press, 2001.

Miller, Seumas. "Individual Autonomy and Sociality." In *Socialising Metaphysics: Nature of Social Reality*, edited by F. Schmitt, 269–300. Lanham, MD: Rowman & Littlefield, 2003.

Miller, Seumas. "Institutions, Collective Goods and Individual Rights." *ProtoSociology: An International Journal of Interdisciplinary Research* 18–19 (2003): 184–207.

Miller, Seumas. "Collective Moral Responsibility: An Individualist Account." *Midwest Studies in Philosophy*, vol. XXX, edited by Peter A. French, 176–93. 2006.

Miller, Seumas. "Civilian Immunity, Forcing the Choice and Collective Responsibility." In *Civilian Immunity*, edited by I. Primoratz, 137–66. Oxford: Oxford University Press, 2007.

Miller, Seumas. "Noble Cause Corruption in Politics." In *Politics and Morality*, edited by I. Primoratz, 92–112. Basingstoke: Palgrave Macmillan, 2007.

Miller, Seumas. "Collective Responsibility and Information and Communication Technology." In *Information Technology and Moral Philosophy*, edited by J van den Hoven and J. Weckert, 226–50. New York: Cambridge University Press, 2008.

Miller, Seumas. *The Moral Foundations of Social Institutions: A Philosophical Study*. New York: Cambridge University Press, 2010.

Miller, Seumas. "Moral Responsibility, Collective Action Problems and the Dual Use Dilemma in Science and Technology." In *On the Dual Uses of Science and Ethics*, edited by B. Rappert and M. Selgelid, 185–206. Canberra: ANU Press, 2013.

Miller, Seumas. "Police Detectives, Criminal Investigations and Collective Moral Responsibility." *Criminal Justice Ethics* 33, no. 1 (2014): 21–39.

Miller, Seumas. "Collective Responsibility for Robopocolypse.," In *Super Soldiers: The Ethical, Legal and Social Implications*, edited by J. Galliott and M. Lotze, 153–66. Aldershot: Ashgate, 2015.

Miller, Seumas. "Design for Values in Institutions." In *Handbook of Ethics, Values & Technological Design*, edited by I. Poel, J. Van den Hoven, and P.Vermaas, 769–81. Dordrecht: Springer, 2015.

Miller, Seumas. "The Global Financial Crisis and Collective Moral Responsibility." In *Distribution of Responsibilities in International Law*, edited by Andre Nollkaemper and Dov Jacobs, 404–33. Cambridge: Cambridge University Press, 2015.

Miller, Seumas. "Joint Epistemic Action and Collective Responsibility." *Social Epistemology* 29, no. 3 (2015): 280–302.

Miller, Seumas. "Joint Political Rights and Obligations." *Phenomenology and Mind* 9 (2015): 136–45.

Miller, Seumas. "Assertions, Joint Epistemic Actions and Social Practices." *Synthese* 1, no. 193 (2016): 71–94.

Miller, Seumas. *Corruption and Anti-Corruption in Policing: Philosophical and Ethical Issues*. Dordrecht: Springer, 2016.

Miller, Seumas. "Cyber-Attacks and 'Dirty Hands': Cyberwar, Cyber-Crimes or Covert Political Action?" In *Binary Bullets: The Ethics of Cyberwarfare*, edited by F. Allfhoff, A. Henschke, and B. J. Strawser, 228–50. Oxford: Oxford University Press, 2016.

Miller, Seumas. *Shooting to Kill: The Ethics of Police and Military Use of Lethal Force*. New York: Oxford University Press, 2016.

Miller, Seumas, "Ignorance, Technology and Collective Responsibility." In *Perspectives on Ignorance from Moral and Social Philosophy*, edited by R. Peels, 217–38. London: Routledge, 2017.

Miller, Seumas. *Institutional Corruption: A Study in Applied Philosophy*. Cambridge University Press, 2017.

Miller, Seumas. *Dual Use Science and Technology, Ethics and Weapons of Mass Destruction*. Dordrecht: Springer, 2018.

Miller, Seumas. "Joint Epistemic Action: Some Applications." *Journal of Applied Philosophy* 35, no. 2 (2018): 300–18.

Miller, Seumas. "Machine Learning, Ethics and Law." *Australasian Journal of Information Systems* 23 (2019): 1–15.

Miller, Seumas. "Collective Responsibility as Joint Responsibility." In *Routledge Handbook of Collective Responsibility*, edited by S. Bazargan-Forward and D. Tollefsen, 38–50. New York: Routledge, 2020.

Miller, Seumas. "Freedom of Political Communication, Propaganda and the Role of Epistemic Institutions in Cyberspace." In *The Ethics of Cybersecurity*, edited by M. Christen, B. Gordjin, and M. Loi, 227–44. Dordrecht: Springer, 2020.

Miller, Seumas. "Joint Rights: Human Beings, Corporations and Animals." *Journal of Applied Ethics and Philosophy* 12 (2021): 1–7.

Miller, Seumas. "Predictive Policing: The Ethical Issues." In *Future Morality*, edited by D. Edmonds, 73–82. Oxford: Oxford University Press, 2021.

Miller, Seumas. "Rethinking the Just Intelligence Theory of National Security Intelligence Collection and Analysis." *Social Epistemology* 35, no. 3 (2021): 211–31.

Miller, Seumas. "Epistemic Institutions: A Joint Epistemic Action-based Account." *Nous-Supplement: Philosophical Issues* 32 (2022): 318–416.

Miller, Seumas, "War, Reciprocity and the Moral Equality of Combatants" *Philosophia*, August 2023. https://link.springer.com/article/10.1007/s11406-023-00678-1.

Miller, Seumas. "Cognitive Warfare: An Ethical Analysis" *Ethics and Information Technology* 25, no. 46 (2023): 1–10.

Miller, Seumas, and John Blackler. *Ethical Issues in Policing*. Aldershot: Ashgate, 2005.

Miller, Seumas, and Terry Bossomaier. "Privacy, Encryption and Counter-Terrorism." In *Counter-Terrorism, Ethics, and Technology: Emerging Challenges at The Frontiers Of Counter-Terrorism*, edited by A. Henschke, A. Reed, S. Robbins, and S. Miller, 139–54. Dordrecht: Springer, 2021.

Miller, Seumas, and Ian Gordon. *Investigative Ethics: Ethics for Police Detectives and Criminal Investigators*. Oxford: Wiley-Blackwell, 2014.

Miller, Seumas, A. Henschke, and Jonas Feltes, ed. *Counter-terrorism: The Ethical Issues*. Cheltenham: Edward Elgar, 2021.

Miller, Seumas, and Pekka Makela. "The Collectivist Approach to Collective Moral Responsibility." *Metaphilosophy* 36, no. 5 (2005): 634–51.

Miller, Seumas, Mitt Regan, and Patrick Walsh, ed. *National Security Intelligence and Ethics* London: Routledge, 2021.

Miller, Seumas, and Michael J. Selgelid. "Ethical and Philosophical Consideration of the Dual-Use Dilemma in the Biological Sciences." *Science and Engineering Ethics* 13, no. 4 (2007): 523–80.

Miller, Seumas, and Marcus Smith. "The Ethical Application of Biometric Facial Recognition Technology." *AI and Society* 36 (2021): 167–75.

Miller, Seumas, and Marcus Smith. "Ethics, Public Health and Technology Responses to COVID-19." *Bioethics* 35, no. 4 (2021): 364–71.

Miller, Seumas, and Marcus Smith. "Quasi-Universal Forensic DNA Databases." *Criminal Justice Ethics* 41 (2022): 238–56.

Miller, Seumas, and Patrick F. Walsh. "NSA, Snowden and the Ethics and Accountability of Intelligence Gathering." In *Ethics and the Future of Spying: Technology, Intelligence Collection and National Security*, edited by J. Galliott and W. Reed, 193–204. Abingdon: Routledge, 2016.

Miller, Virginia. *Child Sexual Abuse Inquiries and the Catholic Church: Reassessing the Evidence*. Firenze: Firenze University Press, 2021.

Morgan, E. M. "The Privilege against Self-Incrimination." *Minnesota Law Review* 34, no. 1 (1949).

Moser, Paul. *Knowledge and Evidence*. Cambridge: Cambridge University Press, 1989.

National Science Advisory Board for Biosecurity Framework for Conducting Risk and Benefit Assessments of Gain-of-Function Research. Washington, DC: NSABB, 2015.

Nissenbaum, Helen. *Privacy in Context: Technology, Policy and the Integrity of Social Life*. Stanford: Stanford Law Books, 2009.

Novossioloval, Tatyana A., Simon Whitby, Malcolm Dando, and Graham S. Pearson. "The Vital Importance of a Web of Prevention for Effective Biosafety and Biosecurity in the Twenty-First Century." *One Health Outlook* 3, no. 17 (2021).

Nye, Joseph. "The Regime Complex for Managing Global Cyber Activities." London: Global Commission on Internet Governance, 2014.

Office of the Director of National Intelligence. *A Guide to Cyber Attribution*. September, 2018. https://www.dni.gov/files/CTIIC/documents/ODNI_A_Guide_to_Cyber_Attribution.pdf. Accessed 31/10/2023.

Olson, Mancur. *The Logic of Collective Action*. Cambridge, MA: Harvard University Press, 1965.

Omand, David, and Mark Phythian. *Principled Spying: The Ethics of Secret Intelligence*. Oxford: Oxford University Press, 2018.

Osiel, Mark. *The End of Reciprocity: Terror, Torture and the Law of War*. Cambridge: Cambridge University Press, 2009.

Ostrom, Elinor. *Governing the Commons: The Evolution of Institutions for Collective Action*. Cambridge: Cambridge University Press, 1990.

Ostrom, Elinor. *The Future of the Commons: Beyond Market Failure and Government Regulation*. Indianapolis: IEA, 2012.

Pearson, Graham. "The Web of Prevention." In *Preventing Biological Threats: What You Can Do. A Guide to Biological Security Issues and How to Address Them*, edited by S. Whitby, T. Novossiolova, G. Walther, and M. Dando, 136–59. Bradford, UK: Bradford Disarmament Research Centre, University of Bradford, 2015.

Perry, David L. *Partly Cloudy: Ethics in War, Espionage, Covert Action and Interrogation*. Lanham, MD: Scarecrow Press, 2009.

Popper, Karl. *Objective Knowledge*. Oxford: Clarendon Press, 1972.

Prunckun, Henry. "The Rule of Law: Controlling Cyber Weapons." In *Cyber Weaponry: Issues and Implications of Digital Arms*, edited by H. Prunckun, 87–100. Dordrecht: Springer, 2018.

Qiang, Xiao. "The Road to Digital Unfreedom: President Xi's Surveillance State." *Journal of Democracy* 30, no. 1 (2019): 53–67.

Rappert, Brian, and Caitriona McLeish. *A Web of Prevention: Biological Weapons, Life Sciences and the Governance of Research*. London: Routledge, 2012.

Rappert, Brian, and Michael Selgelid, ed. *On the Dual Uses of Science and Ethics*. Canberra: ANU Press, 2013.

Rawls, John. *A Theory of Justice*. Cambridge, MA: Harvard University Press, 1971.

Raz, Joseph, "Rights-Based Moralities." In *Theories of Rights*, edited by J. Waldron, 180–93. Oxford: Oxford University Press, 1984.

Reaume, Denise G. "Individuals, Groups and Rights to Public Goods." *University of Toronto Law Journal* 38, no. 1 (1988): 438–68.

Redmayne, Mike. "Rethinking the Privilege against Self Incrimination." *Oxford Journal of Legal Studies* 27, no. 2 (2007): 209–32.

Reed, Alastair, and Henschke, Adam. "Who Should Regulate Extremist Content On-line?" In *Counter-Terrorism, Ethics, and Technology: Emerging Challenges at the Frontiers of Counter-Terrorism*, edited by A. Henschke, A. Reed, S. Robbins, and S. Miller, 175–98. Dordrecht: Springer, 2021.

Regan, Mitt and Poole, Michele. "Accountability of Covert Action in the United States and the United Kingdom." In *National Security Intelligence and Ethics*, edited by Seumas Miller, Mitt Regan, and Patrick F Walsh, 232–48. London: Routledge, 2021.

Ressa, Maria. *How to Stand Up to a Dictator*. New York: Harper Collins, 2023.

Rid, Thomas. *Cyber War Will Not Take Place*. New York: Oxford University Press, 2013.

Robbins, Scott. "Bulk Data Collection, National Security and Ethics." In *Counter-Terrorism: The Ethical Issues*, edited by Seumas Miller, Adam Henschke, and Jonas Feltes, 89–105. Cheltenham: Edward Elgar, 2021.

Robillard, Michael. "Counter-Terrorism and PSYOP." In *Counter-Terrorism: The Ethical Issues*, edited by Seumas Miller, Adam Henschke, and Jonas Feltes, 23–34. Cheltenham: Edward Elgar, 2021.

Roff, Heather. "Cyber Perfidy, Ruse and Deception." In *Binary Bullets*, edited by F. Allhoff, A. Henschke, and B. J. Strawser, 201–27. Oxford: Oxford University Press, 2016.

Rousseau, Jean-Jacques. *The Social Contract*. Translated by H. J. Tozer. Ware: Wordsworth Editions, [1762] 1998.

Rowe, Neil C. "Perfidy in Cyberwarfare." In *Routledge Handbook of Ethics and War: Just War in the 21st Century*, edited by F. Allhoff, N. Evans, and A. Henschke, 394–404. Abingdon: Routledge, 2013.

Sanger, David E. *Confront and Conceal: Obama's Secret Wars and Surprising Use of American Power*. New York: Broadway Books, 2013.

Schauer, Frederick. *Free Speech: A Philosophical Inquiry.* Cambridge: Cambridge University Press, 1981.

Schmitt, Michael M., ed. *Tallinn Manual on the International Law Applicable to Cyberwar.* Cambridge: Cambridge University Press, 2013.

Scudder, Nathan, Dennis McNevin, Sally Kelty, Christine Funk, Simon Walsh, and James Robertson. "Policy and Regulatory Implications of the New Frontier of Forensic Genomics: Direct-to-Consumer Genetic Data and Genealogy Records." *Current Issues in Criminal Justice* 31, no. 2 (2019): 194–216.

Scudder, N., D. McNevin, S. Kelty, S. Walsh, and J. Robertson. "Massively Parallel Sequencing and the Emergence of Forensic Genomics: Defining the Policy and Legal Issues for Law Enforcement." *Science and Justice* 58 (2019): 153–58.

Searle, John R. *Making the Social World.* Oxford: Oxford University Press, 2011.

Selgelid, Michael. "Gain of Function Research: Ethical Analysis." *Science and Engineering Ethics* 22, no. 4 (2016): 923–64.

Shannon, Claude. *A Mathematical Theory of Cryptography*—Case 20878, Alcatel-Lucent MM-45-110-92 (1945).

Singer, Peter W., and Allan Friedman. *Cybersecurity and Cyberwar: What Everyone Needs to Know.* Oxford: Oxford University Press, 2014.

Skerker, Michael. *The Moral Status of Combatants.* London: Routledge, 2020.

Sloss, David. *Tyrants on Twitter.* Stanford: Stanford University Press, 2022.

Smith, Marcus. "Universal Forensic DNA Databases: Balancing the Costs and Benefits." *Alternative Law Journal* 43, no. 2 (2018): 131–35.

Smith, Marcus, Monique Mann, and Gregor Urbas. *Biometrics, Crime and Security.* Abingdon: Routledge, 2018.

Smith, Marcus, and Seumas Miller. "The Ethical Application of Biometric Facial Recognition Technology." *AI and Society* 37 (2021): 167–75.

Smith, Mark P. *Review of Selected Los Angeles Police Department Data-Driven Policing Strategies.* Los Angeles: Los Angeles Police Commission, 2019.

Smith, N. V., ed. *Mutual Knowledge.* Cambridge, MA: Academic Press, 1982.

Solove, Daniel. *Understanding Privacy.* Cambridge, MA: Harvard University Press, 2008.

Soni, S. L., and C. P. Bhargav. *Cyber Security and Cyber Law.* New Delhi: Prashant Publishing, 2016.

Sorell, Tom. "Privacy, Bulk Collection and Operational Utility." In *National Security Intelligence and Ethics*, edited by Seumas Miller, Mitt Regan, and Patrick F Walsh, 141–55. London: Routledge, 2021.

Sparrow, Robert. "Killer Robots." *Journal of Applied Philosophy* 24 (2007): 63–77.

Steinhoff, Uwe. "Killing Them Safely: Extreme Asymmetry and Its Discontents." In *Killing by Remote Control: The Ethics of an Unmanned Military*, edited by B. J. Strawser, 179–210. Oxford: Oxford University Press, 2013.

Stout, Lynn. *The Shareholder Value Myth.* Oakland, CA: Berrett-Koehler, 2014.

Sundberg, Kelly W., and Christina M. Witt. "Undercover Operations: Evolution and Modern Challenges." *Journal of the AIPIO* 27, no. 3 (2019): 3–17.

Sunstein, Cass. *#Republic: Divided Democracy in the Age of Social Media.* Cambridge, MA: Harvard University Press, 2017.

Sutton, Robbie, and Karen Douglas. "Conspiracy Theories and the Conspiracy Mindset: Implications for Political Ideology." *Current Opinion in Behavioral Sciences* 34, no. 118 (2020): 118–22.

Szigeti, Andras. "Are Individualist Accounts of Collective Moral Responsibility Morally Deficient?" In *Institutions, Emotions and Group Agents: Contribution to Social Ontology*, edited by A. Konzelmann-Ziv and H. B. Schmid, 329–42. Dordrecht: Springer, 2014.

Tapscott, Don, and Alex Tapscott. *Blockchain Revolution: How the Technology Behind Bitcoin Is Changing Money, Business and the World*. London: Penguin, 2016.

Taylor, Michael. *The Possibility of Cooperation*. New York: Cambridge University Press, 1987.

Thaler, Richard, and Cass Sunstein. *Nudge*. London: Penguin, 2009.

Theil, S., O. Butler, K. Jones, H. Moynihan, C. O'Regan, and J. Rowbottom. "Response to the Public Consultation on the Online Harms White Paper." Bonavero Report No. 3, 2019, 1–10. Oxford: Bonavero Institute of Human Rights, 2019.

Thompson, Dennis F. "Moral Responsibility and Public Officials: The Problem of Many Hands." *American Political Science Review* 74, no. 4 (1980): 259–73.

Tucker, Jonathan, ed. *Innovation, Dual Use, and Security*. Cambridge, MA: MIT Press, 2012.

Tung, Liam. "Google Accused of Leaking Personal Data to Thousands of Advertisers." ZDNet. Accessed July 6, 2020. https://www.zdnet.com/article/google-accused-of-leaking-personal-data-to-thousands-of-advertisers/.

Turing, Alan M. *Turing's Treatise on Enigma*. Unpublished Manuscript, 1939.

United Kingdom Home Office, *National DNA Database Statistics*. Accessed 2 February 2021. https://www.gov.uk/government/statistics/national-dna-database-statistics.

Urkup, Cagan, et al. Customer Mobility Signatures and Financial Indicators as Predictors in Product Recommendation. *PLOS ONE* 13 (July 2018): 1–5.

van den Hoven, Jeroen, Seumas Miller, and Thomas Pogge, eds. *Designing-in-Ethics*. New York: Cambridge University Press, 2107.

van der Bruggen, Koos, Seumas Miller, and Michael Selgelid. *Report on Biosecurity and Dual Use Research*. The Hague: Dutch Research Council, 2011.

Vayena, Effy, and Lawrence Madoff. "Navigating the Ethics of Big Data in Public Health." In *The Oxford Handbook of Public Health*, ed. A. Mastroianni, J. Kahn, and N. Kass, 370–80. Oxford: Oxford University Press, 2019.

Waldron, Jeremy. *The Harm in Hate Speech*. Cambridge, MA: Harvard University Press, 2012.

Walsh, Patrick F., and Seumas Miller. "Rethinking 'Five-Eyes' Security Intelligence Collection Policies and Practices Post 9/11/Post-Snowden." *Intelligence and National Security* 31, no. 3 (2016): 345–68.

Walzer, Michael. "Political Action: The Problem of Dirty Hands." *Philosophy and Public Affairs* 2 (1973): 160–80.

Warnat-Herresthal, S., et al. "Swarm Learning for Decentralised and Confidential Clinical Machine Learning." *Nature* 594 (2021): 265–70.

Warren, Samuel, and Louis Brandeis. "The Right to Privacy." *Harvard Law Review* 4, no. 5 (1890): 193–220.

Wenar, Leif. "Rights." *Stanford Encyclopedia of Philosophy*. 2020. https://plato.stanford.edu/entries/rights/.

Williamson, E., et al. "Factors Associated with COVID-Related Death Using OPENSafely." *Nature* 584 (2020): 430–36.

Woolley, Samuel, and Philip Howard, eds. *Computational Propaganda*. Oxford: Oxford University Press, 2019.

Xafis, Vicki, G. Owen, Markus Schaefer, K. Labude, Iain Brassington, et al. "An Ethics Framework for Big Data in Health and Research." *Asian Bioethics Review* 11, no. 1 (2019): 231.

Zeleznikow, John. "Can Artificial Intelligence and On-line Dispute Resolution Enhance Efficiency and Effectiveness in Courts." *International Journal for Court Administration* 8, no. 2 (2017): 30–45.

Zimmerman, Michael. "Sharing Responsibility." *American Philosophical Quarterly* 22, no. 2 (1985): 115–22.

Zuboff, Shoshana. *The Age of Surveillance Capitalism*. London: Profile Books, 2019.

Index

For the benefit of digital users, indexed terms that span two pages (e.g., 52–53) may, on occasion, appear on only one of those pages.

Figures are indicated by *f* following the page number